THE CHEMICAL ELEMENTS
AND MAN

Publication Number 1021

AMERICAN LECTURE SERIES®

A Monograph In

The BANNERSTONE DIVISION *of*
AMERICAN LECTURES IN LIVING CHEMISTRY

Edited by

I. NEWTON KUGELMASS, M.D., Ph.D., Sc.D.
Consultant to the Departments of Health and Hospitals
New York, New York

THE CHEMICAL ELEMENTS AND MAN

Measurements · Perspectives · Applications

By

E. I. HAMILTON, B.Sc., D.Phil., F.G.S.

"Woodchurch"
Crapstone, nr Yelverton
Devon, United Kingdom

CHARLES C THOMAS · PUBLISHER
Springfield · Illinois · U.S.A.

Published and Distributed Throughout the World by

CHARLES C THOMAS • PUBLISHER
Bannerstone House
301-327 East Lawrence Avenue, Springfield, Illinois, U.S.A.

This book is protected by copyright. No part of it may be reproduced in any manner without written permission from the publisher.

© 1979 by CHARLES C THOMAS • PUBLISHER
ISBN 0-398-03732-9
Library of Congress Catalog Card Number: 78-5409

With THOMAS BOOKS careful attention is given to all details of manufacturing and design. It is the Publisher's desire to present books that are satisfactory as to their physical qualities and artistic possibilities and appropriate for their particular use. THOMAS BOOKS will be true to those laws of quality that assure a good name and good will.

Library of Congress Cataloging in Publication Data

Hamilton, E. I.
    The chemical elements and man.

    (American lecture series; no. 1021)
    Includes index.
    1. Chemicals—Physiological effect. 2. Body composition. 3. Health. 4. Chemical elements—Analysis. I. Title.
QP531.H28   612'.0152   78-5409
ISBN 0-398-03732-9

Printed in the United States of America
C-1

To all those prepared to accept the need for generalities and holism in the pursuit of the health and well-being of man.

PREFACE

THIS MONOGRAPH is largely based upon a series of investigations carried out between 1967 and 1971 at the late Medical Research Council and Ministry of Health Radiological Protection Service (RPS). The data were obtained to describe, to a first order of approximation, the distribution of the chemical elements in man and to identify major source materials and routes of entry of the elements. From knowledge gained through a study of the stable elements of the Periodic Table, this was considered useful in order to identify sites of deposition and retention for a great variety of radionuclides derived from nuclear technology now released into the natural environment. As a by-product, several investigations were carried out in order to describe relations between the chemical elements and human morbidity.

From planning stages to the final retrieval of data, considerable attention was paid to the development of sampling techniques, to chemical analysis, establishing clean working conditions, and engaging in multidisciplinary cooperative programmes of research in order to reach defined objectives. The enlightened management of the RPS provided an ideal environment for such work, free from the shackles of limited vision imposed by historical demands of many individual disciplines.

The text describes and provides fundamental information essential in trying to understand associations between elements and man. Methods of analysis are presented in brief, designed to identify suitable methods, identify problems of analysis, and provide an overall viewpoint of those methods which are currently available from commercial sources.

No attempt is made to discuss matters in any great detail; while I am acutely aware of the dangers of generalisations, the substance of the text is amply illustrated, providing data for the abundance and distribution of the elements in man and various parts of his environment, and also has a nonmathematical description of analytical methods.

The text is not written for analytical chemists but rather for nonchemists who find a pressing need to understand the capabilities of modern chemical techniques in order to make progress in their own fields of interest. Historical precedent has often produced a barrier between chemistry and biology which is most obvious in those biologists (e.g. naturalists) who now move forward under the flag of environmental biologists and ecologists and become enraptured with the novelties of modern science, especially computer sciences, and include analytical chemistry as a black box technology with most unfortunate consequences. Until education has linked biology and chemistry it is difficult to draw both chemists and ecologists from their ivory towers in order to face up to some of the pressing problems of society today.

The classical approach to biochemistry and study of disease is mainly preoccupied by a consideration of organic systems; this monograph identifies the essential role of the chemical elements, in particular many minor and trace elements, and seeks to gain some perspective in the overall interplay of organic and inorganic systems set against a background of evolution and technological progress for a species, namely man, who originated from a natural environment and progressed rapidly into one which has become radically altered as a result of his actions.

This monograph is directed to those with an interest in associations between the chemical elements present in the natural environment and human morbidity, in particular those of the medical profession who are central to the topic and who could, by careful observations, illuminate the grey areas which link geography, occupation, and social status with the chemical elements and disease.

E. I. HAMILTON

INTRODUCTION

IN STUDIES CONCERNED with the health of man, the elements of the Periodic Table have been classified according to their essential or nonessential role in biochemical processes; as our knowledge of the chemical elements improves, elements hitherto classified as nonessential are shown to have some degree of essentiality and so the number of elements in this category increases, although initially the recognition of essentiality is often obtained from animal experiments. Ignoring those elements whose role in biochemical processes is still uncertain, the present classification of the elements essential in maintaining health, together with some elements which are toxic, is given in Table I. The major elements of the human body, together with those elements associated with enzymes, have received considerable attention compared with the remaining elements of the Periodic Table. This state of affairs undoubtedly does reflect the lack of importance of a large number of elements, as far as man is concerned, but some such as silicon, essential for the development of chicks, and tin, an essential dietary requirement of rats, may eventually also be shown to be essential for man. Although all the chemical elements at some level of acute or chronic intake of exposure are toxic to man, toxicological investigations have identified, by direct or indirect means, maximum permissible levels and the associated characteristic patterns of disease which accrue if they are reached or exceeded. In recent years, attention has been focussed upon understanding dose-response relationships for exposure to subtoxic levels with special reference to acute, chronic, somatic, genetic, and both antagonistic and synergistic effects for exposure to elements in defined chemical forms. Prior to about 1950 each element tended to be studied in an isolated manner and little attention was directed to the consequences of interelement effects, which is understandable when the great difficulties of studying a single element in sufficient detail are con-

TABLE I
SOME ESSENTIAL AND TOXICOLOGICAL ELEMENTS

Essential Elements	Toxicological Elements
Hydrogen	Lithium
Boron	Beryllium
Carbon	Nickle
Nitrogen	Arsenic
Oxygen	Bromine
Fluorine	Silver
Sodium	Cadmium
Magnesium	Antimony
Silicon*	Barium
Phosphorus	Mercury
Sulphur	Thallium
Chlorine	Lead
Potassium	Bismuth
Calcium	Radioactive elements,
Vanadium*	e.g. U, Th Series
Chromium	Transuranium Elements
Manganese	
Cobalt	
Copper	
Zinc	
Selenium	
Molybdenum	
Tin*	
Iodine	

* Confirmed for animals for defined types of experiment.

sidered. However, it should be noted that in terms of participation in chemical reactions no single element is really unique and adjacent elements of the Periodic Table share certain similarities as defined by the group classification of the elements presented in Table II.

In order to investigate levels of elements present in trace quantities in human tissues, very sophisticated techniques of analysis are required; prior to their development, analytical chemistry relied heavily upon wet chemical methods supplemented by some instrumental methods, such as colorimetry. It was not until the dawn of the nuclear era, and the associated technological advances, that it became possible to have available sufficiently sensitive and specific methods of analysis and thus render it possible, in theory at least, to determine any element of the

TABLE II

ELEMENT GROUPINGS BASED UPON THE PERIODIC TABLE OF THE ELEMENTS*

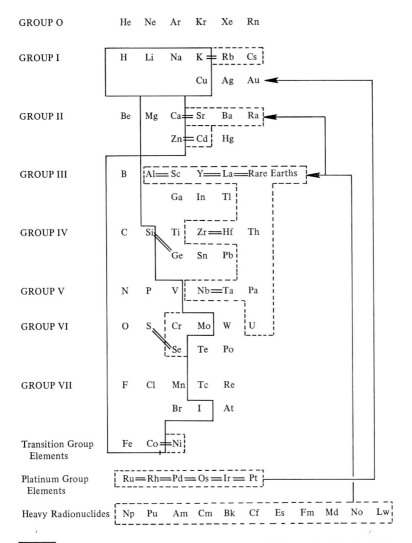

*Elements enclosed by a continuous line are considered to be essential, those enclosed by a broken line represent elements having similar chemistry and amendable to isomorphous substitution. The double continuous lines represent major links between a major and a minor element such that associations are retained in the passage of these elements from the natural environment to man.

Periodic Table even when present in trace quantities. However, one very important fact which is often ignored is that the availability of sophisticated instrumentation does not automatically provide a solution to analysis of the chemical elements. The availability of such instrumentation can overcome many problems, but the major problem lies with the chemical matrix of the sample and the need to develop skills able to isolate the signal from an element from interferences arising from other constituents present in the sample. It is for this reason that the realisation of some form of black box technology, universally applicable to element analysis, lies very much in the future. Other problems concern an ability to present to a detector an aliquot containing elements of interest free from contamination, to accomplish this without loss or gain during processing, and to ensure, as far as possible, that the mass of the element determined is representative of the active state of the element present in living biological systems. Initially the total mass of an element has to be determined, but in the future far more attention needs to be paid to the biologically active phase, otherwise, any relationship between the mass of an element present in a sample and biological effect is likely to remain obscure. Limitations, imposed by the vast analytical load required to sample human populations adequately, impose restrictions on the number of analyses which can reasonably be performed; even if this were no obstacle there remains the problem of adequately sampling a very variable system.

The onset of the nuclear era marked an unprecedented demand for the analysis of a very large number of elements for which virtually no data existed; matters were further complicated by the additional demand to carry out quantitative assays for a series of new elements, such as plutonium, americium, and curium, present in ultra-trace amounts in human materials and environmental samples. Furthermore, the addition of ionizing radiation, associated with man-produced radionuclides in nuclear fission and activation processes, added a new dimension to the health of man; associations between the naturally occurring radionuclides, such as thorium and uranium, and disease were already appreciated such as through studies of uranium miners and individuals exposed to radium in various types of therapy. The very sensitive analytical techniques developed for radionuclides made assay a relatively simple matter, provided that a representative proportion of the radiation could be presented to a suitable detector. Emphasis was now placed upon assay for the stable elements of the Periodic Table on the basis that, if the distribution and metabolic pathways for the stable elements were known, such information could be used to predict environmental and metabolic pathways for the equivalent radionuclides and hence identify sites of deposition, retention, and clearance in order to estimate radiation dose. The tacit assumption underlying this approach, and one which is not always acceptable, is that with few exceptions the radionuclides would be intimately mixed with the stable elements and would enter the human body in the same chemical form, hence knowledge of the manner in which a stable element participated in body processes would provide data for the behaviour of the radioelements. Some elements of interest, such as plutonium, have no stable equivalents and it was necessary to consider their distribution in the body by selecting stable elements having similar chemical properties, for example, one of the rare-earth group elements. As research progressed it soon became apparent that radiological toxicity was in part related to the chemical form of the radionuclides; in one form an element would not cross the gastrointestinal tract or lung, while in another form it would readily pass into the body; in one form an element would not be retained within the body and would be rapidly excreted, while in another form it was avidly retained.

The importance of the chemical elements in metabolic processes was highlighted by the publication in 1959 of the Report of Committee II of the International Commission on Radiological Protection (1) who presented data for the abundance of forty-six elements in most of the tissues of the human body. This, in itself, was a remarkable achievement and represented a detailed scouring of the world's literature for data, very widely disseminated in many journals and reports and brought together for the first time. Later the Report of the Task Group on Reference Man, International Commission on Radiological Protection Publication 23 (2), updated the 1959 report which contains even more details for the distribution of the elements. In dealing with the problems of exposure of large populations to ionizing radiation the accuracy of the data, with some exceptions, is usually adequate and at the same time provides useful guidelines of value to a wide variety of disciplines. Nuclear technology also provided the availability of a large number of radionuclides, in a variety of chemical forms, which opened up new fields of clinical investigation and became widely used to trace metabolic pathways of the stable elements in the human body. Apart from the impetus given to medical research, the nuclear era also gave rise to a new generation of analytical instruments as a result of advances made in the electronic industry. Many non-radioactive-based analytical techniques, such as atomic absorption spectroscopy, gas chromatography, and electrochemical methods, owe much to the fall-out of the nuclear era because of the development of highly stable electronic circuits, improved amplifiers, and better discrimination between instrumental noise levels and sample signals. By mid-1950 nuclear techniques were well established and were applied to a wide variety of materials including human tissues; however, well-planned objective studies for analysis of human tissues were quite rare and such materials were often used to illustrate the versatility of a particular technique with an emphasis on sensitivity of detection. A considerable amount of data was obtained for such exotic elements as scandium, indium, and gold, not because of any real interest in these elements but simply because they were fairly easy to analyse by nuclear techniques. In the field of tissue analysis, nuclear techniques in common use were not suitable for analysis for some of the essential major and minor elements and the tendency was to use other methods, usually with excellent results, but the techniques often required a considerable amount of skill, time, and experience.

In 1970, mankind became sensitive to questions of pollution and contamination of the natural environment, also to associations between disease and the chemical elements, in particular for industrial exposure. By now, very elegant nuclear techniques were firmly established and further advances in instrumentation arose because of developments in space technology. By 1976 with the mass of advanced analytical equipment available and widely used in many disciplines, it would be reasonable to expect that the question of chemical analysis had been solved and the magic black-box era had arrived in which a sample could be presented to an instrument, a button pressed, and an accurate answer emerge. In practice this has not happened, and while detectable signals can now be obtained for most elements, over the picogram to gram range, the quality of much of the data leaves much to be desired. The problem remaining rests with the development of acceptable protocols for defined matrixes, and it will still be many years before real progress has been achieved. Today, the need is to develop acceptable techniques of analysis so that data obtained from different laboratories in many parts of the world may be compared without the need to consider problems of analytical bias or operator skill. The future for gross tissue assay of the chemical elements is limited except in programmes of surveillance, monitoring, and the establishment of regional variations in element abundances. Today, more atention should be directed towards determining the gross levels and distribution of elements in cellular and subcellular fractions of tissues; analytical techniques are available but are often very demanding. It is in such materials that the active elementbiochemical processes take place and it is where the first signs of many diseases are likely to be detected. Apart from now being able to determine levels of elements in very small amounts of sample, instrumentation is also available which enables the distribution of the elements on cell surfaces to be determined without destruction of the sample.

This monograph presents personal impressions of the role and potential of studying levels of elements in human tissues; general discussions are presented for other materials to which man has access and through which the elements enter the human body. In order to obtain some feeling for the significance of the chemical elements in man it is essential to understand the mutual relationships between different elements and, to a first order of approximation using such information, to predict the approximate levels of elements in materials to which man is exposed. My early experiences as a geologist-geochemist have proved to be invaluable in formulating general rules governing the distribution of elements in man; later when I was concerned with medical and radiation chemistry, in particular relationships between stable and radioactive materials in the human body, I found these general rules were still applicable although modifications were required because of the particular modifying influences generated by biochemical processes. Today there is no real lack of suitable instrumentation or techniques for studying the chemical elements in man and related materials; often expert skills are not immediately available but can be obtained given a little time, patience, optimum working conditions, and an enlightened management. However, even if all these requirements were realised, the full potential of element studies in man will not be realised because solutions to problems, or advances in knowledge, do not rest with analytical chemists but require the cooperation of many individuals in different disciplines who can be brought together and their combined expertise focussed upon well-defined problems. The problem of the future is the development of adequate multidisciplinary cooperative research both in the fundamental and applied fields of study. An objective of this monograph is to illustrate some of the requirements for modern research in the field of human element chemistry and biology and to place the essential role of analytical chemistry in perspective.

REFERENCES

1. Report of Committee II on Permissible Dose for Internal Radiation: ICRP Publication 2, Recommendations of the International Commission on Radiological Protection, 1959. Oxford, Pergamon Press, 1959.
2. Report of the Task Group on Reference Man: ICRP Publication 23, 1974. Oxford, Pergamon Press, 1975.

ACKNOWLEDGMENTS

As this monograph represents an extramural activity its completion would not have been possible without the assistance I have received from many individuals and organisations who offered help. It is not possible to mention all but special thanks are sincerely extended to Dr. W. Goody for encouragement; Dr. R. Masironi (W.H.O., Geneva) and Dr. G. V. Iyengar (Kernforsuchungsanlage, Jülich) for library researches; Mr. P. Crawley (Jeol, U.K.) for access to electron microscopy. To the Elsevier Scientific Publishing Company for permission to reproduce figures and tables of my papers published in *Science of the Total Environment*. To those who wish to remain anonymous but provided funds to travel; and to Mrs. M. Chaplin for help in preparing the final stage of the manuscript. To Dr. J. Vennart of the late RPS and to those of my staff who participated in the RPS work, Miss M. J. Minski, Mr. J. J. Cleary, Mr. and Mrs. N. Priestley, and Mrs. V. Dow.

Finally, to the tolerance of my family in accepting my absence from family life during the evenings for the past nine months.

E.I.H.

CONTENTS

	Page
Preface	vii
Introduction	ix
Acknowledgments	xvii

Chapter

		Page
1	Origins	3
2	Interfaces, Elements, Environment—Elements, Man	19
3	Air	26
4	Water	68
5	Diet	95
6	Methods—Preparatory Techniques	130
7	Instrumental Methods	190
8	Elements and Man	363
9	Disease	411
10	Biochemical Considerations	426
Index		469

THE CHEMICAL ELEMENTS
AND MAN

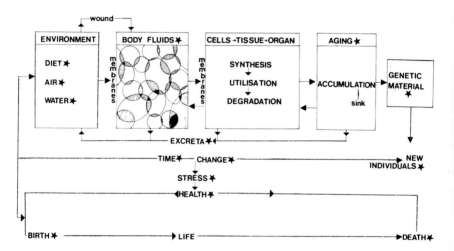

General relationships between the elements, the environment, man and time.

Each box represents a complex system of interacting reactions and pathways which change with time. Key points for monitoring are indicated by an asterisk.

Chapter 1

ORIGINS

KNOWLEDGE OF THE ORIGIN of the chemical elements has a distinct bearing upon our understanding of their presence and levels in biological systems. Charles Darwin (1, 2) demonstrated that evolution in the biosphere can be explained in terms of continuous natural selection through processes of adaption to particular environments. The origin and evolution of the chemical elements cannot be strictly described as a series of chance phenomena, but rather a logical development following fairly well-defined physical characteristics associated with nuclear processes. However, while we can explain how the elements were formed we do not understand how all the nuclear components of the atom were formed.

There are many theories to account for the origin of the chemical elements but that proposed by Burbridge, Burbridge, Fowler, and Hoyle (3) is the most complete. Because of the immense span of time during which bodies in the sky have been in existence, it is impossible to start at the true beginning of the creation of matter, we have to be content, for the purpose of this monograph, with the origin of the elements of which Earth is formed. One of the ubiquitous constituents of space is stellar dust, which is considered to consist of a mixture of hydrogen, graphite, silicates, and ice; its presence in space gives rise to several types of phenomena, such as the microwave background radiation observed on Earth while Earth collects cosmic dust by deposition at a rate of $\approx 2 \times 10^8$ tons. Until recently the density of the dust was estimated to be 10^{-26} g/cm^3, while Wesson and Lerman (4) have calculated a density of less than 10^{-46} g/cm^3; therefore the dust is unlikely to present a barrier to the continued expansion of the Universe. A convenient starting point, in order to consider the origin of the elements, is a volume in space into which the lightest element, hydrogen, gradually accumulated through processes of diffusion. The origin of hydrogen lies in the realms of Albert Einstein's equation $E = mc^2$—where E = energy, m = mass, and c = the velocity of light—through which energy can be converted into matter. A source of energy, such as that created by the detonation of chemical explosives, releases approximately 2×10^3 calories per gram of explosive, but matter is not created as the energy released is derived from the breaking of chemical bonds. On the other hand nuclear explosions release energy from within the atoms of the exploding material and as a result the total mass of the products is less than the mass of the starting materials because of the conversion of matter into energy released in the form of heat.

In order to create matter on a large scale a vast source of energy is required; within the solar system this condition is fulfilled in hot stars which may be considered nuclear furnaces. The constituents of hot stars, such as our sun, will at an early stage be clouds of gas and dust which aggregate in space and are gradually whirled into pockets of high density through forces of gravity. As the hydrogen atoms become crowded together, gravitational energy becomes converted into heat and, although some energy is lost by radiant processes, the internal temperatures of such protostars gradually increase until at temperatures of $5\text{-}10 \times 10^{6}°K$ nuclear fusion occurs. In the fusion process, two protons collide to form a deuterium atom and as a result the original hydrogen atoms lose their electrons. The products of fusion carry a positive charge and therefore do not travel very far before being annihilated by striking an electron; a further product, the neutrino with no charge or mass, passes to the surface of the protostar and is lost into space. The newly formed deuterium atoms in the cloud of hydrogen soon react with one another and give rise to a new element helium, ^3He, consisting of two protons and a neutron which is more massive than the original proton. In the gas cloud, the hydrogen is slowly used up to form helium, which is accompanied by a drop in temperature, followed by a period of renewed heating as the helium in turn is subjected to gravitational forces.

Fusion of two helium nuclei now takes place to form a variety of products depending upon the particular requirements for various nuclear reactions. At temperatures of $1\text{-}2 \times 10^8 \,°K$, helium nuclei react exothermically by (α, γ) reactions to form ^{12}C, while at still higher temperatures successive helium captures result in the formation of ^{16}O, ^{20}Ne, and perhaps some ^{24}Mg. The rate of energy release by helium reactions slows down as the ash of the process of helium burning accumulates in central regions of the protostar. The next stage involves a period of contraction, followed by one of reheating but at a far more vigorous rate than before; for example, reactions between two ^{12}C nuclei occur to form ^{24}Mg ($^{12}C, \gamma$) and ^{23}Na ($^{12}C, p$). As the process of burning, accumulation of ash, contraction, and reheating at even higher temperatures, e.g. $> 10^9 \,°K$, is repeated, elements continue to be formed up to a mass of about atomic weight 56, when an equilibrium process becomes dominant and the iron group of elements are formed, for which the nuclear binding energy per nucleon in this region is maximal.

Above $2 \times 10^8 \,°K$, free neutrons are formed and neutron capture processes take place very similar to those of modern nuclear reactors, with the result that heavier elements are formed. The actual processes are complicated, but the production of further elements is terminated by the spontaneous fission of the very heavy elements. Although at temperatures of about $5 \times 10^9 \,°K$ the final product should be ^{56}Fe, the ultimate fate of the star is influenced by its mass. For a small star most of the core region will consist of iron and the star will cool down; for more massive stars, as the core temperatures approach the critical value of $5 \times 10^9 \,°K$, their cores may be converted into iron in a few seconds; as a result of the rapid change in density, instability sets in, followed by a catastrophic implosion and the explosion of the star to give rise to a supernova.

While weather can have a physiological effect upon man, more severe extraterrestrial events, such as supernova, could have very significant effects on all living matter. Supernova occur in single regions of galaxies which contain lanes of dust; as a result of forces of compression star clusters could form, among which would be some heavy stars subject to rapid evolution when supernova occur. Clark and Stephenson (5) discuss the possibility that the radiation emitted by such events would destroy Earth's ozone layer and increase the radioactive bombardment of Earth's surface by a factor of thirty. In the absence of the ozone layer, the protective greenhouse effect would be lost and an ice age could be initiated. The time for transit of the ejecta would be about two thousand years, hence the reader could visualise an ice age in transit. However, the Milankovitch theory of ice ages argues that they could arise from small perturbations of Earth's orbit caused by other planets, resulting in fluctuations in the distance between Earth and the sun, thus giving rise to periods of warming and cooling. Supernova can occasionally be observed in the skies; one recorded in the year 1054 AD, observed from Earth, actually took place some four thousand years before and the light emitted took forty centuries to reach Earth. Although supernova are rather rare events, they are important as during the explosion rapid neutron capture processes take place and new heavy elements are formed terminating with the production of ^{254}Cf which has a half-life for radioactive decay of fifty-five days, which is also the rate of decay of light emitted by supernova. Following the explosion, matter is flung into space where the debris may condense in the clouds of gas to form new stars or be captured by older stars. As a consequence of the Burbridge, Burbridge, Fowler, and Hoyle theory, we can account for the abundances of the elements in the solar system as illustrated in Figure 1, for example:

- The exponential decrease in the abundance of elements from $A = 1$ to $A = 100$ reflects a fairly young solar system in which is present a major mass of hydrogen and only a small degree of conversion to heavier elements.
- The approximate constant abundances for elements above $A = 100$ reflects the rapid formation of heavy elements by neutron capture processes.
- The relative low abundance of deuterium, lithium, beryllium, and boron is because of their rapid consumption by nuclear processes taking place at high temperatures.
- The high abundance of ^{16}O occurs because of the high stability of these nuclei towards further nuclear reactions and be-

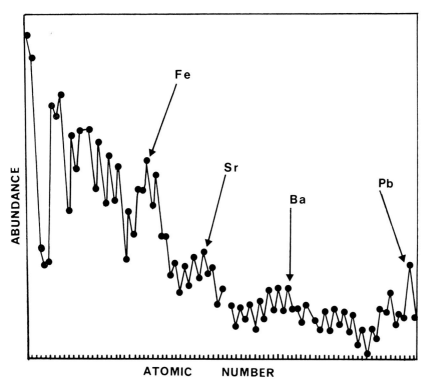

Figure 1: The cosmic abundance of the chemical elements; atomic concentrations relative to Si = 10^6.

cause of the efficiency of the nuclear capture processes at temperatures where neutron production is negligible.
- The ^{56}Fe peak is because the nuclear binding energy is maximal in this region.
- The low abundance of proton-rich nuclides such as ^{74}Se, ^{92}Mo, and ^{130}Ba is probably because of the rare occurrence of proton capture compared with neutron-capture.

One of the surprising outcomes of this theory is the very rapid rate at which some elements are formed during exceedingly short periods of time. Time is a very important factor in terrestrial evolution and the following simple calculation is revealing.

Radionuclides present in the crust of Earth today undergo radioactive decay at a fixed rate and therefore can be considered as a type of clock which is slowly running down at a fixed rate determined by the characteristics of radioactive decay for individual radionuclides (see page 272). If we assume that the isotopes of uranium, namely ^{234}U, ^{235}U, and ^{238}U were all produced at the same time, and in equal amounts, then the initial ratio of unity, at the time the elements were synthesized, will be different today and a measure of the differences in the rates of decay of each of the uranium isotopes. From such relationships it is possible to calculate that some 13×10^9 years have passed since the elements of our galaxy were synthesised, that the oldest stars in our galaxy are about 6.5×10^9 years old. By radiometric dating of meteorites, using relationships between the parent/daughter abundances of naturally occurring radionuclides and their stable end products, it can be shown that the age of the solar system is no more than about 4.5×10^9 years old, and the age of Earth, more precisely, is $4.55 \pm 0.07 \times 10^9$ years (Patterson, Brown, Tilton, and Inghram, 6; Hamilton and Farquhar, 7; Tilton, 8; Nunes, Tatsumoto, and Unruh, 9), which is approximately the same for the age of meteorites and the moon.

Meteorites, which contain the same type of elements present in Earth, are fragments of minor planets which travel around the sun between the orbits of Mars and Jupiter. From the general discussion so far the Burbridge, Burbridge, Fowler, and Hoyle theory provides a logical account of how the nuclides were synthesised and how the chemical elements were formed. We also note that throughout the different processes giving rise to solid bodies, overall composition is a chance phenomena, yet the overall picture is one of some degree of uniformity; where differences are observed simple explanations are forthcoming, for example a planet rich in iron would be accounted for by the fact that when it was formed iron-rich fragments derived from a disintegrating star were more available. By study of the characteristics of the spectrum of light emitted by stars, a high abundance of particular elements can be observed.

In the rocks of Earth today uranium is the heaviest element

which is still undergoing the radioactive decay processes initiated during synthesis of the elements and is decaying very slowly to give rise to stable isotopes of lead. In nuclear physics laboratories, there is considerable interest in bombarding heavy elements with highly energetic particles in order to create super-heavy elements; to date the Soviets have created element 107, a nucleus containing 107 protons and 154 neutrons, which should have chemical properties similar to rhenium. The radioactive half-life of these super-heavy nuclides is very short, possibly fractions of a second, and hence they are not likely to be present in crustal rocks today, although if they could have been incorporated into minerals at an early stage of nuclear synthesis they should be detected by study of radiation damage structures in very old unaltered minerals. In fact a recent controversy has originated as a result of the description by Gentry (10) of radiation damage (pleochroic haloes) features preserved in some very old minerals which could originate from isomers of known elements or from super-heavy elements, although several other investigations by several different techniques do not provide evidence of the existence in the past of super-heavy elements. However some primordial transuranium elements such as ^{244}Pu (Alexander, Lewis, and Reynolds, 11) and ^{247}Cm (Cherdyntsev and Mikhailov, 12) have been reported.

The heaviest naturally occurring element in crustal rocks is plutonium but, unlike uranium, this is being formed today by neutron irradiation processes in uranium minerals. Apart from man-made nuclear explosions such events can also occur by natural processes provided that sufficient uranium is present in a confined space and other essential conditions are met; for example the Oklo Phenomenon (Lancelot, Vitrac, and Allegre, 13; I.A.E.A., 14) was an event which occurred 1.78×10^9 years ago in Gabon, Africa; a natural nuclear explosion involving some 500 tons of uranium, releasing about 1×10^{10} kwh, which gave rise to an integrated neutron flux of more than 1.5×10^{21} neutrons/cm^2; as a result the natural abundance of ^{235}U, as measured today, was depleted at the site from 0.72 percent to 0.027 percent; fortunately such events are rare.

The chemical properties of the elements are a reflection of their nuclear structure and for subsequent discussions it is important to recognise the systematic manner in which the nuclides, and hence the elements, were formed. Although we can now describe processes through which the chemical elements were formed, we are still unable to describe the initial materials, in particular a precise description of the internal structure of the proton. Physics has brought into play several very powerful machines providing very high sources of energy but the real problem lies in being able to simulate the universe in which the fundamental processes took place. Recent investigations of the structure of protons have revealed the presence of quarks, of which there are four in number (called up, down, strange, and charm) held together by glueons and associated with unusual force fields which decrease in strength as quarks approach one another and increase as they part. Further mysteries concern antimatter, that is matter symmetrically opposite to matter as we know it; while it cannot occur on Earth, as particles annihilate one another the instant they meet, here is evidence that antimatter does exist in the universe. If antimatter is accepted, then how did it become separated from matter? Certainly strange features are present of which perhaps the least understood is gravity and its component the gravitons. At sites of black holes, gravity overcomes all forces and not even light can get out, but as matter falls into such spaces an intense emission of X-rays takes place. Such matters are of little concern to elements and biology, but it is worth noting that through physics the simple and beautiful form of nature is revealed.

ROCKS AND THE EARTH

From our previous discussion of the origin of the elements we are presented with a proto-earth consisting of fragments of stellar debris which coalesced to form the planet Earth. It seems probable that at an early stage the elements were fairly evenly distributed, although an iron-rich central core is possible. The next major change which occurred, possibly by radioactive decay processes or heat generated by exothermic chemical processes, resulted in the separation of a central iron-rich core and an outer mantle which underwent further fractionation and differentiation, physically and chemically, to give rise to a thin outer crust upon which life evolved. These initial processes took place between about 4.5 and 3.5×10^9 years ago and gave rise to the proto-continents and hydrosphere. The fractionation processes were controlled by chemical characteristics prevailing under conditions of defined temperature and pressure; the elements became segregated into phases governed by such features as melting point (high temperature phases crystallised first), ionic radius of the atoms, and ionic charge to give rise to a series of mineral phases linked by well-defined chemical bonds.

The continental masses consisted predominantly of silicate, oxide, and sulphide phases; the hydrosphere contained those elements which could be carried in various solutions, while insoluble phases precipitated to the bottom of the early oceans; the atmosphere contained mainly gaseous compounds together with volatile phases which could return to the surface of the crust by condensation processes and be recycled. The segregation of the elements into these three main phases and the major forms are illustrated in Figure 2.

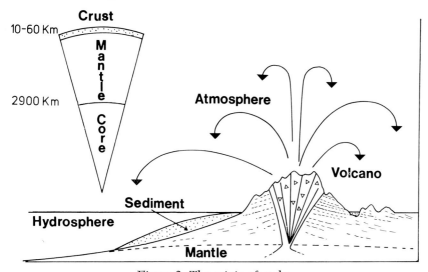

Figure 2: The origin of rocks.

At some early period in the history of Earth, the surface must have been very similar to that of the modern volcanic fields of Iceland. Gradually, areas distant from volcanic activity would cool down and extensive lava fields would become dominant features. As local areas became more stable the action of condensed phases would initiate cycles of erosion accelerated by climate and the first true sediments would be deposited on the land and in the seas. As a result of weathering processes the chemical elements became fractionated, soluble fractions being transported to the seas and the more resistant phases accumulating on the land or in littoral waters; hence the chemistry of these soils and sediments would reflect the chemical composition of the basement rocks from which they were formed, modified by solution processes. Below the surface of Earth, cooling processes were also taking place with the result that elements and minerals of high temperature types would crystallise with conditions of falling temperature and sink to the lower regions of the molten masses of rock called magmas. Chemical differentiation of the elements also occurred as a result of transport by hot solutions and through transfer by volatile phases. As differentiated silica-rich phases, together with volatiles, accumulated above the solidifying magma they would be periodically released to the surface through volcanic activity and along fracture planes. As hot rocks were extruded through the already cooled surface materials, local reheating would take place giving rise to a further series of rock type, the metamorphics. Molten magmas need not have always breached the surface; if emplaced below the surface, where temperatures were low enough they would solidify to form large volumes of igneous rock which, through processes of surface erosion and mountain building, would often become exposed at the surface and be subjected to the cycle of weathering, giving rise to silica-rich sediments as illustrated in Figure 3. A final element-rich phase formed as a result of these processes would be the ore minerals which represent segregations of elements, usually as oxides or sulphides, during very early or late phases of crystallisation, differentiation, and fractionation sequences. In very simple terms we have now considered the major processes of the evolution of the early rocks and their constitu-

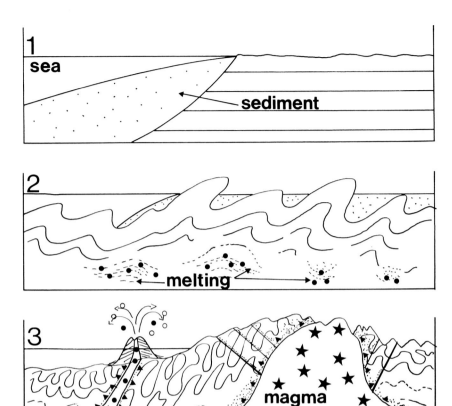

Figure 3: The development of rocks.

1. Erosion of the proto-crust gave rise to the early sediments which were deposited in the seas and lakes.
2. As a result of crustal movements, the early horizontally deposited sediments were deformed and with an increase in temperature with burial at depth the lower zones became molten.
3. High hydrothermal pressure at depth, associated with melting, resulted in fracturing of overlying rock and the intrusion of molten magma to upper horizons associated with surface volcanic activity. As a result of erosion at the surface of the earth, solidified magmas (igneous rocks) were exposed. At the contact of the magmas to the sediments local heating and chemical changes took place to give rise to the metamorphic rocks often associated with the deposition of ore minerals, derived from cooling magmas which would also penetrate into the country rock along zones of fracturing. (N.B. Deformation of either igneous or sedimentary rocks by pressure, for example during episodes of mountain building, also give rise to metamorphic rocks, e.g. clays → shales → slates → schists, and granites → gneisses.)

ents, the minerals which still continue today in the direction of attaining some degree of equilibrium. Areas of the original crust are no longer preserved and have been subjected to complex sequences of reworking, however some of the oldest sediments ($\approx 3 \times 10^9$y old) which are preserved within stable areas still retain the basic chemical composition of the early magmatic rocks.

LIFE

It is clear from the previous discussion that the chemical composition of surface rocks would vary over the surface of the developing crust according to the particular chemical nature of the exposed rocks and their derived sediments. Before embarking upon a very brief discussion of the origin of living matter it is worth noting that one class of meteorites, the carbonaceous chondrites, contain carbon ($\approx 6\%$) and water ($\approx 20\%$) together with organised structures which have been interpretated as bacterial materials; bulk chemical and mineral analyses of these types of meteorites also provide evidence for the presence of organic compounds which could indicate a stellar origin for living materials, although there is always the question of contamination during entry of Earth's atmosphere or after impact.

A search in materials exposed at the surface of Earth today has shown that algal and bacterial bodies are present in rocks which are at least 3×10^9 years old. They are preserved in sedimentary rocks which were deposited or associated with thermal springs; preservation has usually taken place because of the replacement of the original organic structures by hydrated forms of silica, only the external form of the organisms is preserved and any internal structures have been destroyed by crystallisation and diffusion processes. In rocks younger than about 2.0×10^9 years there is no difficulty in tracing the general progress of organic evolution as recorded by the fossil record. However, it is of interest to consider rocks older than 3.0×10^9 years and seek evidence of organic materials which could conceivably be the precursors of life. In order to embark upon this search we seek the presence of fossil organic chemical compounds which can be extracted from rocks. A large variety of such compounds have been found even in rocks which contain no visible fossil remains. The validity of this approach is strengthened by laboratory experiments which have shown that simple mixtures of hydrogen, methane, carbon monoxide, and ammonia, invariably present in gas phases of volcanic regions, can, under the stimulus of light, electrical discharge, and ionizing radiation, give rise to most of the essential building blocks of living matter. It is of considerable interest to note that the amino acids can be formed by such processes and through processes of thermal polymerisation give rise to proteins, including DNA and RNA. Many of these chemical processes require a reducing environment, which existed during the early stage in the development of Earth. Later, the gradual disappearance of hydrogen and an increase in oxygen and carbon dioxide resulted in a change from anaerobic life forms to aerobic forms and the evolution of photosynthetic processes. In relation to living materials, this change possibly represents the greatest episode of terrestrial pollution Earth has ever experienced. Undoubtedly the process was very gradual and there was ample time for biological species to adapt. While we can be fairly certain that the initial forms of life proliferated under reducing conditions, the gradual change to oxidising conditions and the appearance of some form of chlorophyll can be shown by the presence of isoprenoid chain alkanes as fossil molecules formed by the degradation of chlorophyll. Whether or not one particular set of environmental conditions was essential for the establishment of living materials is questionable, but in order for their steady evolution rather constant chemical environments are required; this is distinct from the rapid replication of bacterial bodies which could take place in small puddles and isolated systems which contained the essential chemical and physical ingredients for some forms of life.

Modern thermal springs contain abundant microscopic forms of life and temperatures can reach $90°C$, compared to $12°C$ for a large part of the surface of Earth today which supports a great variety of life forms. In high temperature environments, interest centres about the stability and efficiency of biological processes with particular reference to the stability of proteins

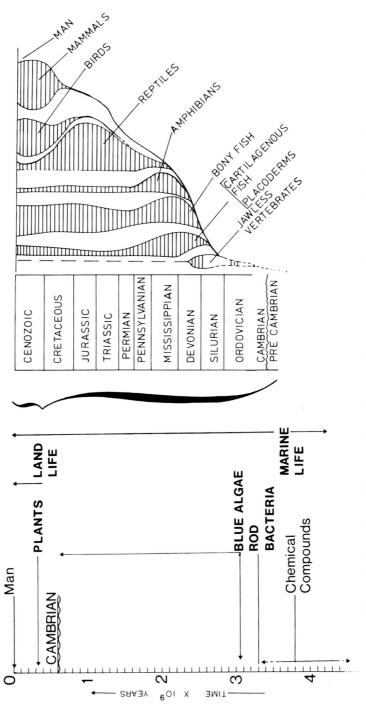

Figure 4: Organic evolution— relationships in time of organic evolution and chordate evolution.

and enzymes. There is considerable interest in the types of enzymes which existed in the primeval environments, as most certainly there was an abundance of suitable elements such as iron, zinc, cobalt, and manganese in reduced forms capable of providing a matrix, through chemical combination with major cations and anions, to form essential electron transfer molecular-systems.

One of the great unknowns is the length of time which was required before recognisable life-forms first formed. There appears to be no problem in the formation of simple forms at a very early stage but no fossil records exist for either structural components or chemical molecules. However, in relation to geologic time man is a very recent addition; the most primitive primates were in existence some seventy-five million years ago and the oldest near-human remains date back about three million years. The agricultural revolution started some ten thousand years ago and man has progressively divorced himself from a close association with nature and is now sustained by technological achievements. The geologic record in relation to chordate evolution is illustrated in Figure 4; a cursory glance of the available details of evolution may suggest that we have a complete record for examination but this is far from the truth. It is estimated (Grant, 15) that today some 4.5 million plants and animals exist while only about 130,000 fossil species have been described and named; this amounts to about 8.7 percent of the number of known living species. Simpson (16) has estimated that the average life span of a single species is 2.75 million years from origin to extinction; Raup and Stanley (17), using this estimate, conclude that the number of species which have lived in the past six hundred million years, since the beginning of the Cambrian Period, is about 982 million of which only about 0.013 of 1 percent have been recognised in the fossil record.

REFERENCES

1. Darwin, C.: *The Origin of Species by Natural Selection.* London, John Murray, 1859.
2. Darwin, C.: *The Expression of the Emotions in Man and Animals.* London, John Murray, 1872.

3. Burbidge, E. M., Burbidge, G. R., Fowler, W. A., and Hoyle, F.: Synthesis of the elements in stars. *Rev Modern Phys, 21*:625, 1957.
4. Wesson, P. and Lerman, A.: Limits on cosmic dust density by observations of the microwave background. *Astronomy and Astrophysics, 53*:383, 1975.
5. Clark, D. H. and Stephenson, F. R.: Frequency of nearby supernovae and climatic and biological catastrophes. *Nature, 265*:318, 1976.
6. Patterson, C. C., Brown, H., Tilton, G. R., and Inghram, M.: The concentration of uranium and lead and the isotopic composition of lead in meteoritic material. *Phys Rev, 92*:1234, 1953.
7. Hamilton, E. I. and Farquhar, R. M.: *Radiometric Dating for Geologists.* London, Wiley, 1968.
8. Tilton, G. R.: Isotopic composition of lead from basalt and ultramafic rocks. *Ann Rept Dir Geophys Lab Carnegie Inst.* Washington, D.C., 1955-56.
9. Nunes, P. D., Tatsumoto, M., and Unruh, D. M.: U-Th-Pb and Rb-Sr systematics of Apollo 17 boulder 7 from the N. Massif of the Taurus-Littrow Valley. *Earth Planetry Sci Lettrs, 23*:445, 1974.
10. Gentry, R. V.: Giant radioactive halos: indicators of unknown radioactivity? *Science, 169*:670, 1970.
11. Alexander, E. C., Lewis, R. S., Reynolds, J. H., and Michel, M. C.: Plutonium-244: confirmation as an extinct radioactivity. *Science, 172*: 837, 1971.
12. Cherdyntsev, V. V. and Mikhailov, V. F.: A primordial transuranium isotope in nature. *Geochemistry, 1*:1, 1963.
13. Lancelot, J. R., Vitrac, A., and Allegre, C. J.: The Oklo natural reactor: age and evolution studies by U-Pb and Rb-Sr systematics. *Earth Planetry Sci Lettrs, 25*:189, 1975.
14. *The Oklo Phenomenon.* Vienna, The International Atomic Energy Agency (Symposium Publication) 1975.
15. Grant, V.: *The Origin of Adaptation.* New York, Columbia U Pr., 1963.
16. Simpson, G. G.: How many species? *Evolution, 6*:342, 1952.
17. Raup, D. M. and Stanley, S. M.: *Principles of Paleontology.* San Francisco, W. H. Freeman, 1971.

Chapter 2

INTERFACES, ELEMENTS, ENVIRONMENT— ELEMENTS, MAN

THE ELEMENTS, present in all living forms, originated from rocks and their by-products; the evolution of living matter took place both on the land and in the seas and it is generally accepted that this medium constitutes the most suitable matrix for continuous evolution and is supported, where available, by the fossil record. In the early stages of evolution it seems inevitable that life forms proliferated in the presence of large amount of elements which today are considered to be toxic to life. It seems reasonable to consider early unicellular forms of life as having the requisite chemical or metabolic systems which discriminated against harmful materials, even though they became limited through adaptation for survival to well-defined chemical environments. While it is possible to describe organic evolution as the development of unicellular to multicellular forms, together with the development of structural and genetic changes, it is also possible to overcome some of the difficulties of this approach by one which requires that individuals with well-defined requirements for existence came together, the marriage offering certain advantages for survival or proliferation in more alien environments. Symbiotic unions could offer distinct advantages for more complex organisms, and it is possible that mitochondria, and similar bodies present in modern cells, once existed as free living organisms equipped with a respiratory machinery; other cells of the human body, such as cilia, may also once have been free living but this is a matter for speculation. By accumulating such small bodies in a larger body, certain essentials for a free living existence would no longer be required and protection against the toxic effect of the chemical elements would be taken over by other specialised organs. Hence unusual elements essential for enzyme processes may owe their presence in man to their selection from primitive environments where they would be readily available for incorporation into developing life forms which required special chemical systems. Certainly essential enzyme elements such as iron and manganese would be rare in oxygenated seas, although they would be present in bottom reduced sediments and available to microorganisms.

In tracing the incorporation of elements present in early crustal materials into living forms, it might be expected that some relict element abundance pattern of rocks could be observed even in an advanced species such as man; indeed to a degree this is true as illustrated in Figure 5, comparing the relative abundances of elements present in crustal rock with those in human blood, which is only one chemical system of the human body.

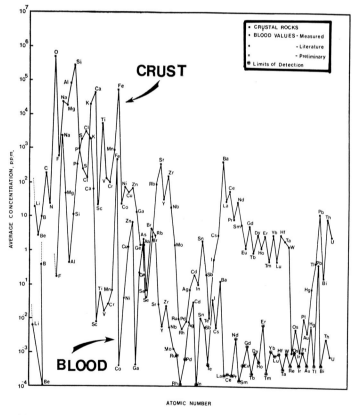

Figure 5A: A comparison between the concentration of the chemical elements in human whole blood and crustal rocks.

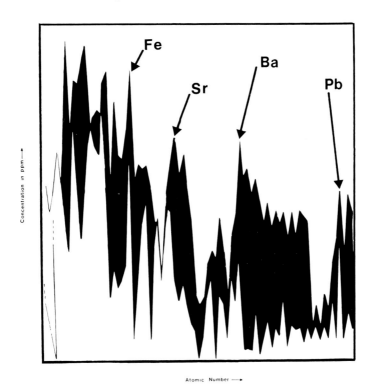

Figure 5B: The area shaded (in black) represents the difference between levels of elements in blood and crustal rocks.

Similarly, the same association is found if we compare the relative abundances of elements present in sea water and human blood as illustrated in Figure 6. The striking feature illustrated in these figures is the similarity between both sets of profiles. In particular the fluctuations of both sets of profiles mirror one another. There are of course exceptions, such as the high abundance of silicon and aluminium in rocks and the high abundance of hydrogen, carbon, and oxygen in blood.

Although perhaps only a small item, illustrations such as this serve to highlight that life cannot be considered as an independent chemical system and one which has developed totally free from the chemistry of the natural environment. It therefore follows that if we add elements or chemicals to today's environment we can expect both adaptation and change to biochemical systems. It should also be noted in these illustrations that the differences in absolute abundance of the elements are large in the case of the rock-blood pair but greatly reduced for seawater-blood, which supports the case for organic evolution from marine environments. These associations, which hold for most of the elements of the Periodic Table, should not be taken as illustrating that man has evolved from rocks, but rather that knowledge of the abundance of the elements in different rocks

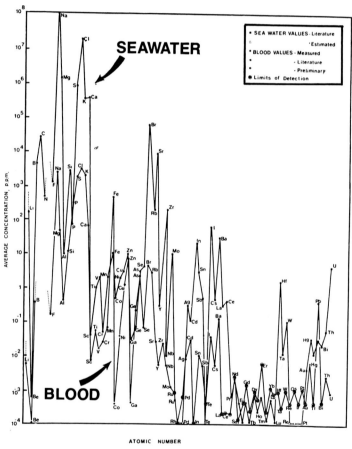

Figure 6A: A comparison between the concentration of the chemical elements in human whole blood and sea water.

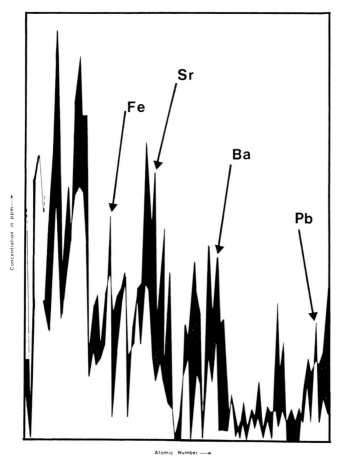

Figure 6B: The area shaded (in black) represents the difference between levels of elements in blood and seawater.

can be useful in understanding levels found in man; biological systems are able to discriminate between similar elements, and also some parts of the body contain high concentrations of some elements which do not show classical characteristic geochemical associations because of the development of very special biochemical pathways.

Associations between the inert and the living can be traced through the element interfaces, namely the soil, air, and vegetation through which elements are made available to man. It therefore follows that the chemical composition of man should, to some degree, reflect that of his environment and the type of foods he consumes; the strongest correlations would be expected in isolated populations which are restricted to defined geographical areas and far removed from modern technology. It is not surprising, as will be shown later, that the chemical composition of man is related to geography and also some forms of disease.

In recent years considerable concern has been expressed for levels of lead found in human tissues arising from contamination of the natural environment; present day normal values undoubtedly contain a proportion of lead derived from the combustion of leaded car fuels. However, as Figure 5 shows, lead is an abundant element in Earth's crust and it is to be expected that it will also be a natural constituent of the human body; it is quite pointless considering a lead-free man. However it is far more important to determine the degree of variability of levels of lead in man residing in different parts of the world as a result of both natural and technological exposures. With the exception of occupational exposure, and those arising as a result of accidents, the generally accepted philosophy of determining mean levels of lead and other elements in man, and establishing maximum permissible intakes or tissue levels, only affords protection to theoretical large populations. The case developed for maximum levels of exposure to ionizing radiation cannot be compared with those for the stable elements. For the former, and accepting a linear dose response relationship, the difference between levels giving rise to effect or no effect is large, for the latter differences can be quite small and are often very dependent upon an individual's susceptibility or threshold to exposure.

With the exception of occupational exposure, the major source of the elements to man is through the food chain; elements may be biologically incorporated into items of food, trapped on external surfaces, or added as a result of manufacturing or preparing foods for consumption; biological availability of the elements to man is often very dependent upon chemical form. The natural pathways through which elements present in the natural environment reach man contain systems which discriminate against a large number of the potentially harmful elements, such as that afforded by root hair systems of plants and G.I. tract transfer in animals. In the case of elements derived from technological processes, the natural protective systems are often bypassed, either as a result of direct entry through processing into food or because of the chemical form of the element which is foreign to natural systems. However, it would be quite wrong to create an impression that exposure to high environment levels of such toxic elements as lead and arsenic, will always be associated with recognised types of morbidity. For example, in southwest England, the county of Cornwall is littered with large and extensive ore tips as a result of past mining activity; waste tips penetrate into towns with locally dense populations. Often the surface of soils consists of a matt of arsenious sulphides together with dispersed grains of galena (PbS), sphalerite (ZnS), and copper ores, but patterns of morbidity attributable to the presence of these minerals, formed of very toxic elements, are not observed. In general, vegetation is prolific on the mine tips and the few areas which are devoid of plant life usually arise because of the development of acid soils as a result of the weathering of sulphide minerals. Within the area there are plenty of potential routes of entry for the elements to man apart from airborne debris: There is a large local production of vegetables and meat which are consumed within the county, milk is often obtained from local farms, while most of the rivers draining the area contain significant levels of a large number of elements.

Chapter 3

AIR

ASSOCIATIONS BETWEEN disease and levels of gaseous compounds in the air we breathe are well known, but data relating levels of the chemical elements present in the human lung with geography as a result of intake of airborne particulates are scant, except for occupational exposure. Throughout the span of a human life it is reasonable to expect that small amounts of dust containing some insoluble materials will slowly accumulate in the human lung. A very limited study was undertaken at the Radiological Protection Service laboratory at Sutton, to determine the overall chemical composition of the human lung for a variety of the chemical elements for individuals who had been exposed to the local urban atmosphere for between fifty to sixty years. The abundance of the chemical elements in these lungs were very similar to that available in the literature for a wide variety of geographical localities; the Sutton data, together with comparative data, are presented in Table III. It would therefore appear that the lung is a remarkably efficient organ in preventing entry of particulate matter. Considerable amounts of material do in fact momentarily enter the lung but are removed through ciliary action to the gastrointestinal tract. Areas of study worthy of consideration concern lung retention of particulate matter in relation to degeneration of the ciliary escalator, such as occurs in chronic bronchitics and asthmatics, also the chemical composition of debris trapped in nasal cavities and the deep lung; clearly entry of particulates into the respiratory system is governed by the different way individuals breathe, that is by nose or mouth.

The mass of particulate matter, water vapour, and various gases present in the air are of considerable importance to man because such constituents absorb incoming solar radiation and hence affect air temperature, a fundamental factor instrumental in major climatic changes. The atmosphere is a relatively stable

TABLE III*
THE ABUNDANCE OF SOME ELEMENTS IN HUMAN LUNG FROM AN URBAN AREA†

(μg/g wet wt.)

Elements	Lung (R.P.S. Values)		No. of Samples	Mean (Anspaugh et al.)	No. of Samples	Range
U	0.001 ±	0.0005	1	ND§		
Th	0.01 ±	0.007	11	ND		
Bi	0.01 ±	0.001	11	<0.05		
Pb	0.4 ±	0.05	11	0.7	141	0.2—1.5
Ba	0.03 ±	0.008	11	0.1	14	0.03—0.4
I	0.07 ±	0.03	11	ND		
Sb	0.06 ±	0.005	11	1.0	3	
Sn	0.8 ±	0.2	11	1.2	1	
Cd	0.48 ±	0.1	4	<0.8		
Ag	0.002 ±	0.0001	11	0.005	1	
Nb	0.02 ±	0.001	11	1.6	2	0.6—2.7
Mo	0.12 ±	0.01	8	ND		
Zr	0.06 ±	0.009	11	3.5	4	1.2—7.5
Y	0.02 ±	0.001	11	ND		
Sr	0.2 ±	0.02	11	0.1	155	0.05—0.5
Rb	3.5 ±	0.4	11	ND		
Br	7.5 ±	0.7	11	ND		
Se	0.1 ±	0.02	11	0.2	4	
As	0.02† ±	0.01	4	ND		
Ga	0.005 ±	0.002	11	ND		
Zn	10.0 ±	0.5	11	15.4	141	9.7—19.8
Cu	1.1 ±	0.1	11	1.4	141	1.0—2.0
Fe	293 ±	47	11	318.7	120	120.9—517.4
Mn	0.08 ±	0.01	11	0.3	141	0.09—0.5
Cr	0.5 ±	0.07	11	0.2	141	0.04—0.5
V	0.1 ±	0.02	11	0.1	1	
Ti	3.7 ±	0.9	11	7.7	1	
Ca	123.0 ±	10.3	11	120.9	119	70.3—175.8
K	2000 ±	110	11	1978	110	1538—2637
Cl	266 ±	48	11	ND		
S	1241 ±	36	11	ND		
P	1000 ±	100	11	1099	111	769.2—1538
Si	43 ±	13	11	ND		
Al	18.2 ±	9.7	11	27.5	141	8.6—43.9
Mg	135.0 ±	28.5	11	100.0	119	68.1—142.9
F	0.04 ±	0.009	11	ND		
Li	0.06 ±	0.01	11	ND		

* From Hamilton, E. I., Minski, M. J. and Cleary, J. J.: The concentration and distribution of some stable elements in healthy human tissues for the United Kingdom. *Sci Total Environ*, 1:354, 1973.

† Methods of analysis: neutron activation, X-ray (crystal) fluorescence, and remainder spark source mass spectrometry.

‡ Data from Anspaugh, L. R., Robison, W. L., Martin, W. H., and Lowe, O. A.: Compilation of published information on elemental concentrations in human organs in both normal and diseased states. University of California, Livermore, Lawrence Livermore Laboratory, UCRL-51013, Pt. I-III, 1971.

§ ND = not determined.

system; solar radiation must be exactly balanced by the heat emitted by infrared radiation. A reduction of only about 2 percent would, according to theory, reduce the mean surface temperature of Earth by 2°C which would be sufficient to produce an ice age. Natural processes, such as volcanic eruptions and dust storms, can eject a considerable mass of particulate material into the atmosphere and together with condensed water provide a veil of dust which acts as an insulating layer around Earth and prevents entry of solar radiation. The consequences of such natural events can be very serious and give rise to dramatic changes of climate which in turn affect the health of man. The magnitude of these natural processes may be judged from the following examples.

In 1883 the Indonesian volcano Krakatoa erupted, the sound of which could be heard some 5,000 km distant, and several cubic kilometers of ash and rock debris were ejected into the stratosphere. The eruption occurred on August 26 and by September 10 the dust cloud had circled the world. Much is to be learned from this event, as more recently it has been shown that the debris of nuclear explosions also take about fourteen days to circle the world (see p. 41). The volcanic eruptions of Surtsey from 1963 to 1965, and Hekla in 1947 in Iceland, produced about 10^8 and 10^9 metric tons of dust while the Mount Agung eruption on Bali in 1963 produced about 10^{10} metric tons of dust (Matthews, Kellogg, and Robinson, 1). No part of the globe is free from the fallout of material injected into the troposphere and stratosphere including the polar ice caps and the deepest parts of the oceans.

Although reliable data are lacking, it has been estimated that apart from natural products, in 1968, anthropogenic sources of atmospheric particles amounted to some 300×10^6 metric tons of material, compared with a natural production rate of small particulate matter of 1300×10^6 tons/year. As a result of the release of natural and man-produced materials into the atmosphere, specific elements become more widely distributed than others; for example, natural volcanic activity can release considerable amounts of mercury and oxides of sulphur, while combustion products of automobile fuels have been responsible for the world-wide distribution of lead.

Changes in weather and climate can have significant effects on human health which has been discussed by Tromp (2, 3). The principal structure in the brain, through which changes in weather and climate affect the body, is the hypothalamus and as a result of meteorological factors changes take place in physiological and pathological processes in the human body. The chemical elements are not directly concerned, but if the general quality of health changes then it is conceivable that normal balances of the chemical elements in the body will be disturbed and so contribute to impaired health.

In order to consider the general composition of air particulate matter, a survey was carried out in 1969 at the Radiological Protection Site, Sutton, United Kingdom, and described by Hamilton (4). Air particulate matter was collected throughout 1969 on 0.45 μm Millipore® filter paper, over periods of sixty-five hours during the weekend in order to reduce local contamination arising from car and pedestrian traffic on the site. The site was on the edge of the London conurbation and situated some two hundred yards from a minor road, surrounded on three sides by tall trees and facing onto extensive areas of grass in which a hospital was situated. Sampling was terminated in December 1969 because new construction work on the site had a profound effect on the levels of elements present in the airborne particulate material. The natural bedrock of the area was chalk overlaid by a thin clay soil. The prevailing wind was from the southwest and local airborne debris is likely to be derived mainly from chalk soils, physical and chemical ablation of building materials, especially clay bricks, effluent from the burning of domestic fuels, and automobile exhaust products. As the site was situated at the southern extremity of the London Basin, fog and smog accumulating in the basin during winter months would often pass over the lip of the basin and engulf the sampling site; under such conditions the air particulate debris became black in colour compared to a pale brown to grey colour common for summer months when the wind came from the southwest across agricultural land. Results for the concentration (ng/m^3) for the elements in air for Sutton are given in Table IV, compared with data obtained for a site at Chilton, Berkshire, reported by Peirson, et al. (5). The Chilton site is situated in a rural area consisting of open glasslands, distant from any major source of industry, while the bedrock and top soil is the same as that found at Sutton. A comparison of both sets of data illustrates remarkable similarities in the levels of many elements; the main differences concern levels of Br and Pb, enhanced at Sutton and possibly derived from locally produced fossil fuels and automobile combustion products. Further data are presented in Table V for the concentrations (μg/g dust) of samples collected at Sutton illustrating changes in the abundance of the elements in dust with different weather conditions. After prolonged periods of rain there is a significant reduction in the total mass of particulates collected and also levels of contained elements. The prevailing southwest winds are maritime in origin but, because of slight changes in wind direction, they can often contain debris from the southern extremities of London. When the wind is from the northeast, London air is carried over the site and is enriched in many elements but with a noticeable reduction in levels of Na and Cl; winds from the east will mainly contain London debris while those from the southwest will have passed over fairly open agricultural country.

Examination of the particle size distribution of the dust showed that about 60 percent of the dust was collected on the first stage of a May Cascade Impactor, implying that the mean particle size at collection was greater than 4 μm; however, detailed examination of the impacted particulate material failed to reveal particles of 4 μm mean diameter and the presence of very fine particulate material indicates disintegration of large particles upon impact. The particulate material consists predominantly of carbonaceous debris, soil debris, chalk particles, various types of organic matter, iron oxides, and both brown and clear spheres of silicate glass, also recorded by Bertine (6), which possibly originate from fly ash. The rapidity with which changes in weather can alter the composition of particulate

TABLE IV*

A COMPARISON BETWEEN THE MEAN CONCENTRATION (ng kg^{-1}) FOR THE MASS OF THE CHEMICAL ELEMENTS IN AIR PARTICULATES FOR THE YEAR 1969 COLLECTED AT THE RADIOLOGICAL PROTECTION SITE, SUTTON, WITH THOSE OBTAINED BY PEIRSON ET AL. (5)

The mean concentration of SO_2 in air for Sutton is for Nov.-April 130 ± 30 $\mu g/m^3$ and for May-Oct. 65 ± 15 $\mu g/m^3$.‡

Element	RPS, Sutton (ng kg^{-1})	Chilton, Berks (ng kg^{-1})	Element	RPS, Sutton (ng kg^{-1})	Chilton, Berks (ng kg^{-1})
U	0.015 ± 0.010	—	Zr	0.53 ± 0.2	—
Th	0.030 ± 0.010	0.06	Y	0.17 ± 0.07	—
Bi	0.20 ± 0.04	—	Sr	1.7 ± 0.6	—
Pb	260 ± 74	98	Rb	2.8 ± 1.0	0.3
Tl	0.048 ± 0.020	—	Br†	180 ± 40	48
Hg†	1.0 ± 0.8	0.15	Se†	0.12 ± 0.04	1.5
Au	0.001 ± 0.001	0.003	As†	6.6 ± 2.1	6.5
W	0.27 ± 0.04	0.3	Ge†	2.2 ± 1.0	—
Lu	0.003 ± 0.001	—	Zn	200 ± 50	194
Yb	0.02 ± 0.01	—	Ga	0.18 ± 0.09	—
Tm	0.003 ± 0.001				
Er	0.02 ± 0.01	—	Cu	26.4 ± 5.0	22
Ho	0.01 ± 0.01				
Dy	0.03 ± 0.02	—	Ni	8.6 ± 2.2	6.4
Tb	0.008 ± 0.003	—	Co	1.1 ± 0.4	0.41
Gd	0.035 ± 0.02	—	Fe	400 ± 70	380
Eu	0.01 ± 0.01	—	Mn	14.3 ± 7.0	21
Sm	0.10 ± 0.02	—	Cr	4.9 ± 1.3	3.6
Nd	0.22 ± 0.10	—	V	32.1 ± 11.0	22
Pr	0.05 ± 0.02	—	Ti	31.9 ± 16	—
Ce	0.48 ± 0.10	0.40	Sc	0.2 ± 0.1	0.10
La	0.23 ± 0.07	0.3			
Ba	4.5 ± 1.0	—	Ca	330 ± 160	550
Cs	0.16 ± 0.06	0.30	K	980 ± 520	—
I†	3.6 ± 0.15	0.1	Cl†	2,800 ± 900	2,040
Te	0.14 ± 0.10	—	S†	2,480 ± 600	—
Sb	1.8 ± 0.6	2.9	P	140 ± 78	—
Sn	1.2 ± 0.9	—	Si	800 ± 300	—
In	0.14 ± 0.02	0.03	Al	370 ± 130	370
Cd	0.4 ± 0.10	< 4	Mg	290 ± 145	—
Ag	5.2 ± 0.20	—	Na	428 ± 214	650
Rh	0.21 ± 0.07	—	F†	1.2 ± 0.5	—
Ru	0.21 ± 0.10	—	B	2.7 ± 1.3	—
Mo	0.38 ± 0.20	—	Be†	3.0 ± 0.2	—
Nb	0.09 ± 0.03	—	Li	1.8 ± 0.5	—

* From Hamilton, E. I.: The chemical elements and human morbidity—water, air and places—a study of natural variability. *Sci Total Environ*, 3:41, 1974.

† Data considered to be low because of incomplete retention on filters; the data are therefore biased towards involatile debris.

‡ Location of sites given in Figure 7.

TABLE V*
THE ABUNDANCE OF SOME ELEMENTS IN AIR PARTICULATES ($\mu g/g$) COLLECTED AT THE RPS SITE, SUTTON, IN RELATION TO RAINFALL AND WIND DIRECTION

Element	Sample Number† 2323	2487	2334	2389	2319	2473	2424
U	0.1	0.01	0.01	0.01	0.1	0.01	0.01
Th	0.2	0.02	0.02	0.02	0.3	0.02	0.02
Bi	1.0	0.4	0.6	0.01	4.0	1.0	0.4
Pb	1,347	476	2,128	29	7,733	2,165	172
Tl	3.0	0.8	2.0	0.07	6.0	2.0	0.7
Hg	10.0	3.0	7.0	1.0	14.0	8.0	4.0
Au	<0.05	<0.05	<0.05	<0.05	<0.1	<0.05	<0.001
Pt	0.001	0.001	0.001	0.001	0.001	0.001	0.001
Ir	0.001	0.001	0.001	0.001	0.001	0.001	0.001
Os	0.001	0.001	0.001	0.001	0.001	0.001	0.001
Re	0.001	0.001	0.001	0.001	0.001	0.001	0.001
W	1.0	1.0	1.0	1.0	3.0	1.0	1.0
Ta	0.09	1.0	1.0	1.0	1.0	1.0	1.0
Hf	0.1	0.1	0.1	0.1	0.1	0.1	0.1
Lu	0.03	nd	nd	nd	0.3	nd	nd
Yb	0.2	nd	nd	nd	0.3	nd	nd
Tm	0.03	nd	nd	nd	0.04	nd	nd
Er	0.2	nd	nd	nd	0.2	nd	nd
Ho	0.1	nd	nd	nd	0.1	nd	nd
Dy	0.3	nd	nd	nd	0.4	nd	nd
Tb	0.05	nd	nd	nd	0.07	nd	nd
Gd	3.0	nd	nd	nd	3.0	3.0	nd
Eu	0.2	nd	nd	nd	0.2	nd	nd
Sm	0.3	nd	nd	nd	0.4	nd	nd
Nd	3.0	nd	nd	nd	3.0	nd	nd
Pr	0.4	0.02	nd	nd	0.5	0.8	0.04

Air

Ce	5.0	0.5	1.0	0.03	5.0	5.0	0.4
La	2.6	0.3	0.6	0.01	3.0	2.0	0.3
Ba	53.0	13.0	14.0	0.8	102.0	114.0	9.0
Cs	12.0	0.9	2.0	0.03	28.0	12.0	2.0
I	2.0	0.3	0.9	1.0	3.0	0.8	0.3
Te	0.2	0.1	0.1	0.1	1.0	0.1	0.1
Sb	26.0	9.0	8.0	0.2	91.0	26.0	3.0
Sn	23.0	16.0	14.0	0.3	157.0	39.0	7.0
In	0.6	0.2	0.2	0.1	4.0	0.7	0.2
Cd	16.0	3.0	6.0	1.0	47.0	8.0	2.0
Ag	2.0	2.0	2.0	0.5	11.0	2.0	0.1
Pd	0.001	0.001	0.001	0.001	0.001	0.001	0.001
Rh	0.001	0.001	0.001	0.001	0.001	0.001	0.001
Ru	0.6	0.3	0.1	0.1	0.6	2.0	0.3
Mo	7.0	3.0	3.0	0.2	8.0	12.0	0.9
Nb	0.9	0.2	0.4	0.1	0.5	2.0	0.1
Zr	5.0	5.0	1.0	0.1	9.0	0.6	0.8
Y	2.0	0.3	0.4	1.0	2.0	0.3	0.2
Sr	49.0	63.0	16.0	1.0	56.0	7.0	13.0
Rb	5.5	351.0	17.0	0.2	251.0	8.0	28.0
Br	510	130	680	7.0	2,100	530	22.0
Se	4.0	9.0	2.0	1.0	8.0	3.0	2.0
As	35.0	67.0	11.0	0.5	80.0	10.0	7.0
Ge	24.0	8.0	2.0	1.0	28.0	4.0	1.0
Ga	32.0	16.0	9.0	0.1	61.0	4.0	3.0
Zn	2,200	698.0	687.0	9.0	2,500	308.0	223.0
Cu	381.0	306.0	301.0	8.0	1,000	67.0	49.0
Ni	101.0	65.0	10.0	0.4	232.0	14.0	21.0
Co	3.0	9.0	2.0	0.06	31.0	1.0	1.0
Fe	4,740	759.0	374.0	103.0	13,600	3,350	2,430
Mn	152.0	326.0	80.0	2.0	382.0	72.0	130.0
Cr	57.0	184.0	90.0	4.0	131.0	81.0	59.0
V	235.0	367.0	74.0	6.0	540.0	81.0	88.0

Element	2323	2487	2334	2389	2319	2473	2424
Ti	299.0	576.0	47.0	4.0	207.0	42.0	31.0
Sc	0.8	0.4	0.1	0.04	0.02	0.6	0.2
Ca	5,710	5,490	2,700	620.0	19,600	2,020	1,460
K	1,080	86,600	8,520	353.0	1,240	1,910	5,520
Cl	8,160	1,820	10,100	120	130	920	95
S	7,030	11,300	5,540	153.0	8,070	2,480	3,590
P	286.0	460.0	452.0	13.0	329.0	203.0	147.0
Si	5,570	17,900	2,630	121.0	6,400	1,180	2,850
Al	2,260	8,020	1,789	54.0	186	1,770	1,643
Mg	1,110	3,570	1,750	483.0	2,550	1,570	1,130
Na	1,060	6,810	6,700	462.0	487.0	1,500	1,090
F	5.0	6.0	28.0	0.2	6.0	4.0	0.4
B	25.0	4.0	10.0	0.3	29.0	4.0	2.0
Be	3	8.0	4.0	2.0	3.0	0.3	6.0
Li	10.0	0.7	1.0	0.2	0.4	8.0	0.05

* From Hamilton, E. I.: The chemical elements and human morbidity—water, air and places—a study of natural variability. *Sci Total Environ*, 3:43-44, 1974.

† Sample No. 2323, March, 1969, sample wt. 45.0 mg, rain 0.16 in, cloud 1, wind SW
Sample No. 2487, June 1969, sample wt. 10.7 mg, rain 0.0 in, cloud 6, wind SW
Sample No. 2334, April, 1969, sample wt. 9.1 mg, rain 0.06 in, cloud 5, wind SW
Sample No. 2389, May, 1969, sample wt. 9.3 mg, rain 0.41 mg, cloud 7, wind SW
Sample No. 2319, February, 1969, sample wt. 32.0 mg, rain 1.06 in, cloud 7, wind NE
Sample No. 2473, June, 1967, sample wt. 19.7 mg, rain 0.08 in, cloud 8, wind E
Sample No. 2424, May, 1969, sample wt. 8.5 mg, rain 0.32 in, cloud 7, wind SE

[See Table IV concerning quantitative collection of some elements.]

phases in the air is amply illustrated by the sudden appearance at Sutton of a thin grey-white deposit on all out-of-door materials on July 1, 1968. The deposit consisted of rounded and drikanter-faceted grains of quartz often encased in a thin ferruginous layer; the particles originated from the Sahara desert, and were carried north by a well-recognised wind system; Stevenson (7) estimated that on this occasion about 10^{12} g of material were deposited over southern England and Wales. Chester and Johnson (8) collected atmospheric dust off the Sahara coast in the Atlantic Ocean and estimated an air concentration of 10 μg dust/m³. Prospero and Carlson (9) found Sahara-type dust at an altitude of between 1.5 and 3.7 km at a concentration of 61 μg dust/m³ compared with a sea level concentration of particulates of 22 μg dust/m³. The authors also calculated that between 25 and 37 $\times$ 10^6 tons of dust are transported through the longitude of Barbados each year, which is sufficient to supply all the needs for pelagic sedimentation across the entire northern equatorial zone of the North Atlantic Ocean; airborne dust collected from ships in the North Atlantic Ocean usually tend to be grey or dark grey in colour because of the presence of industrial carbon and fly ash, illustrating the transport of fossil fuel debris over considerable distances.

Junge (10) has noted that in general terms about 85 percent of the troposphere is filled with a rather uniform aerosol which is supported by the chemical composition of air particulates sampled throughout the world. Rancitelli and Perkins (11) have determined the mass of elements present in air particulates collected at an altitude of 0.3-15 km over the United States; the abundance of Fe, Co, and Sc was compatible with the injection of materials from Earth's natural crust, while levels for Ag, Cr, Sb, and Zn were too high and could have been derived from pollutants, although in the case of Ag an origin related to the artificial seeding of clouds to form rain was a possibility.

Today the tendency is for a decrease in both the amount of deposited dust and the concentration of smoke in the air; this may reflect a cleaner atmosphere but with a tendency to high stack releases this may be an illusion. A better index of cleaner air would be the observation that visibility downwind of cities and industrial complexes is increasing.

THE SIGNIFICANCE OF THE ASSOCIATION OF THE CHEMICAL ELEMENTS IN AIR PARTICULATES

From my training as a geochemist my natural inclination is to consider the extent to which the chemical composition of terrestrial rocks and their products can account for the compositions of air particulates. Such an approach is logical because, prior to the geologically very recent addition of man to the terrestrial ecosystem, particulates present in the air would have been derived from natural constituents of Earth's surface supplemented by acute injections associated with volcanic activity. A simple geochemical classification of the elements was proposed in 1922 by V. M. Goldschmidt, the father of modern geochemistry:

1. Siderophilic (iron associations), e.g. Co, Ni, Mo, Mn
2. Chalcophilic (sulphide associations), e.g. Cu, Zn, Cd, As, Sb, In, Hg
3. Lithophilic (silicate associations bound to oxygen), e.g. Li, Rb, Cs, Ca, Mg, Sr, Al, Sc
4. Atomophilic (occur as gaseous compounds in the atmosphere), e.g. He, Ar, Ne

This type of general classification can also be expressed in relation to the electron structure of the atom. Atomic size is a very important parameter, as it controls the structure of a mineral and also the degree to which atoms of similar size may substitute for one another in crystal lattices. A detailed description of these processes is beyond the scope of this monograph and the reader is referred to standards texts on the subject, for example Ahrens (12), Shaw (13), Mason (14) and McIntire (15).

Data presented in Table VI provide comparisons for characteristic geochemical ratios for elements present in the air at Sutton with other localities and common crustal materials. Geographical localities for the United Kingdom sites are given in Figure 7. Potassium is an abundant crustal element, while rubidium only occurs in trace amounts in most minerals, and rubidium-pure minerals are unknown in nature; the K/Rb ratio for most crustal rocks is about 250 and when significant deviations do occur they point to the operation of very particular geochemical processes; the Rb/Tl ratio of 58 may reflect the enhancement of thallium in the air as a result of selective volatilisation, possibly as a result of the burning of fossil fuels. Germanium is a typical chalcophile element but is very volatile compared to that of silicon, the Si/Ge ratio of 364 found for Sutton air could once more reflect an origin from fossil fuels. In common crustal minerals and rocks, the U/Th ratio shows little change which also applies to the ratio Zr/Hf; most of the Zr and Hf in rocks occurs in the mineral-resistant zircon, which also contains the bulk of U and Th.

An interesting geochemical pair of elements is selenium and sulphur; the ionic radius of S^{6+} is 0.34 Å and Se^{6+} is 0.3-0.4 Å. The Se/S ratio for many of the common crustal rocks is 1×10^{-4}, which may be compared with a value of 0.5×10^{-4} for air collected at Sutton (see Table IV). Glover and Chir (16) have discussed the value of using the Se/S ratio as a means of establishing a value for the maximum allowable concentration of

TABLE VI*

THE ABUNDANCE RATIOS FOR SOME ELEMENTS PRESENT IN AIR AT SUTTON COMPARED WITH THOSE FOR AVERAGE CRUSTAL ROCKS AND SHALES

Element Ratio	Ionic Radius (Å)				Ratio for Air Particulates (Sutton)	Ratio for Average Crustal Rocks	Ratio for Average Shales
Potassium/Rubidium	K,	1.33	Rb,	1.45	350	233	190
Potassium/Cesium			Cs,	1.67	6,125	7,000	5,320
Rubidium/Thallium			Tl,	1.44	58	200	100
Magnesium/Lithium	Mg,	0.65	Li,	0.68	167	1,000	260
Aluminium/Gallium $\times 10^3$	Al,	0.50	Ga,	0.57	2.1	5.5	4.2
Silicon/Germanium	Si,	0.42	Ge,	0.47	364	188,000	45,625
Thorium/Uranium	Th,	1.02	U,	0.97	2.0	3.3	3.2
Zirconium/Hafnium	Zr,	0.79	Hf,	0.78	53	55	57
Calcium/Strontium	Ca,	1.01	Sr,	1.18	194	112	74

* From Hamilton, E. I.: The chemical elements and human morbidity—water, air and places—a study of natural variability. Sci Total Environ, 3:46, 1974.

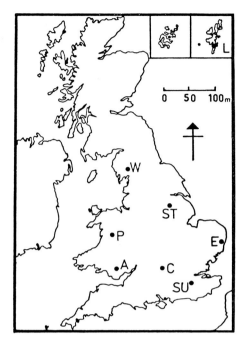

Figure 7: Locations for United Kingdom air particulate collecting sites. L = Lerwick, W = Wraymires, ST = Styrrup, E = Leiston, A = Trebanos, C = Chilton, SU = Sutton.

selenium in urine for industrial workers and rural populations; man excretes about 3 g of sulphate in urine over a period of 24 hours which gives rise to a Se/S ratio of about 0.5×10^{-4}. For industrially exposed workers engaged in grinding selenium metal, urine levels of 0.34 mg Se/l were associated with air levels of 0.33 mg Se/m^3. Other workers engaged in the selenium industry, but not exposed to the metal, had urine levels of 0.06 mg Se/l associated with air levels of 0.04 mg Se/m^3. In some natural seleniferous areas of the world, urine levels as high as 1.12 mg Se/l are quite common. In developing baseline data for the abundance of the chemical elements in man and associations of supposed elevated levels with toxic effects, it is possible, for exposed workers, to predict the natural expected level in relation to defined geographical environments; also for normal members of the population living in areas where high but natural concentrations of the elements occur. Apart from its ubiquitous association with sulphur, selenium also participates in several important biochemical processes of which the detoxification of methyl mercury compounds is important and illustrates the selectivity of some elements in participating in very special biochemical processes; Wood (17) has shown that during this detoxification reaction dimethylselenide is ventilated, indicating that methyl groups are transferred into the less toxic methylselenium cycle.

The rare-earth group of elements offer interesting possibilities as tracers in biochemical processes. The two characteristics of interest are the gradual decrease in ionic radius with increase in atomic number offering many possibilities for studying replacement phenomena in relation to size substitution processes and difference in oxidation state. An example of the coherence between the rare earths in natural materials is illustrated in Table VII for natural rocks and airborne particulates.

Finally data presented in Table VIII provide evidence for the coherence between various groups of elements for air particulates from several localities and common terrestrial rocks.

It is clear from the evidence presented here that the atmosphere provides a medium for the rapid transfer of elements

TABLE VII*

A COMPARISON BETWEEN ELEMENT ABUNDANCE RATIOS FOR VARIOUS RARE-EARTHS FOR ROCK AND AIR PARTICULATES

Source	La/Nd	Pr/Sm	Ce/La	Ce/Eu	Ce/Sm	La/Eu	La/Sm	Eu/Sm	$\dfrac{\Sigma ppm\, La \to Eu}{\Sigma ppm\, Gd \to Lu + Y}$
European shales†	1.1	1.5	2.0	65	13	32	6	0.2	3.2
Russian shales‡	1.1	1.5	2.0	48	10	24	5	0.2	3.0
Russian sandstones‡	1.0	1.1	2.0	47	9	24	5	0.2	3.0
Russian limestones‡	0.9	1.2	1.6	—	5	—	3	—	2.8
Average igneous rocks†	1.1	1.3	1.9	54	11	28	6	0.2	3.1
Sutton air	1.1	1.7	2.1	48	16	23	8	0.3	2.7

* From Hamilton, E. I.: The chemical elements and human morbidity—water, air and places—a study of natural variability. *Sci Total Environ*, 3:46, 1974.

† Data from Herrmann, A. G.: Yttrium and the rare-earths. In Wedepohl, K. H. (Ed.): *Handbook of Geochemistry*. Berlin, Springer-Verlag, II 2, 39, 57-71, 1970.

‡ Data from Ronov, A. B., Balashov, Y. A., and Migdisov, A. A.: Geochemistry of the rare-earths in sedimentary cycles. *Geochemistry International*, 4:1-17, 1967.

TABLE VIII*

ELEMENT RATIOS FOR THE CONCENTRATION OF SELECTED ELEMENTS IN AIR PARTICULATES FOR THE UNITED KINGDOM AND OTHER AREAS COMPARED WITH THOSE FOR VARIOUS COMMON ROCKS, COAL, AND SEAWATER

Ratio	Sutton	Chilton†	Leiston†	Lerwick†	Styrrup†	Waynuirest†	Plynlimon†	Trebanos†	Shales	Sandstones	Limestones	Mean Crustal Rocks	Coal	Seawater
Na/Cl	0.2	0.3	0.5	0.6	0.2	0.5	0.4	0.4	53	330	3	150	2	0.6
Na/Br	16	14	48	191	9	28	36	2	16000	66000	5000	9600	—	160
Br/Cl	0.06	0.02	0.01	0.003	0.02	0.02	0.01	0.02	—	—	—	—	—	0.003
Br/Pb	0.7	0.5	0.3	0.4	0.3	0.2	0.3	0.2	0.2	0.1	0.7	0.2	—	2243333
I/Br	0.02	0.002	0.003	0.006	0.001	0.01	0.009	0.009	2	0.6	5	5	—	0.00095
Al/Sc	1850	3700	4831	4684	4469	2778	4469	1765	6154	25000	4200	886	2000	250
Fe/Sc	2000	3800	6169	5368	5233	5019	5000	2953	3631	98000	3800	2445	2000	850
Co/Sc	6	4	7	5	4	4	6	28	2	0.3	0.1	1	1	100
Zn/Sc	1000	1940	3674	3368	2767	2444	3656	1776	7	16	20	3	10	1250
Mn/Sc	72	210	427	226	327	278	256	118	65	—	1100	43	10	100
Fe/Zn	2	2	2	2	2	2	1	2	497	613	190	800	200	0.7
Fe/Mn	28	18	15	24	16	18	20	25	56	—	4	49	200	8.5
Fe/In	2857	12700	18300	3400	26200	90000	5300	5000	—	—	—	560000	—	—
Cu/Zn	0.1	0.1	0.08	0.02	0.1	0.1	0.2	0.1	0.5	—	0.2	0.8	0.3	0.2
Cu/As	4	3	3	0.4	3	4	9	6	4	—	4	31	3	0.4
Ni/Co	8	16	9	12	14	22	26	11	4	7	200	3	3	17

* From Hamilton, E. I.: The chemical elements and human morbidity—water, air and places—a study of natural variability. *Sci Total Environ*, 3:50, 1974.
† Data from Peirson et al. (5).

contained in particulate materials. Because of the global transport time of about fourteen days for suspended particulates, significant industrial emissions are not only of national concern but also have to be considered in an international context. It is undoubtedly true to say that a person standing at the most westerly part of England, and facing the sea, would be inhaling air particulate matter containing lead derived from Los Angeles which in all probability had passed round the world at least once. In considering matters related to the release of particulate matter into the air, from point and regional sources, it is essential to acknowledge the rapid global transfer of such materials. An example is given in Figure 8 for the global (northern hemisphere) transfer of ^{131}I released from a true point source following the detonation of a nuclear device in the Far East; the iodine 131 cloud passed over Sutton, Surrey, United Kingdom, every fourteen days and was reduced in intensity by diffusion and radioactive decay. A further example is provided by the suggestion that acidic matter, mainly associated with the formation of sulphur dioxide from the burning of fossil fuels, produced in the United Kingdom is transferred across the North Sea to Scandanavia where it falls as acid rain and is considered to be detrimental to fish in freshwater rivers and lakes; nevertheless, it is worth noting that in affected areas, such as southern Norway, natural processes may lead to the production of acid waters enhanced perhaps by man's activities.

In any natural system it is important to consider the extent of natural variability in the levels of the chemical elements; each geographical area will have its own intrinsic types of variability which can be described in changes taking place gradually over tens of years to those which take place within minutes. Today our knowledge of the biological significance of the levels of elements present in the air we breathe is small and it is a matter of some urgency to define levels which cannot reasonably be ascribed to those of a natural system, acknowledging however that not all natural systems are associated with acceptable levels.

In order to consider this, in very crude terms, the abundance of the chemical elements in the air sampled at the Sutton site

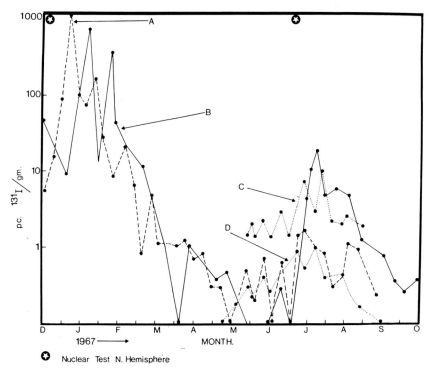

Figure 8: The ^{131}I content of sheep thyroids. Origin of samples: A = Birmingham, United Kingdom; B = Nashville, Tennessee, United States; C = Winchester, United Kingdom; D = Isle of Man, United Kingdom. Data for A and B provided by Professor L. van Middlesworth from unpublished data. In open pastures, sheep readily take up iodine 131 from herbage, which can be detected in the thyroid gland. Considering the geographical distances between sites and the variability in rainfall, all curves show very similar features. The Nashville maximum occurs some two weeks before being recorded in the United Kingdom, the delay reflecting atmospheric transfer rates in the northern hemisphere.

were considered in terms of those expected for the natural state; this analysis was accomplished by taking the abundance of silicon for crustal rocks as unity and comparing the abundance of all the elements to the level of silicon. The silicon-normalised data for air particulates was then subtracted from the normalised crustal rock values; the results of this calculation are given in Table IX and are considered to be valid within a factor of about 10 in order to allow for regional variability. A value of zero denotes no major difference between the relative abundances of the elements in both air particulates and rocks; a value of 10 indicates slight enrichment of the element in air particulates. It should be noted that in Table X relative enrichment factors greater than ×1000 are for volatile elements, while those for factors greater than ×10,000 are for highly volatile elements which could be produced by the combustion of fossil fuels or as a result of high temperature thermal processes in industry. High enrichment factors for chlorine and bromine can reasonably be attributed to fossil fuels together with a marine influence; silver and lead could also be derived from fossil fuels but from other evidence the greater mass of lead appears to be derived from the combustion of leaded automobile fuels. A source for silver is not obvious but this element has often been reported to be elevated in air particulates and it has been assumed that the seeding of clouds provides a convenient source; this is not likely to be applicable to the Sutton site unless of course sufficient silver has been released to give rise to gross global contamination of the air. A well-defined source for antimony is not known but is present in moderate amounts in air particulates from many sites and is possibly derived from the burning of fossil fuels. In order to consider these matters further and to illustrate that levels of elements in the air require considerably

TABLE IX*

The 'RELATIVE ENRICHMENT' (F) FOR THE ABUNDANCE OF THE ELEMENTS IN AIR PARTICULATES SAMPLED AT THE RPS SITE RELATIVE TO SILICON NORMALISED TO THE SILICON CONTENT OF MEAN CRUSTAL ROCKS

F = Mass of elements in air: Si in air./Mass of element in rocks: Si in rocks.

F	Elements
~ 0	U, Th, RE's, Ba, Y, Fe, Cr, Ti
~ × 10	Be, Na, Al, Ca, Sc, Mn, Co, Ga, Rb, Sr, Nb, Cs
~ × 100	Li, B, V, Ni, Cu, Mo, In, Tl
~ × 1000	P, S, K, Zn, Ge, As, Se, Zn, Cd, Sn, I, Hg, Bi
~ × 10000	Cl, Br, Ag, Sb, Pb

* From Hamilton, E. I.: The chemical elements and human morbidity—water, air and places—A study of natural variability. *Sci Total Environ*, 3:60, 1974.

TABLE X*

ESTIMATES FOR THE ABUNDANCE OF SOME OF THE CHEMICAL ELEMENTS IN CAR EXHAUST DEBRIS AND GASEOUS EFFLUENT

Element	Sample Number† 3768	3767A	3767B
Bi	8.0	1.0	—
Pb	58,900	35,100	458.0
W	2.0	—	—
Ba	2,555	2,610	33.0
Cs	0.5	—	—
I	0.5	—	—
Sb	38.0	—	—
Sn	655.0	—	—
In	3.0	—	—
Cd	4.0	—	—
Ag	6.0	—	—
Mo	1,940	33.0	0.5
Nb	4.0	—	—
Zr	7.0	—	—
Y	2.0	—	—
Sr	141.0	—	—
Rb	30.0	—	—
Br	21,000	12,500	160.0
Se	105.0	—	—
As	298.0	—	—
Ge	13.0	—	—
Ga	47.0	—	—
Zn	624.0	107.0	1.4
Co	391.0	—	—
Ni	2,910	33.0	0.4
Fe	34,000	—	—
Mn	7,290	—	—
Cr	2,460	—	—
V	14.0	—	—
Ti	221.0	—	—
Sc	18.0	—	—
Ca	4,100	—	—
K	3,870	880.0	12.0
S	>10,000	>1,700	21.0
P	1,230	—	—
Si	7,990	—	—
Al	3,590	245.0	3.0
Mg	871.0	—	—
Na	Major	—	—
F	8.0	—	—
B	2.0	—	—
Be	3.0	—	—
Li	1.0	—	—

* From Hamilton, E. I.: The chemical elements and human morbidity—water, air and places—a study of natural variability. Sci Total Environ, 3:54, 1974.

† Sample No. 3768: Carbon deposit removed from internal zone of a 1,000 cc car exhaust, 2 cm from the end. Element concentrations, $\mu g/g$ dry deposit.

Sample No. 3767A: Exhaust effluent from car collected over a period of 5 min after the engine had been running for 5 min. Element concentration $\mu g/g$ dry deposit; total weight of deposit 3.2 mg.

Sample No. 3767B: As 3767A but data expressed in terms of $\mu g/m^3$ of air.

Elements Present in the Exhaust Products of Cars

Data are presented in Table X for the concentrations of some elements present in the exhaust products of a 1,000 cc engine run from the cold on 94 octane fuel; the data were obtained by spark source mass spectrometry and are not corrected for relative sensitivity factors (see page 342) which are unknown for this matrix. The data clearly illustrate that high concentrations of many elements are present in car exhaust products which eventually must be released onto roadways; preliminary data for exhaust products emitted from a car after running on the road for one hour indicate the loss of significant amounts of Pb, Sb, Sn, As, Ge, Br, F, and some Be. Much of the mass of the exhaust products will be in the form of large particles which will fall out onto the surface of the road, although some must rapidly become airborne. Material deposited onto the roads is likely to be efficiently removed by rain and through surface run-off find its way to rivers, estuaries, and eventually the sea. With the presence of high concentration of elements on roadways, intake of elements by children by inhalation and finger-sucking could be significant as described by Duggan and Williams (18).

Uranium and Thorium

Industrial processing of uranium and thorium is mainly concerned with the nuclear industry for which working conditions are usually very carefully controlled; however these elements are also used in the pigment and ceramic industries while fossil fuels provide a dispersed source at least for uranium. At the Sutton site a concentration of 0.015 ng U/m^3 air and 0.03 ng Th/m^3 air has been determined over a sampling period of one year. Variability in the abundance of uranium in air, using mean weekly data, are presented in Figure 9, which illustrates a twofold increase in air uranium levels in the winter months which could be related to the burning of fossil fuels. In order to determine the extent to which uranium in air over continental land masses extends over the oceans, the mass of uranium in air was

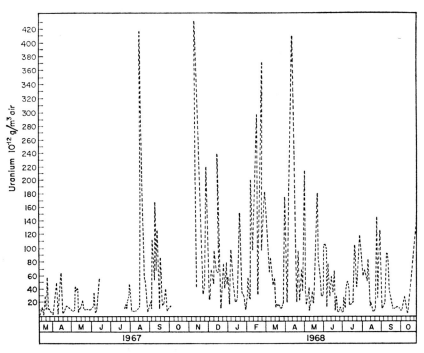

Figure 9A: The daily concentration of uranium in air at Sutton.

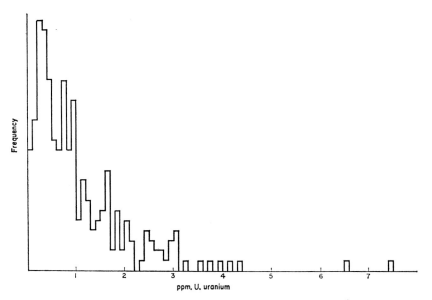

Figure 9B: Frequency distribution for the concentration of uranium in particulate matter in air at Sutton.

determined by Hamilton (19) for a series of stations from Antwerp, Belgium to the Antarctic. A concentration of 0.004 ng U/m^3 was determined for air sampled over the North Atlantic. In the South Atlantic Ocean, a concentration of 0.002 ng U/m^3 air was detected for one sample out of a total of four; for air over the Antarctic pack ice, seventeen samples were analysed and a mean concentration of 0.001 ng U/m^3 was detected for three samples; for air over the land mass of Antarctica, a mean concentration of 0.003 ng U/m^3 was determined for a total of eighteen samples and no uranium was detected in a further two samples. As a result of this study it was concluded that uranium present in the air consists of local debris together with distant debris transported as a result of tropospheric mixing processes; there appeared to be little mixing between the Northern and Southern Hemispheres. An average concentration of between 0.01 and 1.47 ng U/m^3 air has been reported by McEachern, Myers, and White (20) for rural and urban air within New York State, United States, and has been attributed to naturally occurring uranium sources and non-nuclear industrial activity.

Coal

For almost all land and many marine stations at which air particulates are collected, the most obvious visible material is carbon and at least 90 percent of the carbonaceous-organic material by weight has not been identified. In the United Kingdom, in recent years, smokeless forms of fuel have been widely used for domestic heating, while oil has become a popular type of fuel. Domestic chimneys are still the major source of air pollution while the change from steam to diesel and electric locomotives has eliminated a significant source of pollution. In the United Kingdom there has been a long history to reduce air pollution which has shown a significant improvement since the first Clean Air Act became law in 1956. Very little data has been published for the levels of elements in coal but some are provided in Tables XI and XII.

On the assumption that in urban areas domestic hearths are a major source of coal debris, an attempt was made to determine the budget for one element, namely lead, under normal conditions of coal burning. A house was selected which had been newly built; the flue was made of standard clay bricks and high grade anthracite coal was burnt in a closed glass fronted appliance. This type of coal, partially depleted in volatiles, was burnt throughout the winter months for 24 h/day over a period of three years before representative samples were removed for chemical assay. The results of this study showed that about 50 percent of the lead in the coal was retained in the ash and most

TABLE XI

A. THE ABUNDANCE OF THE CHEMICAL ELEMENTS (ESTIMATES FOR AVERAGE COMPOSITION) IN COAL AND COAL ASH FOR THE UNITED KINGDOM

Element	Concentration (ppm)
U	1
Th	4
Bi	5
Pb	16
Hg	<0.1
Ce	20
La	15
Ba	300
Sb	10
Sn	1
Cd	0.2
Ag	2
Mo	3
Nb	5
Zr	<50
Y	15
Sr	280
Rb	160
Zn	120
Cu	50
Ni	<20
Co	9
As	<10
Mn	100
Cr	40
V	40
Mg	2000
B	10
Li	30
Be	3

[Data obtained from bulked samples of coals representative of United Kingdom coalfields.]

B. THE ABUNDANCE OF THE CHEMICAL ELEMENTS IN COAL AND COAL ASH FOR TWO SELECTED SITES IN THE UNITED KINGDOM*

Element	Barnsley† Vitrain Seam (ppm ash wt.)	Uskmouth‡ Pulverised Coal Ash (ppm ash wt.)
Pb	467	300
Te	—	10
Ba	—	100
Sb	150	<10
Sn	110	—
Mo	127	180
Rb	—	100
Zr	300	<30
As	—	200
Ge	633	200
Ga	177	—
Zn	600	300
Cu	867	300
Ni	1333	300
Co	217	—
Mn	1383	400
Cr	617	100
V	633	60
Ti	4667	4600
P	4000	—
B	1133	100
Be	100	30

* From Hamilton, E. I.: The chemical elements and human morbidity—water, air and places—a study of natural variability. *Sci Total Environ* 3:63, 1974.

† Data from Horton, C. and Aubrey, K. V.: The distribution of minor elements in vitrains from the Barnsley Seam. *J Soc Chem Ind*, 69 (Supp issue No. 1):841, 1950.

‡ Data from Smith, A. C.: Determination of trace elements in pulverised fuel ash. *J Appl Chem*, 8A:636, 1958.

of the remainder was trapped in the flue system, hence very little lead was vented to the atmosphere. In urban areas the discharge of cinder waste and soot onto gardens is therefore likely to enhance levels of many elements present in garden soils. Although the abundance of the chemical elements in coal will be related to geochemical characteristics of source materials, for United Kingdom coals, compositions tend to be rather similar and for several elements only slight regional variations occur as illustrated in Table XIII for potassium, thorium, uranium, and lead; one of the controlling factors accounting for variability in many of the pollutant elements will be the differences in the mass of ore sulphide phases ubiquitously present in most coals.

TABLE XII*

CONCENTRATION OF ELEMENTS IN FOSSIL FUELS

Element	Fossil-Fuel Concentration (ppm)	
	Coal	Oil
Li	65	
Be	3	0.0004
B	75	0.002
Na	2,000	2
Mg	2,000	0.1
Al	10,000	0.5
P	500	
S	20,000	3,400
Cl	1,000	
K	1,000	
Ca	10,000	5
Sc	5	0.001
Ti	500	0.1
V	25	50
Cr	10	0.3
Mn	50	0.1
Fe	10,000	2.5
Co	5	0.2
Ni	15	10
Cu	15	0.14
Zn	50	0.25
Ga	7	0.01
Ge	5	0.001
As	5	0.01
Se	3	0.17
Rb	100	
Sr	500	0.1
Y	10	0.001
Mo	5	10
Ag	0.5	0.0001
Cd		0.01
Sn	2	0.01
Ba	500	0.1
La	10	0.005
Ce	11.5	0.01
Pr	2.2	
Nd	4.7	
Sm	1.6	
Eu	0.7	
Gd	1.6	
Tb	0.3	
Ho	0.3	
Er	0.6	0.001
Tm	0.1	
Yb	0.5	
Lu	0.07	

Air

TABLE XII *(continued)*

Re	0.05	
Hg	0.012	10
Pb	25	0.3
Bi	5.5	
U	1.0	0.001

* From Bertine, K. K. and Goldberg, E. D.: Fossil fuel combustion and the major sedimentary cycle. *Science, 173*:233, 1971. Copyright 1971 by the American Association for the Advancement of Science.

Lead

Air-borne lead as an environmental toxin has recently been reviewed by Hicks (21). In spite of a considerable body of data for the mass of lead in air from a large number of sites scattered throughout the world, there is no consensus of opinion for the level in air which is detrimental to health. For many years there has been considerable disagreement between those who

TABLE XIII*

THE CONCENTRATION OF POTASSIUM, THORIUM, URANIUM AND LEAD IN REPRESENTATIVE COALS USED IN ELECTRICITY GENERATING POWER STATIONS IN THE UNITED KINGDOM

Area	Number of Samples	$K(\%)(1)$	$Th\,(ppm)(1)$	$U\,(ppm)(2)$	$Pb\,(ppm)(3)$
N. Scotland	10	0.5 ± 0.3	4.3 ± 3.7	1.3 ± 0.8	13.9 ± 5.1
S. Scotland	15	0.2 ± 0.1	4.5 ± 4.2	1.6 ± 0.4	22.3 ± 10.3
N. Durham	6	0.8 ± 0.8	4.3 ± 4.1	1.8 ± 0.5	19.7 ± 6.2
S. Durham	8	0.4 ± 0.1	2.6 ± 0.7	1.4 ± 0.2	18.5 ± 5.4
Northumberland	4	0.4 ± 0.2	4.3 ± 1.3	2.3 ± 0.5	20.0 ± 2.0
N. Yorkshire	25	0.7 ± 0.2	4.7 ± 1.7	1.6 ± 0.3	10.8 ± 3.0
S. Yorkshire	9	0.4 ± 0.1	2.2 ± 1.3	1.2 ± 0.3	14.8 ± 2.0
Barnsley	6	0.5 ± 0.2	4.0 ± 1.2	1.5 ± 0.2	7.7 ± 4.5
Staffordshire	8	—	—	—	19.0 ± 8
S. Nottinghamshire	12	0.5 ± 0.2	4.1 ± 1.8	1.6 ± 0.7	22.8 ± 7.7
E. Midlands	6	0.3 ± 0.2	2.4 ± 0.7	0.9 ± 0.2	12.8 ± 2.6
Kent	3	0.8 ± 0.8	2.9 ± 0.9	1.4 ± 0.4	11.7 ± 2.3
E. Wales	5	0.5 ± 0.3	4.4 ± 2.4	1.1 ± 0.4	6.8 ± 2.2
W. Wales	3	0.5 ± 0.5	0.6 ± 1.0	1.6 ± 0.9	17.0 ± 4.6
Average	120	0.4 ± 0.3	4.3 ± 3.7	1.4 ± 0.5	15.8 ± 7.4

* From Hamilton, E. I.: The chemical elements and human morbidity—water, air and places—a study of natural variability. *Sci Total Environ, 3*:63, 1974.

† Methods of assay: 1 = γ-ray spectrometry; 2 = delayed neutron counting; 3 = X-ray fluorescence (crystal).

TABLE XIV*
THE CONCENTRATION OF LEAD IN AIR PARTICULATES COLLECTED AT SUTTON BY ELECTROSTATIC PRECIPITATION

Sample No.	Period Collected	Wt. of Dust Particulates (g)	Total Vol. Air Sample m^3	μg Particulates/ m^3 Air	Rainfall (in.)	μg Pb/g Dust Particulates	μg Pb/m^3 Air
NEP. 20A	1961 Dec. 4-11	5.36	13.0×10^4	41.2	0.88	4,076	0.17
NEP. 24A	1962 Jan. 3-9	4.96	11.0×10^4	45.1	1.67	5,895	0.27
NEP. 30A	1962 Feb. 14-20	4.92	11.3×10^4	43.5	0.20	7,304	0.32
NEP. 40A	1962 Apr. 30-May 7	5.16	13.2×10^4	39.1	0.00	5,150	0.20
NEP. 46A	Bulked aliquots	—	—	—	0-1.0	7,420	0.16
NEP. 50A	June-August						
NEP. 54A							
NEP. 78A	1963 Jan. 21-28	10.82	13.4×10^4	80.7	n.r.	3,829	0.31
NEP. 82A	1963 Feb. 18-25	8.86	12.8×10^4	69.2	n.r.	4,962	0.34
NEP. 86A	1963 Mch. 18-25	4.18	12.9×10^4	32.4	n.r.	5,919	0.19
NEP. 94A	1963 Mch. 27-Jne. 4	9.96	10.7×10^4	93.1	0.40	7,213	0.41
NEP. 101A	1963 July 15-29	7.68	19.4×10^4	40.4	0.28	5,364	0.22
NEP. 103A	1963 Aug. 7-12	1.54	9.3×10^4	16.6	0.00	9,481	0.27
NEP. 104A	1963 Aug. 13-28	3.24	28.1×10^4	11.5	2.35	7,872	0.09
NEP. 106A	1963 Sept. 4-12	4.98	14.4×10^4	34.6	2.10	10,618	0.37
NEP. 107A	1963 Sept. 13-20	9.24	13.6×10^4	67.9	0.10	7,998	0.54
NEP. 108A	1963 Sept. 20-30	8.00	18.3×10^4	43.7	1.10	7,184	0.31
NEP. 110A	1963 Oct. 7-14	5.44	13.5×10^4	40.3	0.90	9,954	0.40
NEP. 113A	1963 Nov. 4-18	8.11	25.8×10^4	31.7	3.00	7,372	0.23
NEP. 115A	1963 Nov. 25-Dec. 3	5.78	15.4×10^4	37.5	0.80	8,702	0.33
NEP. 116A	1963 Dec. 3-9	9.76	11.1×10^4	87.9	0.00	5,743	0.51
NEP. 120A	1964 Jan. 27-Feb. 10	16.22	26.1×10^4	61.2	0.00	10,572	0.68
Mean				46.0 ± 4.4		$7,131 \pm 454$	0.32 ± 0.03

* From Hamilton, E. I.: The chemical elements and human morbidity—water, air and places—a study of natural variability. *Sci Total Environ, 3*:64, 1974.
Millipore 0.45 μm filtration. Jan.-June 1969, $2,400 \pm 710$ μg Pb/g particulates, 0.40 ± 0.14 μg Pb/m^3 air.

have a vested interest in the lead industry and those who regard lead as a very toxic element which should be removed from the environment. The geochemical approach requires that lead be considered as a natural constituent of the human body. In the United States, the federal government considers that exposure to a concentration of 2 μg Pb/m^3 air, averaged over a period of three months, will begin to become associated with a risk at the physiological level and is potentially hazardous to health (22). In the United Kingdom, the Secretary of State for the Environment stated that no evidence was forthcoming to indicate that present lead-air levels were a hazard to health (Walker, 23).

Two separate studies by Hamilton (4) report levels of lead from the Sutton site, presented in Table XIV. Total air particulates were collected by electrostatic precipitation for December 1961, January-May 1962, and January-February 1964; those for 1969 were collected by 0.45 Millipore filtration. The apparent concentration of lead in the particulates is different for the two methods of sampling but the concentrations in the air are similar. Hunt, Pinkerton, McNulty, and Creason (24) determined lead levels in air particulates collected from seventy-seven mid-Western cities in the United States and found 1636 ppm Pb for residential, 1512 ppm Pb for industrial, and 1636 ppm Pb for commercial areas. Recently Archer and Barratt (25) determined levels of 932-2350 ppm Pb for mean levels of lead in roadside dust samples for 1972-1975, 1515-2780 ppm Pb for urban dust, and 990-1270 ppm Pb for a complex motorway interchange; all studies were made at Birmingham, United Kingdom, and the levels reflect general values commonly found for the United Kingdom in 1976.

In the United Kingdom it is very difficult to predict past or future levels of lead in air from data obtained today because of the decrease in the use of coal and the increase in car traffic as illustrated in Figure 10 for the annual sales of coal and petrol in the United Kingdom between 1948 and 1970. Over the past twenty years I have been concerned with a number of studies attempting to associate levels of lead in the environment with definable forms of human morbidity. While it is dangerous to

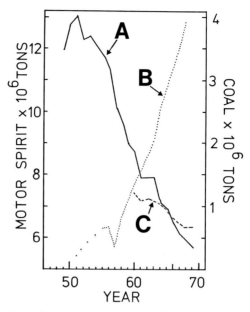

Figure 10: Trends in the consumption of coal and motor spirit for the United Kingdom for the period 1950 to 1969. A = Domestic coal London region, B = Motor spirit, C = Industrial coal London region.

make general comments on present evidence, including that obtained from the vicinity of smelters, there appears to be very little evidence of harm to adult man; the most refined techniques of neurological response often contradict one another when comparative studies are made. I certainly would express concern for exposure of the very young to lead common to many environments today, but would reserve judgement for adult populations until more sensitive techniques of clinical diagnosis for detrimental changes are forthcoming.

The Domestic Home

While considerable effort is expended on determining levels of the chemical elements in external natural environments, very few studies have been made within homes; for many individuals approximately half their life is spent indoors, and levels of airborne dust can be quite considerable. Thomson, Hensel, and Kats (26), Takeuchi, Okamoto, and Emoto (27), and Benarie,

Chuong, and Nonat (28) have described differences between levels of elements present for internal and external environments.

With the present state of the art it is not possible to review the importance of dust present within households or comment upon the significance of the levels of the chemical elements. However, brief comment is made concerning some chemical and physical characteristics of dust common to a household in a rural environment. Data are presented in Table XV for the abundances of some elements in household dust; the sample contained 34 percent of carbon and 5 percent of nitrogen, giving a C/N ratio of 6. The inorganic phase of the dust consisted mainly of very fine mineral debris and about 0.5 percent of carbonate. The levels of zinc and arsenic are high compared with those found for an urban area presented in Table XVI, which is somewhat surprising considering that the house from which

TABLE XV

CONCENTRATIONS OF VARIOUS ELEMENTS IN RURAL DUST FROM A DOMESTIC HOME (n = 4)

Element	Concentration (Dry wt.) in Rural House Dust
Pb	3788 ppm
Sr	252 ppm
Rb	123 ppm
As	141 ppm
Zn	43582 ppm
Cu	857 ppm
Fe	3.87%
Ni	136 ppm
Mn	840 ppm
Cr	66 ppm
Ti	449 ppm
Ca	3.86%
K	2.13%
Cl	1401 ppm
S	<100 ppm
P	2978 ppm
Si	11.03%
Al	3.23%
Mg	2900 ppm
Na	0.99%
C	34.1%
N	4.9%

TABLE XVI*

RELATIVE CONCENTRATIONS (ppm dry wt) FOR SOME ELEMENTS IN HOUSEHOLD DUST FROM AN URBAN SITE TOGETHER WITH DATA FOR PARTICULATE MATERIAL REPRESENTING EXTREMES OF METEOROLOGICAL CONDITIONS

Element	Household Dust 547	Household Dust 546	Urban Air Particulates† 2389	Urban Air Particulates† 2319
U	0.3	1.0	0.01	0.1
Th	1.4	3.0	0.02	0.3
Bi	2.5	8.2	0.01	4.0
Pb	4770	3900	29	7730
Tl	0.6	0.4	0.07	6.0
Au	2.4	2.6	1.0	4.0
W	3.9	2.3	0.001	3.0
Hf	0.8	26.0	0.1	0.1
Yb	5.0	10.0	—	0.3
Er	4.3	6.0	—	0.2
Ho	1.0	2.0	—	0.1
Dy	11.9	26.6	—	0.4
Tb	1.0	6.0	—	0.07
Gd	17.5	68.0	—	3.0
Eu	11.6	13.0	—	0.2
Sm	6.1	13.0	—	0.4
Nd	33.2	108.0	—	3.0
Pr	5.1	36.0	—	0.5
Ce	28.8	125.0	0.03	5.0
La	17.0	55.0	0.01	3.0
Ba	7020	5030	0.8	102
Cs	24.0	52.0	0.03	28.0
I	5	5	1	3
Te	5	5	0.1	1.0
Sb	172.8	112.0	0.2	91.0
Sn	900.0	390.0	0.3	157
Cd	32.0	70.0	1.0	47.0
Ag	36.5	161.0	0.5	11.0
Mo	2	5.0	0.2	8.0
Nb	5.6	7.0	0.1	0.5
Zr	32.2	698.0	0.1	9.0
Sr	602.0	418.0	1.0	56.0
Br	10	10	7	2100
Se	10	20	1.0	8.0
As	20	50	0.5	80.0
Ge	~100	~100	1.0	28.0
Ga	6.4	14.0	0.1	61.0
Zn	4280	6160	9.0	2500
Cu	748.0	809.0	8.0	1090
Ni	1330	2870	0.4	232.0
Fe	31100	33600	103.0	13600
Mn	999.0	1430.0	2.0	382.0
Cr	748.0	488.0	4.0	131.0
V	185.0	134.0	6.0	540.0
Ti	3140	3400	4.0	207.0

TABLE XVI (continued)

Sc	163.0	1170	0.04	0.02
Ca	74800	80900	620.0	19600
K	21300	23000	353.0	1240
Cl	~500	~500	120	130
S	13800	29900	153	8070
P	1180	2930	13.0	329.0
Si	36500	15700	121.0	6400
Al	4920	7080	54.0	186.0
Mg	7280	15700	483.0	2550
Na	2890	9400	462.0	487.0
F	50	100	0.2	6.0
B	49.0	350.0	0.3	29.0
Be	0.3	0.4	2.0	3.0

* From Hamilton, E. I.: Review of the chemical elements and environmental chemistry—strategies and tactics. *Sci Total Environ*, 5:44, 1976.
† Data for extreme meteorological conditions.

the dust was collected is very remote from any urban conurbation or industrial site and is set amid wooded countryside which receives a high rainfall and is noted for its clean uncontaminated air. The dust samples, pale fawn-grey in colour, represent material removed from a room some thirty by eleven feet used as a general lounge and inhabited mainly in the evenings; dust was sampled from the vanes of a specially prepared contamination-free unit used to circulate warm air in the winter and cool air during the summer months and represented material collected over a period of five years. Many common materials in the house can contain very high concentrations of some elements which through processes of general wear become abraded and are distributed throughout rooms. During winter months a coal fire was used but the room was not associated with any significant carbon deposits, hence coal debris are not likely to provide a source for the observed levels of elements unless volatilised species were able to diffuse across the natural draught supply of the fire. If the dust had been mainly derived from coal then it would have had a characteristic coal C/N ratio of about 50. Further, an external origin from air distributed debris from an area of natural mineralisation, some three miles distant, is not considered to be a likely source for most of the elements including lead; however, it is possible that some of the arsenic may be derived from such sources, as this element occurs at high concentrations in algae growing on the bark of local trees facing the direction of the prevailing southwesterly winds.

In order to consider the nature of the dust further, it was subjected to examination by scanning electron microscopy (see page 350) as illustrated in Figure 11. Although it is difficult to identify materials present in household dust with any certainty, Figure 11 does illustrate the diverse nature of the material; most of the white material consists of organic debris and the grey material is inorganic debris, illustrating the physical nature of the material that the human respiratory system has to handle. A preliminary examination of nasal wipes for inhabitants of the household showed material of a similar type. An illustration of the high resolution obtained with this type of instrumentation is provided in Figure 12 for two types of water-soluble material

Figure 11: Scanning electron microscope, electron backscatter picture, illustrating the morphology of dust present in a domestic home. Magnification: ×3000.

believed to consist of sodium sulphate. One small area of the dust was analysed for the distribution of a few elements using back scatter electron imagery: Figure 13 is the normal electron picture of the dust, while Figure 14 represents the distribution of iron, which tends to be rather evenly distributed although discreet areas containing concentrations higher than the general background level are apparent. Figures 15 and 16 show the distribution of silicon and aluminium which describe the distribution of common rock-forming minerals present in the sample; Figure 17 illustrates the distribution of sulphur; Figure 21 depicts copper and Figure 18 lead, neither of which is noticeably associated with major point sources or with the silicate minerals; Figure 19 illustrates the distribution of arsenic which does contain some noticeable point sources but also a disseminated distribution, which is also a feature for zinc in Figure 20 and copper in Figure 21. In the future, more attention should be paid to the abundance of the chemical elements in household dusts, with particular reference to the nature of the materials containing specific elements, and in conjunction with analysis of nasal swabs and lung debris, such as phlegm, this would provide some information concerning the chemical nature of material which enters the respiratory system which may then be compared with the chemical composition of lung tissue from different environments.

Although a wide variety of analytical methods are available for the analysis of air particulate matter, problems of human health require the use of very sophisticated methods of sampling. Although a discussion of the topic is beyond the scope of this monograph, an excellent account of the assessment of airborne particles has been presented by Mercer, Morrow, and Stöber (29). Today there exists a complex global network of air monitoring stations to promote the international exchange of information on levels and trends of air pollution and the use of uniform methods for assessing air quality standards which have been described for air quality in selected urban areas for the years 1973-1974 by the World Health Organisation (30). Long range transport of airborne debris and the problem of trans-

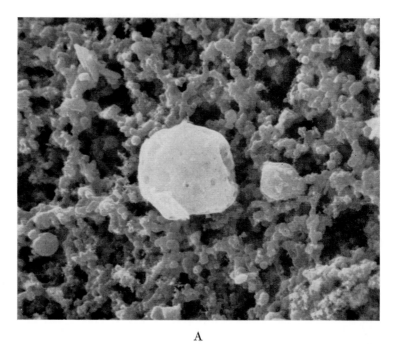

A

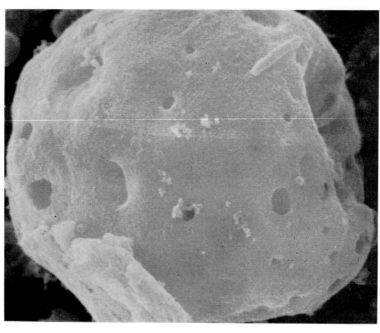

B

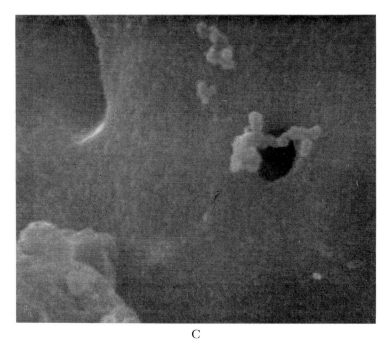

C

Figure 12: Electron micrograph studies illustrating the form of particulate materials in the air collected on Millipore filter paper at Sutton, United Kingdom. All photographs were taken with a JEOL JXA microprobe analyser at 25kv. These illustrate the resolution obtainable for examining surface features of particulate matter. The white body is a particle of ammonium sulphate. Magnification: $A = \times 3000$, $B = \times 10000$, and $C = \times 30000$.

Figure 13: Electron micrograph of air particulates, $\times 1000$.

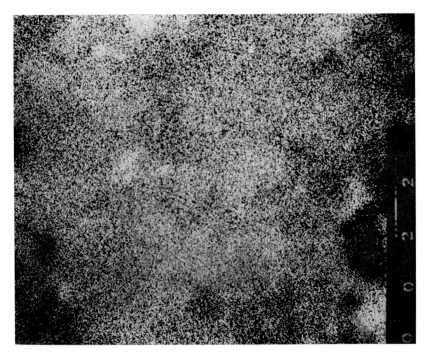

Figure 14: The distribution of iron, Fe K α, in Figure 13, ×1000.

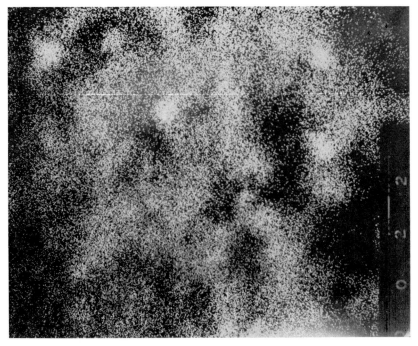

Figure 15: The distribution of silicon, Si K α, in Figure 13, ×1000.

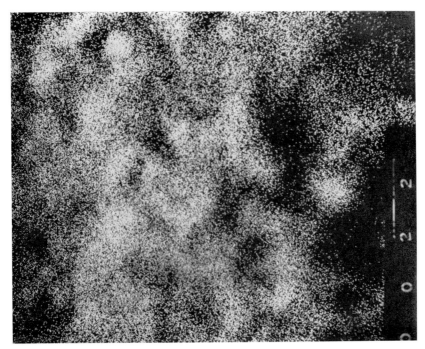

Figure 16: The distribution of aluminium, Al K α, in Figure 13, ×1000.

Figure 17: The distribution of sulphur, S K α, in Figure 13, ×1000.

Figure 18: The distribution of lead, Pb M α, in Figure 13, ×1000.

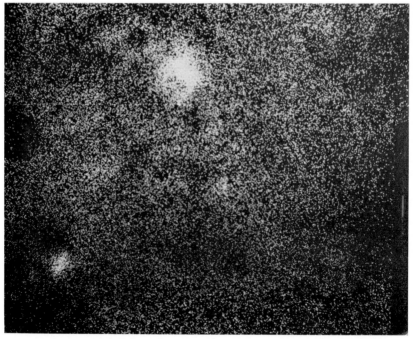

Figure 19: The distribution of arsenic, As L α, in Figure 13, ×1000.

Figure 20: The distribution of zinc, Zn K α, in Figure 13, ×1000.

Figure 21: The distribution of copper, Cu K α, in Figure 13, ×1000

frontier pollution is illustrated for southern Norway and is described by Braekke (31) for acid rain and Lund et al. (32) for organic micropollutants in precipitation.

REFERENCES

1. Matthews, W. H., Kellogg, W. W., and Robinson, G. D. (Eds.): *Man's Impact on the Climate*. Cambridge, Mass Inst Tech, 1971.
2. Tromp, S. W.: The relationship of weather and climate to health and disease. In Melvyn Howe, G. and Lorains, J. A. (Eds.): *Environmental Medicine*. London, William Heinemann Medical Books Ltd., 1973.
3. Tromp, S. W.: *Medical Biometeorology*. Amsterdam, Elsevier, 1963.
4. Hamilton, E. I.: The chemical elements and human morbidity—water, air and places—a study of natural variability. *Sci Total Environ*, 3: 3, 1974.
5. Peirson, D. H., Cawse, P. A., Salmon, L., and Cambray, R. S.: Trace elements in the atmospheric environment. *Nature, 241*:252, 1973.
6. Bertine, K. K. and Goldberg, E. D.: Fossil fuel combustion and the major sedimentary cycle. *Science, 173*:233, 1971.
7. Stevenson, C. M.: The dust storm and severe storms of 1 July 1968. *Weather, 24*:126, 1969.
8. Chester, R. and Johnson, L. R.: Atmospheric dusts collected off the West African Coast. *Nature, 229*:105, 1971.
9. Prospero, J. M. and Carlson, T. N.: Vertical and areal distribution of Saharan dust over the Western North Atlantic Ocean. *J Geophys Res, 77*:5255, 1972.
10. Junge, C. E.: The nature and residence times of tropospheric aerosols. In Matthews, W. B., Kellogg, W. W., and Robinson, G. D. (Eds.): *Man's Impact on the Climate*. Cambridge, Mass Inst Tech, 1971.
11. Rancitelli, L. A. and Perkins, R. W.: Trace element concentrations in the troposphere and lower stratosphere. *J Geophys Res, 75*:3055, 1970.
12. Ahrens, L. H.: *Distribution of the Elements in Our Planet*. New York, McGraw, 1965.
13. Shaw, D. M.: Element distribution laws in geochemistry. *Geochim et Cosmochim Acta, 23*:116, 1961.
14. Mason, B.: *Principles of Geochemistry*, 3rd ed. New York, Wiley, 1966.
15. McIntire, W. L.: Trace element partition coefficients—a review of theory and applications to geology. *Geochim et Cosmochim Acta, 27*: 1209, 1963.
16. Glover, M. A. and Chir, M. B.: Selenium in human urine: a tentative maximum allowable concentration for industrial and rural populations. *Ann Occup Hyg, 10*:3, 1967.

17. Wood, J. M.: Biological cycles for elements in the environment. *Naturwissenschaften, 62*:357, 1975.
18. Duggan, M. J. and Williams, S.: Lead in dust in city streets. *Sci Total Environ, 7*:91, 1977.
19. Hamilton, E. I.: The concentration of uranium in air from contrasted natural environments. *Health Phys, 19*:511, 1970.
20. McEachern, P., Myers, W. G., and White, F. A.: Uranium concentrations in surface air at rural and urban localities within New York State. *Env Sci Tech, 5*:700, 1971.
21. Hicks, R. M.: Air-borne lead as an environmental toxin. A review. *Chem-Biol Interactions, 5*:361, 1972.
22. Anonymous: Health hazards of lead. In Report of the United States Environmental Protection Agency's Bureau of Air Pollution Sciences, Washington, D. C., 1972.
23. Walker, P.: *Secretary of State of the Environment: A Statement.* Hansard, 26th January 1972, Column 433.
24. Hunt, W. F., Pinkerton, C., McNulty, O., and Creason, J. P: *Trace Substances in Environmental Health,* vol IV. Columbia, Missouri, U Missouri, 1971.
25. Archer, A. and Barratt, R. S.: Lead levels in Birmingham dust. *Sci Total Environ, 6*:275, 1976.
26. Thomson, C. R., Hensel, E. G., and Kats, G.: Outdoor-indoor levels of six air pollutants. *J Air Pollut Contr Assoc, 23*:881, 1973.
27. Takeuchi, K., Okamoto, S., and Emoto, Y.: The seasonal variation of metallic contents in suspended particulate matter in indoor air. *Jap Air Cleaning Assoc, 49*:33, 1974.
28. Benarie, M., Chuong, B. T., and Nonat, A.: Pollution particulaire à l'intérieur et à l'extérieur d'un local. *Sci Total Environ, 7*:283, 1977.
29. Mercer, T. T., Morrow, P. E., and Stöber, W. (Eds.): *Assessment of Airborne Particles.* Springfield, Thomas, 1972.
30. *Air Quality in Selected Urban Areas 1973-1974.* WHO Offset Publication No. 30. (ISBN 92-4-170030), 1976.
31. Braekke, F. H. (Ed.): Impact of acid precipitation on forest and freshwater ecosystems in Norway. Research Report SNSF 6/76. NISK, 1432 Ås-NLH, Oslo, 1976.
32. Lunde, G., Gether, J., Gjos, N., and Lande, M-B. S.: Organic micropollutants in precipitation in Norway. Research Report SNSF 9/76. NISK, 1432 Ås-NLH, Oslo, 1976.

Chapter 4

WATER

WATER IS AN ESSSENTIAL constituent of living matter; in spite of the apparent simplicity of its chemical formula, it is a very complex substance. Water is an excellent solvent for many elements in a variety of chemical forms, some of which exist in the true dissolved state while others are present in various forms of complexes. Possibly the greatest problems in the analysis of water are the separation of soluble and insoluble constituents and an identification of the chemical form of an element to be determined. In living systems some form of real or quasi-equilibrium exists for complexed ions, and in many analyses it can be difficult to retain original chemical forms which are subject to rapid change as a result of changes in pH during chemical processing. The efficiency of gut transfer of the chemical elements is very dependent upon chemical form; in practice it is necessary to consider the acidic medium present in the human stomach, which becomes very important when considering the ease of solution of solid particulate phases and the stability of complex ions in relation to times of retention in the gut.

The first analytical problem concerns adequately sampling natural water, followed by the problem of preservation of the sample prior to chemical analysis, which is generally concerned with preventing the breakdown of chemical species, the loss of elements by absorption onto the walls of collection vessels, and the extraneous entry of elements as a result of contamination. An initial stage in the separation of elements present in natural waters is the use of some form of filtration; a commonly used method is to filter the sample through a 0.45 micron pore filter and then to analyse the filtrate; while this may be considered as a practical approach it must be remembered that it is only an arbitrary procedure. If the sample contains an appreciable load of suspended particulates, then the rapidly formed bed of solids on the filtering medium provides an additional filtering aid which can remove solids whose mean diameter is less than 0.45 microns. A problem therefore exists when comparing the level of elements present in samples containing macro and trace amounts of solids.

The chemical quality of water has been implicated in many forms of human morbidity; considerable attention has been paid to the incidence of arteriosclerotic heart disease, which, in many countries, accounts for about 50 percent of all causes of death; the disease is most prevalent in industralised countries but is less frequent in developing countries and almost nonexistent in remote rural areas of such countries. To date, no single or group of elements have been identified as causative agents, but there is sufficient evidence to indicate that there is a water factor beyond that which could reasonably be attributed to statistical chance. Apart from a discussion of a few basic facts, this monograph will not be concerned with the details concerning levels of elements in drinking water and cardiovascular diseases. Suffice it to say that hard water appears to be beneficial and soft water detrimental to health. Most studies which are concerned with particular elements simply repeat overall chemical features associated with the two major types of water which reflect levels of the alkaline earth group of elements, namely calcium and magnesium. Table XVII presents a list of some characteristics of hard and soft waters and, if compared with the

TABLE XVII
SOME CHARACTERISTIC ELEMENTS FOUND IN HARD AND SOFT WATERS

Hard	*Soft*
Barium	Aluminium
Bromine	Arsenic
Boron	Copper
Calcium	Chromium
Fluorine	Iron
Iodine	Lead
Magnesium	Manganese
Nickle	Potassium
Silicon	Rubidium
Silver	Scandium
Strontium	Tin

vast literature on the subject, it can be shown that the "good and bad" elements are clearly defined by the overall degree of water hardness.

In terms of a general overview of the subject it becomes increasingly difficult to identify one element of concern, on the grounds that the total mass of water drunk by individuals, relative to total fluid intake, is very small; most elements are only present in trace qualities and in terms of total human intake the mass is very small compared with that derived from other items of diet. Elements are likely to be more important in the water story for restricted geographical regions where deficiencies or enhanced levels are present and where nutritional disturbances occur. There are also many other rather vague properties of some environments which appear to have a casual relationship with cardiovascular diseases, for example the overall geology of an area and geomorphological history. A general review of trace elements and cardiovascular disease has been presented by Masironi (1), and analytical matters have been described in a technical report published by the International Atomic Energy Agency (2). Before leaving the subject for the present it is worth noting both the data given in Table XVIII for mortality and hardness of water described by Crawford, Gardner, and Morris (3) and the observation by Crawford, Gardner, and Morris (4) that in towns of the United Kingdom where water became harder residents showed a favourable response to cardiovascular disease compared with those towns in which the water became softer; this feature was not observed for non-cardiovascular death rates.

In considering the quality of water and human disease it is possible to identify two major sources of water, namely natural water supplies whose composition is directly related to the geology and the geochemistry of geographical regions (modified by many factors such as climate, weather, and seasons of the year) and domestic waters as supplied from taps to houses. Generalities concerning natural water cannot be made and each area, or source, requires separate consideration; nevertheless, it is relevant to consider the following:

1. surface run-off waters which can be sampled from rivers,

TABLE XVIII*

A. THE CONCENTRATION OF ELEMENTS IN BULKED SAMPLES OF DRINKING WATER IN TOWNS IN BRITAIN, 1962-63

Elements†	9 Towns With Soft Water‡		6 Towns With Hard Water§	
	Mean (μg/l)	Range (μg/l)	Mean (μg/l)	Range (μg/l)
Calcium	8500	7000-10,000	102,000	90,000-114,000
Magnesium	2500	200-3000	9250	450-14,000
Sodium	4500	3000-6000	36,000	17,000-55,000
Manganese	51	43-58	<10	<10
Aluminium	58	30-85	12	<10-19
Boron	15	7-22	123	75-170
Iodide	1	1	5	3-7
Fluoride	40	40	135	90-180
Silica	7600	6600-8500	17,300	16,800-17,800

B. ASSOCIATION BETWEEN COMPONENTS OF WATER HARDNESS (1961) AND DEATH RATES AT AGES 45 TO 64 (1958-64) IN THE COUNTY BOROUGHS OF ENGLAND AND WALES WITH A POPULATION OF 80,000 OR MORE IN 1961

Water Constituents ppm	Correlation With Death-Rates For:					
	Cardiovascular Disease		Bronchitis		All "Other" Causes	
	Male	Female	Male	Female	Male	Female
Total solids	−0.61	−0.58	−0.41	−0.36	−0.23	−0.39
Total hardness:	−0.65	−0.63	−0.45	−0.40	−0.26	−0.40
Temporary hardness (carbonate)	−0.63	−0.62	−0.48	−0.42	−0.29	−0.36
Permanent hardness (non-carbonate)	−0.51	−0.45	−0.24	−0.19	−0.10	−0.38
Electrolytes (as ions):						
Calcium	−0.72	−0.71	−0.55	−0.47	−0.37	−0.44
Magnesium	−0.02	−0.02	+0.06	+0.01	+0.15	−0.06
Sodium	−0.24	−0.28	−0.25	−0.21	−0.11	−0.19

* From Crawford, M. D., Gardner, M. J., and Morris, J. N.: Mortality and hardness of local water supplies. *The Lancet*, 2:827, 1968.

† Analyses by Government Chemist of samples taken at a point in the distribution system.

Only elements showing consistent differences between the soft and hard waters are shown.

‡ Average total hardness, 29 ppm as calcium carbonate.

§ Average total hardness, 310 ppm as calcium carbonate.

streams, lakes, and wells. One of the areas of interest in this field of study concerns interactions with organic debris, such as the humic substances, and a consideration of the presence of biologically active materials such as bacteria and viruses;

2. ground water supplies which accumulate in underground natural aquifers and are available through artesian wells or by pumping. Although there can be considerable variability in the composition of such waters, even from a single aquifer system, their chemistry has been modified as a result of transport through the ground and storage at depth; usually these waters are not subject to rapid changes in chemical composition but once depleted, or contaminated in chemical elements or their compounds, they take a long while to recover.

For man, at least in developed countries, more concern should be directed towards piped domestic water supplies which have not always received the attention they deserve. With the exception of the addition of traps to present the backflow of sewage, the physical method of water distribution has not changed radically for the past 3,400 years. Improvements have been made in the treatment of water and because of the need to conserve water today methods are used whereby waters polluted by sewage and other materials can be rapidly recycled and pronounced fit for human consumption. In many countries, the general public often becomes alarmed if chemicals are added to water supplies in order to improve some aspects of health, for example the addition of fluoride ions to improve the quality of teeth. However it seems probable that many individuals do not realise that almost all domestic supplies of water should really be considered as a manufactured product which has been subjected to considerable technical processing. At most water-undertakings the input water is first filtered, chemicals such as alum are added, and by changing the pH insoluble hydroxides are flocculated followed by their removal through sedimentation, filtration, and finally the addition of chlorine gas to kill off bacteria and viruses; in addition, as local demands decree, there are many other different processes to which particular waters are subjected. The final product is then tested for quality and passed into the general distribution system for household use. The system is not infallible but water-undertakings have an excellent history of providing the public with a safe supply of water. Serious mishaps are rare but slight accidents do occur; for instance I can cite an instance where water supplies to a rural community are supplied by a manual station. Briefly the station deals exclusively with surface run-off water which is purified by coagulation techniques, dosed with chlorine, and then passed into the supply with a residual chlorine level of 1-2 ppm Cl. The chlorine dosing rate is set late at night to provide acceptable residual levels of chlorine before the plant is closed; however, if during the night there is a heavy downpour of rain, organic debris enters the system and the residual chlorine is used up and poorly treated waters can enter the system; when this occurs the domestic water is coloured a faint brown due to the presence of humic substances. No well-defined patterns of morbidity are known to be associated with such episodes but they do tend to occur in the fall and winter when various vague types of stomach upsets are also common.

In order to illustrate the significance of water quality and health a few examples are now given.

Relations Between the Chemical Composition of Water and Human Blood

Results for the chemical composition of water for two households sampled in the morning and evening from both hard and soft water regions are given in Table XIX. The most obvious difference between the two types of water is the greater abundance of the alkaline earths and related elements in the hard water; the abundance of zinc and copper in both supplies could, in part, be due to corrosion from plumbing materials, which have been described by Packham (5) and Obrecht and Pourbaix (6), while the relative enrichment of aluminium in the soft water could be a natural phenomena or derived from chemical

TABLE XIX*

THE RELATIVE ABUNDANCE (μg/l) OF VARIOUS ELEMENTS IN DOMESTIC HARD AND SOFT WATERS FROM S. WALES FOR FIRST (AM) AND EVENING (6 PM) SAMPLES

Element	Hard Water First Flow (AM)	Hard Water Flow (6 PM)	Soft Water First Flow (AM)	Soft Water Flow (6 PM)
U	2	1	<1	<1
Pb	2	2	2	4
Tm	0.008	<0.001	<0.001	<0.001
Dy	0.6	0.5	<0.001	<0.001
Tb	0.2	0.1	<0.001	<0.001
Eu	0.4	0.3	—	—
La	0.6	0.1	—	—
Ba	2,600	1,700	90	67
Cs	0.6	0.1	—	—
I	3	2	3	1
Sb	0.6	0.2	—	—
Sn	0.7	0.7	2	2
Cd	0.4	0.4	2	0.6
Y	0.08	0.04	—	—
Sr	1,440	1,450	74	120
Rb	3	3	3	2
Br	140	71	37	18
Ga	0.5	0.05	0.8	0.2
Zn	200	100	550	51
Cu	910	180	96	23
Ni	10	3	2	3
Fe	83	540	850	210
Mn	3	0.8	39	13
Cr	28	17	43	21
V	2	3	0.2	0.4
Ca	150,000	154,000	9,000	15,000
K	15,800	2,620	4,100	6,720
Cl	20,000	66,000	11,000	11,000
S	3,000	3,000	4,000	3,000
P	1,400	2,800	2,100	700
Si	8,000	8,000	800	160
Al	2,000	700	38,000	31,000
Mg	5,000	5,000	500	500
Na	9,000	7,000	3,000	5,000
B	12	6	0.5	2
Li	10	21	2	0.9

* From Hamilton, E. I.: The chemical elements and human morbidity—water, air and places—a study of natural variability. *Sci Total Environ*, 3:15, 1974.

TABLE XX*

A COMPARISON BETWEEN THE CONCENTRATION OF VARIOUS ELEMENTS IN FRESH HUMAN BLOOD (ppm wet wt) FOR SUBJECTS RESIDING IN BOTH SOFT (HARDNESS, 30-40) AND HARD (HARDNESS, 250) WATER AREAS IN S. WALES, UNITED KINGDOM

Comparative data are also presented for the mass of the elements in average U.K. blood.

Element	Hard Water (n = 38)	Soft Water (n = 38)	U.K. Master Mix (n = 2,500)	U.K. Mean Values (n = 168)
Pb	0.3	0.3	0.3	0.3
Hg	0.0008	0.0007	—	—
Ba	0.07	0.1	0.08	0.1
Cs	0.03	0.03	0.01	0.005
I	0.03	0.03	0.06	0.04
Sb	0.02	0.02	0.002	0.005
Sn	0.02	0.1	0.005	0.009
Cd	0.004	0.005	—	—
Ag	0.009	0.009	0.01	0.008
Mo	0.001	0.001	0.001	—
Nb	0.006	0.006	0.004	0.005
Zr	0.02	0.003	0.01	0.02
Y	0.007	0.007	0.007	0.005
Sr	0.80	0.08	0.02	0.02
Rb†	2.7	2.8	2.8	2.7
Br†	4.3	4.7	4.8	4.6
Se	1.2	0.8	0.08	0.06
As	0.001	0.001	0.001†	0.001†
Ge	< 0.3	< 0.3	—	—
Zn†	7.0	7.0	7.0	7.0
Cu†	1.1	1.1	1.2	1.1
Fe†	497	501	493	491
Mn	0.07	0.04	0.09	0.07
Cr	0.2	0.2	0.07	0.03
Ca†	54	54	60	62
K†	1,900	1,900	1,900	1,860
Cl†	2,700	2,600	3,020	3,000
P†	331	325	335	328
Si	13	12	—	—
Al	0.2	0.2	0.2	0.4
Mg	59	98	49	46
F	0.4	0.8	0.07	0.2
Li	0.006	0.002	0.002	0.006

* From Hamilton, E. I.: The chemical elements and human morbidity—water, air and places—a study of natural variability. *Sci Total Environ*, 8:17, 1974.

† Assay by X-ray fluorescence analysis; remainder by spark source mass spectrometry.

TABLE XXI*
CLINICAL DATA† FOR RESIDENTS IN SOUTH WALES FROM WHICH THE DATA IN TABLE XX WAS OBTAINED

	Hard Water Areas		Soft Water Areas	
Age	46.6 ± 1.46 years	(91)	43.1 ± 1.38 years	(97)
E.C.G.				
QT duration	0.37 ± 0.005 sec	(91)	0.37 ± 0.005 sec	(97)
Extrasystoles	7%	(91)	3%	(97)
R.E.M.P.S.	0.95 ± 0.008	(78)	0.93 ± 0.006	(81)
Cholesterol	224.5 ± 5.38 mg	(73)	220.0 ± 5.74 mg	(70)
B.P. Systolic	139.6 ± 4.64 mm	(72)	134.0 ± 2.14 mm	(67)
Clotting				
O.S.P.T.	12.91 ± 0.14 sec	(23)	12.96 ± 0.45 sec	(23)
Ca C.T.	134.09 ± 5.38 sec	(23)	150.83 ± 5.62 sec	(23)
Thrombofax‡	74.67 ± 3.01	(23)	96.76 ± 3.22 sec	(23)
K.C.C.T.‡	40.67 ± 0.65	(23)	43.35 ± 0.68	(23)

* From Hamilton, E. I.: The chemical elements and human morbidity—water, air and places—a study of natural variability. *Sci Total Environ*, 3:18, 1974.
† Data provided by P. Elwood.
‡ Significant at P = 0.05.

treatment processes. The data illustrate higher levels for Ba, Br, Ga, Zn, Cu, Mn, and Cr for the early morning sample which may reflect the settlement of solids in pipes during the night. The availability of such materials at the tap outlet is very dependent upon the configuration of domestic plumbing systems and water use factors. In some areas lead, as a result of the plumbosolvent properties of some soft waters, can contaminate systems but in others the presence of humic substances in the water can lead to an internal protective coating of pipes. Data are presented in Table XX comparing the abundance of the chemical elements in human blood for the same two hard and soft water areas. No dramatic differences are observed when compared with mean values for blood of United Kingdom residents, except perhaps for levels of calcium, chlorine, and manganese. In a further study, mean clinical data (provided by Dr. P. Elwood) were obtained for the hard and soft water areas in which water and blood had been analysed. The results of this study are given in Table XXI; no significant differences were observed for the two areas in the nine clinical tests.

The Chemical Composition of Kettle Fur as an Index of Water Quality

One of the major obstacles in attempting to study any relationships between domestic water quality and disease is how to obtain an integrated measurement for levels of elements in water over long periods of time; it is of course possible to introduce ion-exchange cartridges associated with flow-meters to taps but for any study to be worthwhile such an exercise would be costly. A preliminary investigation indicated that a partial history of the abundance of elements in water could be obtained, at least for hard water areas, by analysing the deposits found in buried distribution pipes, but these are not easily sampled. Instead a preliminary investigation was carried out on the composition of kettle fur which is easy to obtain and, with some reservations, can be used in both hard and soft water areas. The deposits tend to be white in colour in hard and brown in soft water areas. No attempt was made to investigate the efficiency of precipitation of elements from water or the extent to which deposited materials were leached back into the water, neither was the effect of diffusion in the solid phases considered in any detail. For the area in which this study was carried out, the chemical composition of the kettles was not considered to influence the composition of the deposits as any metal surface was quickly covered with a deposit. The data are presented in Table XXII and can only be considered as representing relative differences as the data were obtained by spark source mass spectrometry and the relevant matrix sensitivity factors (see page 342) for kettle fur are not known. Table XXII also includes data for the composition of blood from the same households as the samples of kettle fur were obtained, but because only two samples were analysed absolute errors are likely to be high. Elements enriched in kettle fur for soft relative to hard water are Pb, Sn, Ta, Zn, Mn, and Cr. In spite of the problems of sampling and analysis, semiquantitative examination of the composition of kettle fur could serve to indicate the presence of an unusual element present in a water supply and in the case of multielement studies the

TABLE XXII*

A COMPARISON BETWEEN THE CONCENTRATION OF THE CHEMICAL ELEMENTS IN BLOOD (ppm wet wt) AND KETTLE FUR (ppm dry wt) FOR BOTH HARD AND SOFT WATER AREAS IN SOUTH WALES, U.K.†

Element	Hard Water Area‡		Soft Water Area§		Kettle Fur Hard Water Area		
	Blood 1,358	Kettle Fur 3,381	Blood 3,161	Kettle Fur 3,380	3,383‖	3,382¶	3,379**
U	—	18	—	5	18	6	—
Th	—	4	—	<1	—	—	—
Bi	—	2	—	<1	—	—	—
Pb	0.1	61	0.09	121	15	60	77
Tl	—	3	—	4	—	—	3
Ta	—	4	—	34	4	41	—
Tm	—	0.03	—	0.1	nd	nd	<1
Er	—	0.4	—	1.0	nd	nd	5
Ho	—	0.03	—	0.8	nd	nd	3
Pr	—	5	—	4	nd	nd	6
Tb	—	0.3	—	0.5	nd	nd	1
Gd	—	0.3	—	0.4	nd	nd	1.7
Eu	—	0.1	—	0.3	nd	nd	0.6
Sm	—	0.3	—	1.5	nd	nd	2
Nd	—	9	—	8	nd	nd	12
Pr	—	1	—	1	nd	nd	<1
Ce	—	3	—	2	nd	0.8	3
La	—	2	—	1	nd	0.7	3
Ba	0.003	5,950	0.02	1,790	7,240	2,920	12,500
Cs	0.01	0.4	0.02	0.4	<1	<1	0.9
I	0.01	20	0.02	4	5	2	4
Sb	0.02	0.9	0.01	2	<1	<1	2
Sn	0.007	8	0.02	314	2	29	3
Cd	0.007	68	0.1	16	7	5	29
Ag	0.003	16	0.008	1	5	1	7
Mo	0.001	<1	0.001	<1	<1	<1	<1
Nb	0.002	0.7	0.003	0.6	0.4	0.5	0.6

Sr	—	0.02	0.003	—	1	0.7	17
Rb	3	4,920	0.12	3,280	24,100	2,670	10,300
Br	—	37	2.4	110	10	3	8
Se	4	123	5	29	6	4	154
As	0.2	73	0.2	49	36	24	51
Ge	0.01	12	0.05	14	9	2	24
Zn	0.07	—	0.2	—	—	—	51
Cu	7.4	182	5.6	4,380	2,679	17,800	15,200
Ni	1.1	9,560	1.3	6,376	2,330	31,200	133
Fe	—	1,020	—	678	2,480	1,660	700
Mn	505.8	Major	477.4	Major	Major	Major	Major
Cr	0.2	3	0.06	170	21	834	36
V	0.08	19	0.05	115	479	156	200
Sc	0.4	2	0.06	1	19	3	5
Ca	0.1	139	0.03	84	35	68	87
K	52.3	Major	57.2	Major	Major	Major	Major
Cl	2,010	Major	1,800	Major	Major	Major	Major
S	2,600	—	3,800	—	—	—	—
Si	1,700	236	1,800	235	287	1,920	1,470
Al	5	2,790	2	560	4,660	4,570	19,500
Mg	0.007	837	0.002	503	209	682	1,750
F	—	6,780	—	678	24,800	3,320	42,600
B	0.05	15	0.2	18	15	—	19
Be	2	2	0.3	0.4	5	1	3
Li	0.003	1	0.001	0.8	0.3	1	3
	0.0006	0.4	0.002	0.5	0.6	0.6	25

* From Hamilton, E. I.: The chemical elements and human morbidity—water, air and places—a study of natural variability. *Sci Total Environ*, 3:18-19, 1974.

† (Chromium plate taps possibly composed of brass; chromium plate on kettles possibly composed of iron.) Assay by spark source mass spectrometry, no relative sensitivity factors applied.

‡ Male blood; kettle fur from same household. Plumbing, mains Fe-Zn; sink, Cu; tap, Cr; Al.

§ Male blood, kettle fur from same household. Plumbing, mains Fe-Zn; sink, Cu; tap, Cr; kettle, Cr plate, brown deposit.

‖ Mains, Fe-Zn; sink, Fe-Zn; tap, Cr; kettle, Cr plate, white deposit.

¶ Mains, Fe-Zn; kettle, Cr plate, white deposit.

** Mains, Fe-Zn; sink, Pb-Cu; tap, Cr; kettle, Cr plate, pink deposit.

The Chemical Composition of Organic Deposits in Household Water Supplies

In hard water areas the calcareous deposits present in pipes and kettles serve to characterise the levels of elements in distribution systems. Problems are usually more severe in soft water areas because of the thin nature of surface deposits. In pursuing this type of investigation it was noted that the header water tank in homes supplied with soft water often contains a brown floclike deposit. In order to study this material further a header tank was lined with a polythene sheet and left for two years, after which the deposit was removed for analysis. The colour and floclike nature of the deposit was similar to that observed for humic substances abundant in the moorland water from which the household supply was obtained. The dried material was carbon rich, contained virtually no carbonate and only trace amounts of nitrogen (less than 110 ppm N). The chemical composition for a few selected elements is given in Table XXIII and XXIV for two different houses, which illustrates a very high iron and lead content. The water and suspended solids sampled in this system are the same as those present in the general mains water drinking supply used by the household. Using iron as an index of water quality, the mains supply was split into two sections, one was left connected to the tap while the other was passed through a filtration system in order to remove particles down to a mean particle size of about 0.45 microns; a comparison of the iron content for both supplies is illustrated in Figure 22; with few exceptions all iron peaks were preceded by heavy rainfall. In a further study a broken mains pipe resulted in the flow of very brown turbid water at the tap; analysis of this material for a few elements is given in Table XXV—while levels of some elements are similar to those found in the header water tank, there are notable differences. In order to probe a little further into the composition of the water floc some preliminary electron-microprobe studies were undertaken. A sample of floc was removed from the tank and, with minimum disturbance,

TABLE XXIII

CONCENTRATION OF SOME ELEMENTS IN HUMIC SUBSTANCES PRESENT AS FLOCS IN A COLD WATER HEAD TANK IN A RURAL DOMESTIC HOUSE*

Element	Concentration (Dry wt)
C†	7.2%
N†	1.1%
Mg	0.52%
Al	3.68%
Si	5.12%
P	2,280 ppm
S	1,673 ppm
Cl	51 ppm
K	2,800 ppm
Ca	2.39%
Ti	1,631 ppm
Mn	5.88%
Fe	21.02%
Ni	86 pm
Cu	1,630 ppm
Zn	85 ppm
Br	211 ppm
Rb	28 ppm
Sr	37 ppm
Pb	1.12%

* Data obtained by inductively coupled plasma spectrometry by Applied Research Laboratories, United Kingdom.
† Combustion/gas analysis.

freeze dried; an element distribution analysis of the sample is illustrated in Figure 23. The floc consists mainly of fibrous materials impregnated by discreet mineral grains and solid materials as yet unidentified.

Electron microscopy element pictures clearly illustrate a very even distribution for copper and lead (perhaps indicating the presence of ionic dissolved species), the segregation of iron in well-defined areas, partial segregation of arsenic (N.B. Arsenic is used as an additive to some types of bronze water fittings), and the presence of well-formed crystals of zinc sulphate exhibiting some zonal and overgrowth features. Further measurements have identified significant levels of arsenic but a natural local source is not obvious although arsenic mineralisation does occur nearby. I have no reason to believe that the results of these

TABLE XXIV

DETERMINATION OF SOME ELEMENTS IN HUMIC RESIDUES OF A DOMESTIC COLD WATER TANK TOGETHER WITH DATA USED FOR STANDARDISATION*

Element	0.1 ppm Standard	Precision (n = 10)	Humic‡ Alkaline Extract	Humic‡ Acid Extract
Al	0.104	0.015	3,086	2,819
As	—	—	25.9	< 0.05
B	0.110	0.004	< 0.003	54.2
Ba	0.142	0.006	0.3	112.6
Ca	0.096	0.002	7.1	2,294
Cd	0.095	0.003	< 0.001	2.8
Co	0.13	0.03	< 0.02	25.4
Cr	0.124	0.005	0.6	41.3
Cu	0.101	0.002	208.3	917.4
Fe	0.096	0.002	28.8	27,857
Hg	0.11	0.04	< 0.02	87.6
La	—	—	< 0.003	14.2
Mg	0.096	0.0008	0.4	583.8
Mn	0.102	0.001	1.1	6,922
Mo	0.10	0.03	3.0	46.1
Ni	0.101	0.008	110.1	121.3
Pb	(0.09)†	0.04	8.3	2,711
Sb	(0.10)	0.09	11.6	542.2
Se	—	—	5.8	31.3
Sn	(0.12)	0.06	13.3	30.2
Zn	0.103	0.005	137.6	12.5

* Data obtained by inductively coupled plasma spectrometry by Applied Research Laboratories, United Kingdom.
† Equal to or below detection limits.
‡ Dry wt, ppm.

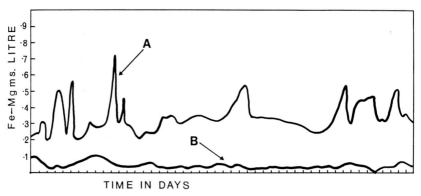

Figure 22: Variability in the levels of iron present in a domestic cold water supply by the visual assay for iron using the o-phenathroline reaction. Frequency of sampling: every 2h from 06 to 24h. Curve A is for the natural domestic supply and curve B for the same supply but filtered through a 0.45 μm filter.

TABLE XXV

SEMIQUANTITATIVE DATA FOR LEVELS OF SOME ELEMENTS IN
DOMESTIC WATER FOLLOWING A PIPE FRACTURE*

Element	Concentration in Solid (dry wt)	Concentration in Water (µg/l)†
Si	2.6%	730
Al	0.3%	90
P	1.6%	470
S	150 ppm	4
Ca	1.0%	290
Mn	1.3%	380
Fe	17%	4,800
Ni	290 ppm	8
Cr	280 ppm	8
Cu	340 ppm	10
Zn	860 ppm	25
Pb	750 ppm	22

* Data obtained by inductively coupled plasma spectrometry by Applied Laboratories, United Kingdom.
† Particle load: 28.6 mg/l.

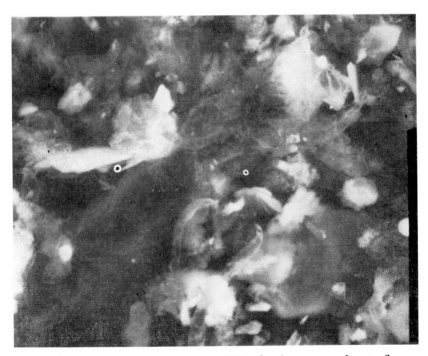

Figure 23A: Electron micrograph ($\times 1000$) of a ferruginous humic floc removed from the cold water header tank of a domestic home. As a result of examining element distributions (depicted in Figure 23B—23G), Figures 23C and 23D provide evidence for the presence of a crystal of zinc sulphate, which can be faintly seen in this figure.

Figure 23*B*: copper

Figure 23*C*: zinc

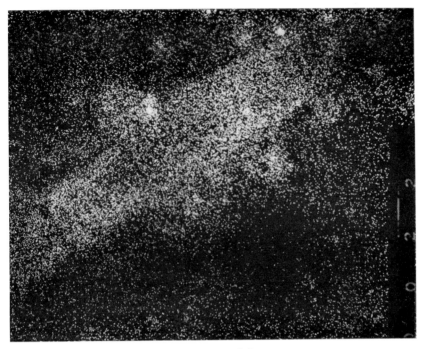

Figure 23*D*: sulphur

Figure 23*E*: lead

Figure 23E: arsenic

Figure 23G: iron

preliminary investigations are in any way unusual and would certainly expect that in some parts of the United Kingdom much higher levels of elements will be present in domestic drinking supplies. However this type of work does illustrate the complex nature controlling element levels, in particular the significance of solid phases. In the house in which this work was carried out it is known that the main road water supply is carried in lead pipes and copper is used within the house. Today in most parts of the world lead and copper plumbing materials are being replaced with plastics but even these materials can contaminate water with elements such as zinc and copper.

Water Quality and Haemodialysis

In haemodialysis the quality of water is an important factor and as Ginn (7) has stated "not all water is good for kidney machines." Whether or not this statement indicates a real cause for concern remains to be shown but nevertheless it deserves serious consideration. It was perhaps fortunate that the original application of the technique took place at a hospital receiving soft water, low in dissolved solids and suspended materials. To place the matter in perspective a normal individual may drink between five hundred and one thousand litres of water each year, while a patient undergoing renal dialysis is exposed to between 30,000 and 60,000 litres of dialysing fluid for the same length of time. The quality of water in hospitals varies considerably and in some areas deserves further study in order to determine the abundance of the chemical elements present in supplies. The problem is potentially more critical for individuals who use kidney machines at home, particularly if mains water is not available and the supply is obtained from leats and wells. Fortunately in recent years this problem has been appreciated and it is common practice for home dialysis units to be equipped with ion-exchange cartridges in order to remove elements present in the water. In 1971 there were 375 centres in Europe performing hospital dialysis; the total number of individuals receiving treatment was 8,666 (Parsons, Brunner, Gurland, and Harlen 8).

The major causes of renal failure are glomerulonephritis or

pyelonephritis and polycystic disease, while most deaths are caused by uraemia, hypertension, and cerebrovascular accident. Apart from effects directly attributable to renal failure other forms of morbidity, particularly uraemic oesteodystrophy (oestoporosis, spontaneous fracture) and various neurological disorders, have been associated with individuals undergoing regular dialysis. Regardless of the quality of water used, present techniques have been successful and are used in a routine manner. However, in recent years it has been noted that the quality of some waters may be detrimental to treatment, but as yet no constituent has been identified as harmful. Some individuals suffering from oesteodystrophy, and associated muscular weakness, have shown a rapid recovery when the normal hospital water has been replaced with deionized water. In Newcastle, United Kingdom, and Plymouth, United Kingdom, a significant number of patients developed disabling bone disorders, for example bone fractures, myopathic weakness, and liability to pathological fracture. Similar features have been noted at Denver, Colorado, United States, and Iowa City, Iowa, United States, while a low incidence rate for such features have been observed for Fulham, United Kingdom, E. Birmingham, United Kingdom, Montreal, Canada, and Rochester, New York, United States. It has been suggested, by some, that the presence of fluorine could be the cause of oesteodystrophy, but seems unlikely as the condition is not always associated with fluorinated water. Parsons et al. (9) have noted that bone from patients with renal failure contains more aluminium than found in normal bone; patients with high levels of aluminium in bone also show signs of severe neuropathy and renal bone disease.

If the chemical elements are considered to be the culprits, then a simple remedy, at first sight, would be the use of ion-exchange systems to remove elements; however such procedures should not be used without very careful monitoring. Hard water areas have been associated with hypercalcemia and hypermagnesia, while in one hospital using copper pipes a case of copper intoxication resulted from the failure of the ion-exchange unit (Schreiner, 10). Although the conductivity of pure water is

$0.0548 \times 10^{-6}/\Omega$ cm at 25°C, it can still contain elements present in solution at the ppb level; a measurement of conductivity will not indicate the presence of colloidal or solid matter. Very little information exists which can even tentatively explain why deionized water can be beneficial to some forms of morbidity following dialysis treatment. As a result of progressive renal failure, very complex biochemical changes take place and if deionized water is implicated the following items may be worthy of consideration:

1. Water contains both organic and inorganic compounds present in solution, as complexed ions and particulate matter. In relation to dialysis each phase, or combination of phases, can pose a particular problem; at some stage of investigation each phase must be considered; the selection of one phase without the evaluation of others may be misleading, for example the removal of one phase may result in the repartition of an element among the remaining phases.
2. With few exceptions most of the chemical elements can be removed from water by passage through an ion-exchange resin bed; for some elements the capacity for removal is limited. In practice any advantage gained by using an ion-exchange system would seem to be lost simply because the chemicals used in the preparation of the dialysing fluid are impure. A commonly used dialysing mixture consists of NaCl, 190 g/l; Na-Acetate, 190 g/l; KCl, 4 g/l; $CaCl_2$, 12 g/l; $MgCl_2$, 6 g/l; and dextrose, 70 g/l; the mixture is then diluted with water in the ratio of 1 : 34. Data are not available for the mass of elements present as impurities in the reagents used in dialysis, but an evaluation of possible levels and their significance can be obtained by using data available for analytical grade reagents.

With the exception of sulphate ions, the reagent impurities can make a significant contribution to the total mass of impurities in the dialysing fluid. It is therefore apparent that levels of elements present in the untreated normal water supply are likely to be present at quite insignificant levels.

The addition of the dialysis salt mixture takes place after the water has passed through the deionizer. As the dialysis-salt solution is common to both normal tapwater and deionized water systems, it is difficult to accept that the process of deionization, in terms of chemical elements, has any significance. Any benefit which may accrue from the use of deionized water could be related to the removal of particulate matter present in water as a consequence of the filtering action of a bed of ion-exchange resin. If an untreated water supply contained quantities of organic humic substances, then it is conceivable that this could impair the permeability of the dialysing membrane.

There are very few papers which attempt to relate patterns of morbidity with the abundance of the elements in samples of tissue for individuals undergoing renal dialysis. Alfrey et al. (11), report a neurological syndrome for patients receiving chronic haemodialysis which was associated with minimal anatomic abnormalities. In this study a common etiology was indicated and it was concluded that the syndrome was a result of a metabolic encephalopathy. Evidence of vascular disease and infection was lacking and the syndrome had an unrelenting course leading to total disability and eventually death over a period of six to seven months. The characteristics of the syndrome were an initial stage of speech abnormality, the patient's speech becoming slow and deliberate, with stuttering associated with dyspraxia, followed by a state of marked tremulousness, myoclonus, asterixis, dyspraxia of movement, impairment of memory, personality changes, and at times overt psychoses manifested by depression and a paranoid state with hallucinations. The authors determined the relative abundance of Fe, Cu, Zn, Br, Sr, Zr, Mo, Pb, Ni, and Co, together with quantitative assay for Rb, Sn, and K in brain samples for both dialyses and normal individuals. They found a depression in the level of potassium for dialysed patients and a substantial increase in the level of tin. The authors note that in animals alkyl-tin can induce encephalopathy which is characterized by muscular weakness, loss of ability to ambulate, and death. Alajouanine, Dérobert, and Thiéffry (12) report cases of tin poisoning as a result of injection of an acne preparation to a large number of people; of 210 cases diagnosed, 110 deaths occurred as a result of tin encephalopathy. Hamilton (13) investigated incidences of severe disorders for patients undergoing renal dialysis in which very characteristic symptoms disappeared when the normal hospital water was replaced with a deionized supply. The hospital was sited in Plymouth, United Kingdom, supplied with soft water originating from a granite area associated with tin mineralisation; tin was only present in trace amounts, about 2 $\mu g/l$, in the raw water of the catchment area, but slightly higher levels were found for the supply to the hospital. A flow diagram illustrating the processing of the hospital water is given in Figure 24 and the chemical composition of the water at various stages of treatment is given in Table XXVI; following passage through the deionizer the levels of all elements were significantly reduced. Prior to the installation of the deionizer, the system contained a bronze filter which resulted in an increase in the concentrations of Zn, Cu, Fe, and Al in the water. Analyses were also performed on postmortem samples of brain and bone but no significant differences, apart from elevated levels of aluminium and iron in the brain, were noted in the levels when compared with normal values for the United King-

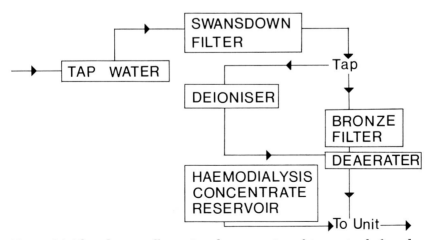

Figure 24: Flow diagram illustrating the processing of tap water before delivery to a renal dialysis machine.

TABLE XXVI*

THE ABUNDANCE OF SOME ELEMENTS (μg/liter) FROM A SOFT WATER AREA IN THE UNITED KINGDOM IN RELATION TO USE IN A HOSPITAL FOR A RENAL DIALYSIS UNIT (RDU)

Element	Filtered Natural Water†	Unfiltered Acidified Natural Water‡	Local Tap Water§	RDU Raw Feed	RDU Water From Coarse Filter Unit	RDU Post Coarse Filter	RDU Post Main Filter and Sintered Bronze Filter	RDU Post Deioniser
Pb	< 0.003	1.3	3	3	3	1	1	< 1
W	—	—	3	1	1	4	1	< 1
Ba	3	17	4	2	4	7	6	< 1
I	1	1	2	1	1	1	1	< 1
Sn	0.4	2	21	10	11	14	10	7
Cd	—	—	2	2	< 1	3	1	< 1
Ag	—	—	—	—	< 1	—	—	< 1
Zr	< 0.0004	0.2	—	—	—	—	—	—
Y	0.3	0.4	< 1	< 1	< 1	1	< 1	< 1
Sr	9	16	16	8	8	11	11	1
Rb	5	36	1	2	3	2	1	< 1
Br	—	—	24	35	18	37	11	< 1
Se	12	15	—	—	—	—	—	—
As	1	3	< 1	< 1	< 1	1	< 1	< 1
Zn	< 0.003	7	24	175	122	16	34	3
Cu	3	9	31	45	16	7	44	1
Ni	0.3	0.7	13	6	7	6	2	< 1
Co	0.5	0.9	1	1	1	2	1	< 1
Fe	26	276	167	56	29	26	55	14
Mn	8	49	83	6	42	19	17	1
Cr	—	—	8	12	4	5	6	< 1
V	0.3	1.4	1	1	1	1	1	< 1
Sc	0.7	1	—	—	—	—	—	—
Ca	6,000	6,000	13,000	6,298	6,585	6,690	6,184	22
K	5,000	5,000	546	1,849	964	839	1,811	19
Cl	2,000	2,000	213	343	359	164	92	7
S	4,000	2,000	2,742	3,094	3,235	2,958	911	23
P	—	—	20	10	10	7	2	5
Si	—	—	4,136	2,861	2,091	910	655	48
Al	24	820	409	297	207	63	194	< 1
Mg	—	—	9,978	2,410	1,681	2,196	2,368	50
Na	4	11	Major	Major	Major	Major	Major	53
F	2	8	9	14	4	4	4	< 1
B	4	12	—	—	—	—	—	—
Li	4	11	—	—	—	—	—	—

* From Hamilton, E. I.: The chemical elements and human morbidity—water, air and places—a study of natural variability. *Sci Total Environ*, 3:24, 1974.

† Water sampled from a local main reservoir passed through a 0.45 μm filter.

‡ Unfiltered water sampled from a local main reservoir and acidified with 10% nitric acid.

§ Domestic tap water from a new house supplied with copper and plastic plumbing.

dom. Although elevated levels of tin were not found in the tissues, this element cannot be ruled out as instrumental in causing the symptoms: in the particular method selected, for the very small samples of tissue which were available, alkyl forms of tin could have been lost by volatilisation processes.

It now seems almost certain that in some hospitals the quality of water is responsible for some of the characteristic disorders associated with renal dialysis; whether or not the chemical elements are implicated remains a mystery. Analytical techniques are now available and hopefully will be used in an attempt to solve the problem and improve the health of patients undergoing renal dialysis.

Finally, it must be stated that the analysis for elements in water is very difficult; Ellis (14) reports the results of an interlaboratory water analysis programme and remarks that while progress has been made in recent years in improving the quality of trace element analysis, in many laboratories the standard of water analysis for many common constituents still leaves much to be desired.

REFERENCES

1. Masironi, R.: Trace elements and cardiovascular diseases. *Bull Org Mond Sante Bull WHO, 40*:305, 1969.
2. WHO meeting of investigators on trace elements in relation to cardiovascular diseases. Joint WHO/IAEA research project, CVD/71.02, 1971.
3. Crawford, M. D., Gardner, M. J., and Morris, J. N.: Cardiovascular disease and the mineral content of drinking water. *Br Med J, 27*:21, 1971.
4. Crawford, M. D., Gardner, M. J., and Morris, J. N.: Changes in water hardness and local death-rates. *Lancet,* August 14: 7720, 1971.
5. Packham, R. F.: The leaching of toxic stabilisers from unplasticized PVC water pipe: Part I—A critical study of laboratory test procedures. *Water Treatment Examination, 20*:108, 1971. Part II—A survey of lead levels in uPVC distribution systems. *Water Treatment Examination, 20*:144, 1971. Part III—The measurement of extractable lead in PVC pipes. *Water Treatment Examination, 20:* 152, 1971.
6. Obrecht, M. F. and Pourbaix, M.: Corrosion of metals in potable water systems. *J Am Water Works Assn,* 59:977, 1967.
7. Ginn, H. E.: Not all water is good for kidney machines. Proc. 5th Int. Water Qual. Symp. Washington, D.C. *The Water Quality Research Council, U.S.A.*:76, 1970.

8. Parsons, F. M., Brunner, F. B., Gurland, H. J., and Harlen, H.: Combined report on regular dialysis and transplantation in Europe I. In Cameron, J. S. (Ed.): *Dialysis and Renal Transplantation.* London, Pitman Medical, 1971.
9. Parsons, V., Davies, C., Goode, C., Ogg, C., and Siddiqui, J.: Aluminum in bone from patients with renal failure. *Br Med J,* 4:273, 1971.
10. Schreiner, G.: *Water Requirements for Kidney Dialysis.* London, Elga Pub, 1969.
11. Alfrey, A. C., Mishell, J. M., Burks, J., Contiguglia, S. R., Rudolph, H., Lewin, E., and Holmes, J. H.: Syndrome of dyspraxia and multifocal seizures associated with chronic hemodialysis. *Trans Am Soc Artif Intern Organs, XVIII:*257, 1972.
12. Alajouanine, Th., Dérobert, L., and Thiéffry, S.: Étude clinique d'ensemble de 210 cas d'intoxication par les sels organiques d'étain. *Rev Neurol,* 98:7, 1958.
13. Hamilton, E. I.: The chemical elements and human morbidity—water, air and places—a study of natural variability. *Sci Total Environ,* 3: 3, 1974.
14. Ellis, A. J.: The I.A.G.C. interlaboratory water analysis comparison programme. *Geochim et Cosmochim Acta, 40:*1359, 1976.

Chapter 5

DIET

THE MAJOR PATHWAY through which the chemical elements enter man is through the food chain. In spite of the availability of suitable methods of analysis, large scale studies describing the total element abundances in diet are seldom reported, although detailed analyses of some items of diet have been carried out in relation to nutritional requirements, or studies related to contamination and pollution. Funds for nutritional research are not very forthcoming, which is rather surprising considering the importance of this type of work and the lack of information for many elements and compounds.

With few exceptions most items of diet, including water, have some relationship with the abundances of the chemical elements in the natural environment from which they are derived, although in transport through the system (soils→vegetation→animals→man), all elements are subjected to some degree of discrimination, usually brought about by different rates of transport and diffusion through biological membranes; in some items of food, secondary enrichment of particular elements occur through the action of specific biochemical systems. In the first instance we need to know the abundance of the chemical elements in natural products, without considering the question of chemical form or biological availability to man. We then need to determine the levels of elements present in materials which have been subjected to some form of technological processing either during growth, such as the addition of agricultural chemicals, during manufacturing processes, or in the preparation of foods prior to consumption. These types of measurements can be concerned with major sources defined by geography or identified manufacturing processes and serve to provide information for the characteristic composition of foods available to large populations. The relevance of such information to the actual composition of food individuals consume is not simple because of individual preferences and methods of preparing foods; however it does provide sufficient information in order to provide baseline data for large populations.

One important item of diet, namely water, rarely receives adequate attention; data for daily intakes are seldom reported in spite of the fact that the quality of water is considered by some as the causative agent for various major forms of human morbidity. This subject has been discussed elsewhere, but it is worth noting that the data which are required concern the chemistry of water sampled at domestic taps, as distinct from that obtained by analysis of natural waters and those altered by processing at water works prior to transport and storage, in a variety of materials, before reaching homes. The total daily intake of fluids by an adult in temperate latitudes is as follows: total fluids 1000-2400 ml, milk 120-450 ml, tapwater 45-730 ml, and water-based drinks 320-1450 ml. Bransby and Fothergill (1) determined a total fluid intake of 1850 ml per day for 270 persons residing in a suburb of London. Data are reported in Table XXVII for two surveys, A and B, carried out in the United Kingdom on normal healthy individuals, described by Hamilton (2). In survey A a total of sixty individuals (males 20-50 years and females 20-60 years of age) residing in a hard water area were asked to record a total inventory of fluid intake over a period of one week during a winter month. In survey B, data obtained over a period of one week are for forty-one males (46-65 years of age) residing in both hard and soft water regions of South Wales, United Kingdom. While recognising the limitations imposed by the small numbers of samples studied, they add support to the previous estimates and show that the intake of tapwater in temperate latitudes is small when compared with the total intake of fluids. In Table XXVII the higher consumption of milk for males in survey A, compared with survey B, is possibly due to the lower mean age of the individuals sampled.

When correlating levels of an element in water with morbidity, it is essential to only use data derived from the analysis of tapwater; water is of course used in the preparation of many items of food but in most circumstances the element composition of the original water is changed considerably. Tapwater is

TABLE XXVII*
TOTAL FLUID INTAKE (ml/day) FOR ADULTS RESIDING IN BOTH HARD AND SOFT WATER AREAS

	Total Fluid Intake	Tea	Domestic Tap Water	Soft Drinks	Beer, Cider	Coffee, Cocoa	Total Milk	Milk Drinks
Survey A—Radiological Protection Service—Sutton, Surrey, Water type, hard								
Males† (27)	1,926±602(27)	816±403(26)	184±119(22)	79± 58(10)	313±502(23)	516±374(27)	580±245(27)	227±148(22)
Females (33)	1,784±760(33)	1,133±780(29)	193±159(23)	120±117(21)	141±115(25)	340±239(31)	433±198(33)	285±166(2)
Survey B—South Wales—Water type, soft and hard areas								
Males (41)‡								
Hardwater	1,869±923(21)	1,479±699(21)	284	426±200	568±560(11)	284±134	261±125(2)	—
Softwater	1,572± 69(20)	1,278±647(20)	715±208	284(2)	490±600(6)	189(2)	221± 97(26)	—

* From Hamilton, E. I.: The chemical elements and human morbidity—water, air and places—a study of natural variability. *Sci Total Environ*, 3:14, 1974.
† Total number of individuals sampled.
‡ Total number of individuals for whom data were recorded.

used in making tea, coffee, and other forms of liquid refreshment but, because of the different methods of preparation and chemical form of the elements in the ingredients, the chemical form and biological availability of the elements present in the original tapwater will inevitably be altered. Shah, Filby, and Davis (3) have shown that tea and ground or instant coffee can contain significant quantities of Na, K, Cr, Mn, Fe, Cu, Zn, Se, Br, Rb, Sb, Cs, and Hg; a proportion of the elements present in the residues of ground coffee and tea infusions will be retained in solid residues, but for instant coffee the total mass of the elements will be ingested. Anderson, Hollins, and Bond (4) have shown that metal and ceramic utensils used for making tea do not result in serious contamination of the infusion. The authors considered relations between the complex polyphenolic flavonoids (daily intake for the United Kingdom $\approx$ 2 g/day) and calcium, also the repression of extractable oxalic and pectinic acids by calcium. Tea infusions contain between 62 and 98 mg of oxalate per litre while the daily intake of oxalate in different parts of the United Kingdom varies inversely with the degree of water hardness. The whole question of water chemistry and associations, if any, with human health is a very complex subject of which very little is known. Many recent studies creating alarm concerning environmental pollution of water have little substance, at least as far as the chemical elements are concerned, simply because the actual mass of raw tapwater is small and the contribution to total intake of elements from other sources is usually large.

In order to obtain a fairly realistic estimate of the total intake of foods by individuals of a large population, specially designed studies are required but often for practical reasons many have to be rather limited in coverage and restricted to small geographical regions. As a result of the transport of foods over large distances, and in the absence of demographic stratification and an appropriate adjustment to allow for the contribution of locality in relation to a recognisable population exposed to an environment containing a characteristic assemblage of elements, mean data obtained from large diet surveys may have very limited value; this is particularly true when considering possible relationships between the mass of elements in diet and human morbidity. Data are presented in Table XXVIII for the composition of diet for major areas of the world and illustrates that for many areas the differences in gross material intake is small.

In order to obtain an estimate for the daily intake of the chemical elements by individuals of the United Kingdom, a survey aimed at reproducing normal household intakes of foods

TABLE XXVIII*

CONSUMPTION OF FOOD (g/Head/Day) FOR DIFFERENT PARTS OF THE WORLD

Food Group	Consumption (g/Head/Day)							
	Far East	Near East	Africa	Latin America	Europe	North America	Oceania	Mean
Milk	51	214	96	240	294	850	574	360 ± 290
Meat	24	35	40	102	111	248	312	126 ± 113
Fish	27	12	16	18	38	26	22	23 ± 9
Eggs	3	5	4	11	23	55	31	19 ± 19
Fats + oils	9	20	19	24	44	56	45	31 ± 17
Sugar + Preserves	22	37	29	85	79	113	135	71 ± 44
Starchy roots	156	44	473	247	377	136	144	225 ± 151
Vegetables and fruits	128	398	215	313	316	516	386	325 ± 127
Cereals	404	446	330	281	375	185	243	323 ± 93
Pulses and nuts	56	47	37	46	15	19	11	33 ± 18
Total	880	1,258	1,259	1,367	1,872	2,204	1,873	1,530 ± 463

	United Kingdom	European Community	United States	
Milk	382	287	508	392 ± 111
Cheese	12	21	19	17 ± 5
Meat	137	118	206	154 ± 46
Fish	21	22	22	22 ± 0.6
Eggs	34	21	47	34 ± 13
Fats	44	63	49	52 ± 10
Sugar + preserves	77	57	69	68 ± 10
Potatoes	202	196	103	167 + 56
Other vegetables	118	180	202	167 ± 44
Fruit	108	114	184	135 ± 42
Cereals	246	346	207	266 ± 72
Total	1,381	1,425	1,611	1,472 ± 122

* From Hamilton, E. I., and Minski, M. J.: Abundance of the chemical elements in man's diet and possible relations with environmental factors. *Sci Total Environ*, 1:383, 1972/73.

was undertaken. Samples of diet for the year 1966-67, obtained principally for pesticide analysis, were provided by the Ministry of Agriculture, Fisheries, and Food. Items of food characteristic for different regions of the United Kingdom containing the bulk of the population were purchased by various domestic science colleges and cooked and prepared by local commonly used methods. The foods were arranged according to the following groups: cereals, meat and fish, fats, fruits and preserves, root vegetables, and milk. After cooking, or preparation for eating, each foodstuff was allocated to one of these categories and monthly aliquots removed, homogenised and stored prior to chemical analysis. The mass of food in each category reflects their relative proportions in the average diet. In order to calculate average daily intakes for various elements, the 1963 United Kingdom consumption and population data were used. The chemical composition of water used in cooking, for making beverages, or for drinking direct from the tap was not studied separately, neither was the possibility investigated that cooking utensils could contribute to the abundance of some elements in the prepared diet. Further, no consideration was given to the loss of elements as a result of cooking or through wastage which, for

TABLE XXIX[*]

TOTAL DAILY INTAKE (μg/day) FOR THE CHEMICAL ELEMENTS IN DIET FOR THE UNITED KINGDOM TOGETHER WITH SOME COMPARATIVE DATA

Element	RPS	ICRP II[†]	Cresta et al.[‡]	Tipton et al.[§]
U	0.99	2.0		
Th	<0.05	400		
Pb	320 ± 150	400		
Bi	< 5			
Tl	< 2			
Hg	<16	20		
Au	< 7			
Pt	< 1			
Ir	< 1			
Os	< 1			
Re	< 1			
W	< 1			
Ta	< 1			
Rare earths	< 5[c]			
Ba	603 ± 225	900		650; 920

TABLE XXIX (continued)

Cs	13 ± 7	10		
I	220 ± 51	200	89.8 ± 16.4	
Sb	34 ± 27			
Sn	187 ± 42	17,000		5,800; 8,800
Cd	64 ± 30			100-220[d]
Ag	27 ± 17	0.08		
Mo	128 ± 34	450		210; 460
Nb	20 ± 4			
Zr	53 ± 34			430; 550
Y	16 ± 7			
Sr	858 ± 144	1,000	763 ± 108	1,900; 2,100
Rb	$4,350 \pm 1,547$			
Br	$8,400 \pm 900$	17,000		
Se	~ 200			
As	< 50			
Ge	367 ± 159			
Zn	$14,250 \pm 1,220$	17,000	$7,545 \pm 1,186$	11,000; 18,000
Cu	$3,110 \pm 760$	3,000	$1,635 \pm 273$	9,500; 17,000
Ni	< 300	400		
Fe	$23,250 \pm 1,120$	27,000	$12,090 \pm 1,940$	15,000; 28,000
Mn	$2,674 \pm 854$	3,100	$3,200 \pm 540$	3,300; 5,500
Cr	320 ± 162	150		200-290
Ti	~ 800	540		
Ca	$1.37 \pm 0.02 \times 10^6$	1.0×10^6	$0.69 \pm 0.17 \times 10^6$	0.94×10^6 2.3×10^6
K	$2.80 \pm 0.03 \times 10^6$	3.0×10^6	$3.0 \pm 0.8 \times 10^6$	1.7×10^6 3.8×10^6
Cl	$5.40 \pm 0.06 \times 10^6$	6.7×10^6		
S	$0.94 \pm 0.06 \times 10^6$	1.3×10^6		
P	$1.87 \pm 0.03 \times 10^6$	1.4×10^6		
Al	$2,330 \pm 1,060$			
Mg	$0.25 \pm 0.02 \times 10^6$	0.53×10^6		0.18×10^6 0.36×10^6
Na	$4.64 \pm 2.39 \times 10^6$			
F	~ 500	1,000		
B	$2,819 \pm 1,554$	6,000		1,200; 2,800
Be	< 15			
Li	107 ± 53	2,000		

* From Hamilton, E. I., and Minski, M. J.: Abundance of the chemical elements in man's diet and possible relations with environmental factors. *Sci Total Environ,* 1:378, 1972/73.

† Data from I.C.R.P. Publication 2, Report of Committee II. *Recommendations of the International Commission on Radiological Protection.* Elmsford, New York, Pergamon Press, 1959.

‡ Data from Cresta, M., Ledermann, A., Garnier, A., Lombardo, E., and Lacourly, G.: *Euratom/CEA,* Eur 4218f, 589, 1969.

§ Data from Tipton, I. H., Steward, P. L., and Martin, P. G.: Trace elements in diets and excreta. *Health Physics,* 12:1683, 1966.

[a] The mean of data reported for dietary intake of 9,000 families in eleven regions of the European Community during the period 1963-1965.

[b] Mean dietary intake of the elements by two men in 1969. Sample period: 50 weeks.

[c] Estimate for each individual rare earth element.

[d] For 63 days only.

[Note data reflecting urban rather than rural communities.]

some foods, could amount to about 15 percent. Methods of analysis were spark source mass spectrometry, X-ray fluorescence spectrometry, various neutron activation techniques, and atomic absorption spectrometry. The data are presented in Table XXIX in terms of total daily intakes of the chemical elements and compared with other published values. Not all the elements present in the samples of diet were determined in this survey and one, cobalt, was not determined because of analytical problems.

Bhat (5) has determined a daily intake of 254 ± 16.5 μg Co for residents of Tarapur in India; Yamagata, Kurioka, and Shimuza (6) a value of 35.7 μg Co for the Japanese; Schroeder, Nason, and Tipton (7) 300 μg Co for United States diet; while Tipton, Stewart, and Martin (8) and Tipton, Stewart, and Dickson (9) determined the United States daily intake of cobalt in the range of 160-470 μg. While surveys such as that reported in Table XXIX have not been repeated in the United Kingdom, data on similar materials has been obtained for lead and cadmium. In a lead survey (MAFF, 10) the intake of lead from 1.5 kg of food consumed daily by the average person in the United Kingdom is in the region of 200 μg, and a further 20 μg of lead per day is ingested from beverages. In a further study Thompson (11) determined an intake of 274 μg Pb/day for normal individuals of the United Kingdom, but also noted that daily intakes varied between 70 and 750 μg Pb; the lead ingested was almost totally eliminated in faeces and urine, only about 10 μg on average per day was retained. The data obtained from large surveys must be treated with caution but they do serve for the protection of large populations. However obtaining such data can only constitute an initial stage of study as it is quite clear that large numbers of a population can have intakes higher or lower than the calculated means; an individual's tolerance to lead, or any other element, can be a very individual feature and until more is known about the dose response relationships for most of the chemical elements complacent attitudes should not be tolerated. It should also be noted that an individual's response to insult by an element may be determined by the general state of health, since various adaptation processes may take place which reduce the extent of toxicity, while body response may alter with age. As far as chemical analysis is concerned it is quite unrealistic to expect that individual data, such as that presented in Table XIX, will be all of the same quality; very severe analytical problems exist for many elements and in this study some problems were reduced by the use of particular methods most suited for defined elements. Finally, the significance of the data has to be considered in the perspective of the natural variability of the chemical elements in foodstuffs which is a combination of sampling problems and geography. Large scale national surveys tend to obscure regional and local levels of elements present in foods which are characteristic of particular areas, the more rural the area and the lower the population the less is the chance that such areas will be identified. Geography and communications are important factors influencing levels of elements in foods; some foods such as cereals from North America may be considered for universal distribution while others are grown and consumed within quite restricted areas. For example in the county of Devon, United Kingdom, between 90 and 95 percent of vegetables are grown and consumed locally which is of considerable interest to studies of human morbidity as the area is heavily mineralised, although well-defined associations of disease and abundance of elements associated with the mineralisation are not obvious, except perhaps low fluorine levels and a higher incidence of dental caries in some areas. Natural distributions of the chemical elements are also disturbed by the use of fertilisers and soil additives, such as phosphate, selenium, nitrogen, magnesium, cobalt, and boron; this also applies to livestock, not only to overcome natural deficiencies, but also to improve the quality or yield of the product.

Intensive livestock rearing involves the use of a large number of additives, including vitamins and antibiotics, which themselves can interact with some elements. The importance of the availability of elements to some animals is often obscure; for example, in the past, battery-bred chickens utilised iron and zinc from galvanised cages as a source of these elements; when they were replaced by plastics various patterns of disease emerged and it became necessary to provide these elements as food additives. Veterinary practice has become an important area of study in which element imbalances have been detected, and some may be relevant to man. In one study (Lloyd, Hill and Meerdink, 12) a molybdenosis-copper deficiency was considered to be due to a borderline copper deficiency which was aggravated by abnormally high levels of dietary molybdenum and sulphate; the excess molybdenum was identified as a contaminant (4-12,200 ppm Mo) in magnesium oxide which was added to the diet at a level of 1 percent in the form of a trace mineral complex.

Analysis for the composition of natural foods may often have little bearing upon dietary intake; parts of some foodstuffs are removed prior to preparation; as a result of cooking some elements (for example many metal-organic complexes) may be lost, while others (for instance silicon and aluminium from cooking utensils) will be added. It is beyond the scope of this monograph to consider details of all the elements in terms of origin and pathways into foods consumed by man; brief comment will be made for a few elements.

Arsenic

It is well known that arsenic is toxic to man but unlike mercury, cadmium, and lead which have caused concern about the health of man, arsenic is seldom studied. Apart from the well-known association between arsenic and skin cancer, exposed workers also show a high incidence rate of lung and lymphatic cancer. The toxicity of organic-arsenic compounds is higher than inorganic forms and they are widely used in garden and farm pesticides, defoliants, wood treatment to prevent rot, and the control of sludge in lubricating oils. The toxicity of arsenic has been reviewed by Vallee, Ulmer, and Wacker (13) and Luh, Baker, and Henley (14). Awareness of the toxic nature of arsenic has resulted in stringent precautions when handling this element and its compounds, yet arsenic is ubiquitous in the natural environment.

Arsenic is a common constituent of crustal rocks and becomes enriched in soils, reaching a maximum in areas associated with sulphide mineralisation where levels of at least 2,000 ppm As are not uncommon. The distribution of arsenic from stream sediments of the United Kingdom is illustrated in Figure 25; Colbourn, Alloway, and Thornton (15) have described the distribution of arsenic and heavy metals in soils with geochemical anomalies in southwest England; arsenic and copper are highly enriched in topsoils and the trace element content of pasture herbage reflects, in part, the degree of soil contamination. In the same area Porter and Peterson (16) note that ecotypes can accumulate arsenic to extreme levels and although arsenic occurs at high levels through a plant, including the seeds, the highest concentrations were recorded in old leaves, for example up to 10,000 ppm of arsenic for individual leaves.

The county of Cornwall, and parts of the adjacent county of Devon, England, provide an opportunity to study possible associations between disease and the chemical elements, particularly arsenic, copper, zinc, and lead. Between 1815 and 1925 very intensive mining was carried out in the area, which is littered with mine tips. The ore minerals were associated with considerable quantities of arsenic for which there was no large commercial market. When the ores were calcined, the arsenic was volatilised and much was trapped in elaborate brick labrynths; the remaining volatile matter was then usually passed into a large tunnel constructed of brick which would pass beyond areas of immediate habitation and vent into a tall stack. Figure 26 illustrates the distribution of mine sites in the vicinity of Camborne, Cornwall, England which clearly follow the lines of mineral lodes below the surface. Figure 26B is a sketch of an arsenic labrynth and associated flue. Mining activity in the area was intimately associated with the development of road and rail transport linking quite large towns and hamlets. Today the whole central area of Cornwall is littered with waste tips containing very large amounts of arsenic, while labrynths and associated buildings can still be found more or less intact. Today, in spite of considerable exposure of people in the area to arsenic in a variety of forms, I know of no identified pattern of morbidity which can be related to the presence of arsenic in the region, a clear illustration that the simple presence of a very toxic element need not always be associated with obvious toxic effects on man.

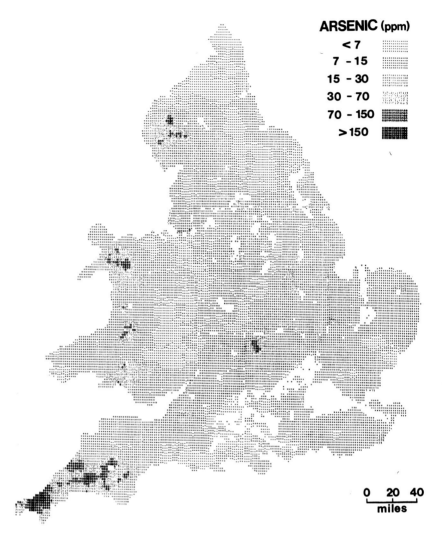

Figure 25: Regional map for the distribution of arsenic in England and Wales. The counties of Cornwall and Devon in Southwest England are associated with extensive ore mineralisation, the presence of a large number of mine waste tips, and many remains of old stacks and associated brick labrynths in which arsenious oxide was removed by sublimation. This reconnaissance forms part of an extended research programme of regional geochemical mapping carried out by the Applied Geochemistry Research Group at Imperial College, London, United Kingdom. Sampling was based upon stream sediments collected at tributary/road intersections at a mean density of one sample per square mile. Reproduction by permission of Professor J. S. Webb.

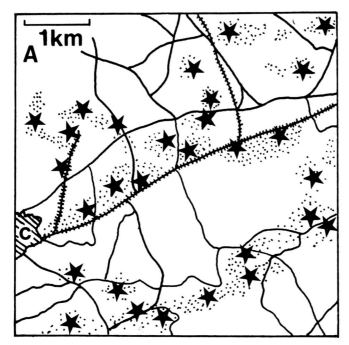

Figure 26A: Mining activity in Cornwall, England. An illustration of the close association between mines (stars), and associated waste tips (following the line of subsurface mineral lodes) with road and rail transport for the outskirts of Camborne (C), Cornwall, England.

Figure 26B: A sketch illustrating an arsenic labrynth at a mine site and the associated flue leading to a distant chimney for dispersing volatiles which were not removed by sublimation in the flue. Today the downwind areas of many dispersal chimneys are often devoid of vegetation and contain high concentrations of several elements; loss of vegetation may in part be due to toxic effects of some elements but also because of acid soils attributable to deposition of oxides of sulphur and associated particulate matter.

Although there are many arsenic-rich areas in the world which are associated with food production, very little is known of the manner in which this element is transferred through the food chain. Of all the foods consumed by man, highest levels of arsenic are found in seafoods where it can occur in organic and inorganic forms. A nation such as Japan relies heavily upon seafoods as a constituent of normal diet and it is of interest to note blood levels of 60 ng As/ml, compared to about 3 ng As/ml for subjects not exposed to seafoods (Iwataki and Horiuchi, 17). In some countries seaweed is often used as a health food; Walkiw and Douglas (18) have described cases of arsenical poisoning associated with symptoms of peripheral neuropathy for individuals who consume such preparations, which can contain between 0.6 and 28 ppm As. In some countries arsenic in food is consumed but it does not appear to impair health, possibly because of the bulk composition of the material or perhaps because the arsenic is in a different chemical form; trivalent arsenic is highly toxic and inhibits enzyme systems by chelation to dithiol groups, while pentavalent arsenic is rapidly excreted in urine by the kidneys.

Mercury

The hazards from mercury have been well described by Friberg and Vostal (19) and O.E.C.D. (20). Apart from the disasters of Minimata and Niigata, where fifty-two people died following eating fish contaminated by methylmercury which originated in the effluent from a plastics manufacturing process, a further 459 hospital deaths occurred in Iraq in 1973 (Bakir et al., 21) following the eating of wheat grain contaminated with about 8 ppm Hg and flour with about 9 ppm of methylmercury. Apart from these exposures to toxic levels of mercury, much is to be learned from studying normal individuals who are exposed through eating habits to anomalous levels of this element but for which well-defined toxic effects are not observed. Hoshino et al. (22) have demonstrated that nationals from Japan who have lived in other countries for more than 1.5 years show a decrease in levels of mercury in hair but an increase in levels upon their return to Japan; it takes between one and two years before levels common to residents of Japan are reached. Suzuki, Matsubara-Khan, and Matsuda (2) report a similar study for Japanese nationals living in Burma and Bangladesh, while Suzuki and Ohta (24) report data for residents in Bolivia and Brazil. In all these studies concerned with Japanese nationals, it is presumed that the main source of mercury was present in seafoods.

Uranium

Although uranium is an ubiquitous constituent of man's natural environment, it does not appear to be an essential element for living matter and in man, at least, is best described as an adventitious element. However, uranium is able to enter body fluids and participate in various biochemical reactions through processes of limited anionic and cationic substitution. By making use of the very sensitive and rapid method of delayed neutron counting (see page 288) described initially by Amiel (25), Hamilton (26) obtained data, presented in Table XXX, for the concentration of uranium in various items of diet together with annual and daily intakes for the United Kingdom. Data presented in Table XXXI are for the concentration of uranium in some other commonly used ancillary items of diet; a concentration of 40 ng U per g salt, with an estimated United Kingdom intake of 3 g each day, provides a significant part of the annual intake of this element, namely about 50 μg U/y. The results of this investigation suggest an intake of about 1 μg per day which may be compared with the results of Welford and Baird (27) for an intake of 1.3, 1.4, and 1.3 μg uranium per day for New York, Chicago, and San Francisco respectively. The concentration of uranium in various items of diet shows very little variation with origin of samples. In relation to the concentration of uranium in soils, the components of man's food chain appear to discriminate against uranium.

Tin and Some Other Elements in Relation to Diet of the Newborn

Of all the human tissues available in sufficient quantity for study, possibly those provided by the newborn are the most accessible but yet, except for a few individuals, they are rarely subjected to analysis for levels of the chemical elements. If we

TABLE XXX*

THE CONCENTRATION OF URANIUM IN VARIOUS ITEMS OF DIET TOGETHER WITH ANNUAL AND DAILY INTAKE

Group	European Diet Intake (g/Person/Day)	No. Samples Analysed	Mean Concentration of Uranium in Diet (ng U/g Sample)	Annual Uranium Intake (µg)	% Annual Intake
Cereals[†]	375	15	0.5	68	19
Starchy Roots[‡]	377	10	1.0	138	38
Sugar	79	5	0.2	6	2
Vegetable and fruit[§]	316	18	0.8	92	26
Meat[‖]	111	14	0.4	16	4
Eggs	23	8	0.4	3	0.9
Fish[¶]	38	4	0.2	3	0.8
Milk	494	6	0.01	2	0.5
Fats, Oils**	44	8	2.0	32	9.0
			Total	360 µg U/yr	0.99 µg U/day

* From Hamilton, E. I.: The concentration of uranium in man and his diet. *Health Physics,* 22:151, 1972.
† Wheat flour, spaghetti, biscuits, rusks, oats, rice.
‡ Potatoes, parsnips, turnips, swedes.
§ Caggage, sprouts, carrots, green beans, apples.
‖ Beef, pork, N.B. lamb $< 1 \times 10^{-10}$ g U.
¶ Herring, cod.
** Butter, margarine.

are to be concerned with environmental levels of pollutants and contaminants to which adults are exposed, then surely we should pay attention to those levels of elements present in human tissues at birth and before infants are exposed to modern environments. The placenta could provide a very useful material in order to monitor levels of elements to which the mother is exposed during pregnancy. Thieme et al. (28) compared levels of trace elements in human placenta from different regions of Bavaria and relate high levels of Sb, Br, La, and Co for Munich with the environmental pollution of a large city. Widdowson et al. (29) examined livers of thirty human fetuses from twenty weeks gestation to term and of five adults for copper, zinc, manganese, chromium, and cobalt; the concentration of copper was ten

TABLE XXXI*
CONCENTRATION OF URANIUM IN VARIOUS INDIVIDUAL ITEMS OF DIET

Items	No. Samples Analysed	Sample (Mean Values) (ng U/g)
Tea	2	5
Coffee	4	6
Parsley (dry)	1	60
Thyme (dry)	1	90
Pepper (red)	1	5
Pepper (white)	2	0.2
Mixed spice	2	30
Paprika	2	20
Ginger	2	40
Cloves	2	8
Taragon	2	30
Beef extract Type 1	1	20
Beef extract Type 2	2	20
Mustard	2	<0.2
Dried fruit	6	3
Salt–table	14	40
cooking	30	40
iodised	1	40

* From Hamilton, E. I.: The concentration of uranium in man and his diet. *Health Physics*, 22:152, 1972.

times as high in the fetal livers than in the adult ones, and the liver at full term had nearly 1.5 times as much copper as in those from adults. The concentration of zinc was higher in the livers of the younger fetuses than of the older ones and all the fetal values were above the adult level. There seemed to be no accumulation of manganese, chromium, or cobalt in the liver before birth.

The chemical (inorganic) composition of the newborn, to some degree, reflects the chemistry of the mother. Prior to birth the placenta provides some degree of protection to the developing fetus, but the efficiency of placental transfer for some elements, known to be toxic to adults, is largely unknown for the newborn. After birth, apart from inhalation of elements present in the air or through skin absorption, either the mother's milk or some manufactured form of milk provides the sole route of entry of the chemical elements. Very little data are available for the levels of elements present in mother's milk, or alternative types of milk, and I know of no comprehensive description for levels of elements in the colostrum. As the newborn develop, the liquid food is gradually replaced by solids, usually specially prepared for infants, the process gradually taking place over a period of about one year. Apart from the availability of postmortem samples of the newborn, it is relatively easy to collect samples of urine and faeces, and also hair, in order to observe changes in levels of elements with growth.

An example of differences in the chemical form of an element, namely zinc, in natural and artificial milk has been reported by Massi et al. (30). Differences have been observed in both the quantitative and qualitative zinc content of human and cow milk, and also colostrum, compared with milk powder preparations commonly used as a baby's food. In unprocessed milk and colostrum, the zinc was mainly present as a casein-dependent element, whereas the remaining zinc was found in the immunoglobulin fraction and only trace amounts in the free ionic form. In milk powders the total zinc concentrations were lower than those found for natural products and were mainly present as free or protein-independent zinc; the immunoglobulin fraction of zinc was practically absent.

In 1971 my attention was drawn to the inside of a tin can containing condensed milk upon which a baby was fed soon after birth; the inner surface of the can was marked by stains which were more prominent along the soldered seam; mother's milk was not available to the infant, whose sole source of food was the condensed milk and water from a domestic supply. This form of diet is not uncommon and as far as the tinned milk is concerned this was advocated as a matter of routine practice in a large maternity hospital in southeast England. The particular brand of milk was contained in unlined tin cans and in a random examination of several all showed signs of staining; as a result of this investigation other baby milks contained in tin cans were examined and the results clearly showed that the staining was a feature of the unlined cans while those which had an internal protective layer of plastic film were not stained. Data are presented in Table XXXII for levels of some elements in normal bottled (glass) cow's milk and are compared with data obtained from evaporated milk from an unlacquered can. Apart possibly from lead, anomalously high concentrations of tin, manganese, and chromium were found. A mean concentration of about 8×10^{-3} µg Sn/ml of milk was found for bottled cow's milk compared with a concentration of about 16 µg Sn/ml for the tinned evaporated milk, when diluted to the equivalent volume of cow's milk. The concentrations of tin in milk from unlacquered and lacquered cans were measured in other samples of evaporated milk and are reported in Table XXXIII. Significant amounts of tin, ranging between 28 and 100 µg Sn/ml of evaporated milk, were only found to be associated with milk from unlacquered cans; higher levels of tin were found in the milk from smaller cans.

In order to determine whether or not the tin present in the milk from the unlacquered cans could cross the gastrointestinal tract, the concentration of tin in the infant's faeces and urine was also determined. The results of these measurements are given in Table XXXIV and show that practically all the tin is excreted in the faeces. By the time this investigation was completed, five weeks after birth, the infant's milk supply was changed to that from a lacquered can; data presented in Table XXXIV clearly illustrate the rapid clearance of tin from the gut. In relation to adult exposure to tin for the United Kingdom the data given in Table XXIX indicate a dietary intake of 187 ± 42 µg Sn/day compared with an intake of 11230 µg Sn/day for the infant receiving milk from an unlacquered can. In a limited study from the United States, an intake of 7110 ± 900 µg Sn/day has been reported by Tipton, Stewart and Dickson (9). A further study by Ratu and Sporn (13) reports the following levels of tin in various foodstuffs: marmalade 16.35 ± 6.8 mg Sn/kg (max 55 mg Sn/kg), tomato paste 4.5-4.7 mg Sn/kg (max 45 mg Sn/kg), and vegetables 4-68 mg Sn/kg. Quite clearly corrosion of the tin lining and the solder used in the seams can contaminate foods; examination of the chemical form of the tin in the infant's milk indicated the presence of insoluble fragments of tin, dis-

TABLE XXXII*

THE CONCENTRATION OF THE CHEMICAL ELEMENTS IN
UNITED KINGDOM COW'S MILK AND THE EQUIVALENT
CONCENTRATION IN DILUTED EVAPORATED
MILK FROM AN UNLACQUERED CAN
(Assay by spark source mass spectrometry)

Element	Mean Values for U.K. Cow Milk (μg/ml) (12 Samples)	Semi Quantitative Data for Evaporated Milk From an Unlacquered Can (μg/ml), Equivalent Normal Milk
U	$< 2.5 \times 10^{-3}$	$< 2.5 \times 10^{-3}$
Th	$< 2.5 \times 10^{-3}$	$< 2.5 \times 10^{-3}$
Bi	$< 2.2 \times 10^{-3}$	$< 2.2 \times 10^{-3}$
Pb	$2.5 \pm 0.8 \times 10^{-2}$	8×10^{-2}
Lu	$< 1.9 \times 10^{-3}$	$< 1.9 \times 10^{-3}$
Yb	$< 5.8 \times 10^{-3}$	$< 5.8 \times 10^{-3}$
Tm	$< 1.8 \times 10^{-3}$	$< 1.8 \times 10^{-3}$
Er	$< 5.3 \times 10^{-3}$	$< 5.3 \times 10^{-3}$
Ho	$< 1.8 \times 10^{-3}$	$< 1.8 \times 10^{-3}$
Dy	$< 6.9 \times 10^{-3}$	$< 6.9 \times 10^{-3}$
Tb	$< 1.7 \times 10^{-3}$	$< 1.7 \times 10^{-3}$
Gd	$< 7.6 \times 10^{-3}$	$< 7.6 \times 10^{-3}$
Eu	$< 3.0 \times 10^{-3}$	$< 3.0 \times 10^{-3}$
Sm	$< 6.0 \times 10^{-3}$	$< 6.0 \times 10^{-3}$
Nd	$< 5.6 \times 10^{-3}$	$< 5.6 \times 10^{-3}$
Pr	$< 1.5 \times 10^{-3}$	$< 1.5 \times 10^{-3}$
Ce	$< 1.7 \times 10^{-3}$	$< 1.7 \times 10^{-3}$
La	$< 1.5 \times 10^{-3}$	$< 1.5 \times 10^{-3}$
Ba	0.2 ± 0.1	0.1
Cs	$6.6 \pm 1.0 \times 10^{-3}$	2×10^{-3}
I	0.4 ± 0.1	4×10^{-2}
Sb	$9.4 \pm 1.7 \times 10^{-3}$	5×10^{-3}
Sn	$7.8 \pm 1.2 \times 10^{-3}$	16
Ag	$< 1.7 \times 10^{-3}$	4×10^{-3}
Cd	$3.4 \pm 0.5 \times 10^{-2}$	7×10^{-3}
Mo	$5.6 \pm 0.7 \times 10^{-2}$	
Nb	$2.5 \pm 0.3 \times 10^{-2}$	1×10^{-2}
Zr	$1.2 \pm 0.2 \times 10^{-3}$	2×10^{-2}
Y	$3.5 \pm 0.4 \times 10^{-3}$	1×10^{-2}
Sr	0.3 ± 0.1	0.3
Rb	2.7 ± 0.2	2
Br	4.2 ± 0.4	7
As	$< 6 \times 10^{-4}$	$< 6 \times 10^{-4}$
Zn	4.4 ± 0.1	4
Cu	0.3 ± 0.1	0.3
Fe	4.1 ± 0.7	7
Mn	$7.4 \pm 0.8 \times 10^{-2}$	2
Cr	$4.7 \pm 0.5 \times 10^{-2}$	0.4
Ca	$1.4 \pm 0.1 \times 10^{3}$	1.4×10^{3}
K	$1.6 \pm 0.1 \times 10^{3}$	1.6×10^{3}

TABLE XXXII (continued)

Cl	$1.3 \pm 0.1 \times 10^3$	1×10^3
S	403 ± 8	330
P	871 ± 11	871
Si	6.0 ± 2	11
Al	0.9 ± 0.1	0.2
Mg	78.8 ± 10.4	74.6
F	$9.1 \pm 1.7 \times 10^{-2}$	1×10^{-2}
B	0.4 ± 0.1	0.9
Li	$1.6 \pm 0.3 \times 10^{-2}$	9×10^{-3}

* From Hamilton, E. I., Minski, M. J., Cleary, J. J., and Halsey, V. S.: Comments upon the chemical elements present in evaporated milk for consumption by babies. *Sci Total Environ*, 1:206-207, 1972.

TABLE XXXIII*
THE CONCENTRATION OF TIN (μg/ml) IN EVAPORATED MILK FROM UNLACQUERED AND LACQUERED CANS
(Assay by X-ray fluorescence analysis)

Sample No.	Type of Can/Volume of Milk	μg Sn/ml of Milk
3834 UJR 31	Raw tin/394 ml	29
3840 UJR 31	Raw tin/394 ml	28
3832 2DR 12	Raw tin/197 ml	81
3831 PPX AAK	Raw tin/197 ml	110
3830 711C 700A	Lacquered tin/197 ml	< 5
3835 63871.1	Lacquered tin/394 ml	< 5

* From Hamilton, E. I., Minski, M. J., Cleary, J. J., and Halsey, V. S.: Comments upon the chemical elements present in evaporated milk for consumption by babies. *Sci Total Environ*, 1:207, 1972.

solved tin, and trace amounts of organic tin compounds. While the real effects of elevated levels of tin in infant's milk upon health, during childhood, and later in life are unknown, it is apparent from this work that the unlacquered cans could be a source of some other toxic elements, such as lead. Boppel (32) determined an intake of 46 μg Pb/day for six-months-old infants compared with an adult intake, measured for four individuals over a period of one week, of 121 μg Pb/day. It is some comfort today to note that about this time the lead problem was recognised and solid foods at least began to be supplied in glass containers.

As the sole supply of food for the newborn is contained in animal or human milk, considerably more attention should be paid to the levels of elements present in such materials. Murthy

TABLE XXXIV*

DAILY INTAKE AND LOSS OF TIN FOR A BABY FIVE WEEKS OLD
FED ON EVAPORATED MILK FROM UNLACQUERED CANS

Sample	Tin Content
Mean concentration of Sn in milk	28.5 µg/ml
Intake of milk/24 h	394 ml
Mass of Sn in milk/24 h	11.23 mg
Weight of faeces in 24 h	54 g
Mass of Sn in faeces/g	197 µg/g
Mass of Sn in faeces/24 h	10.64 mg
Weight of urine in 24 h	~ 300 g
Mass of Sn in urine/ml	0.77 µg/ml
Mass of Sn in urine/24 h	0.23 mg
Total Sn excreted/24 h	10.87 mg
Total mass Sn retained	0.36 mg

EFFECT OF TYPE OF CAN LINING ON EXCRETION OF TIN IN FAECES

Sample No.	Date/Time of Sample (h)	Type of Can Lining	µg Sn/g of Wet Faeces
3838	16/12/71 12:00	Raw tin	197
3829	17/12/71 11:00	Change raw → lacquered tin	4.4
3841	18/12/71 23:30	Lacquered tin	< 4
3843	19/12/71 08:00	Lacquered tin	< 4
3844	20/12/71 03:00	Lacquered tin	< 4

* From Hamilton, E. I., Minski, M. J., Cleary, J. J., and Halsey, V. S.: Comments upon the chemical elements present in evaporated milk for consumption by babies. *Sci Total Environ, 1*:208, 1972.

(33) has discussed levels of Cu, Mn, I, Se, Zn, Mo, Co, F, Cr, Al, Cd, Pb, Ni, As, B, Br, Hg, Rb, Ag, and Sr in human and animal milk and concludes that both types of milk are poor sources of trace elements of nutritional importance; in the future there is a need to emphasise a requirement for nationwide studies, better sampling, and improved quality of chemical analysis. Because of the close association between soils, herbage, and animal milk, variations in the levels of elements in herbage will be reflected by those in milk, together with geographical differences such as described by Shearer, Demetrios, and Hadjimarkos (34) for selenium in milk.

The essential or nonessential nature of the chemical elements for man, at all stages of development, is a subject worthy of urgent consideration. First of all it is essential to develop the optimum methods of analysis and control experiments for both animals and man. The work of Schwarz and his co-workers (Schwarz, 35, 36) serve as excellent examples of how to tackle this difficult and exacting subject (for animal studies); Schwarz set himself some rigid standards which should be fulfilled before claims for essentiality can be made, namely:

1. the experiment should produce highly significant responses;
2. it should be reproducible at will, and in a series of tests over a lengthy interval of time;
3. a dose-response curve should be established and the minimum effective dose level of the element should be determined;
4. several compounds of the same element should be tested and compared in potency;
5. the effect should be physiological, i.e. it must be obtained using amounts which are normally present in foods and tissues.

In order to carry out experimental work on animals, Schwarz and Smith (37) developed a trace-element free isolator system which consisted of the following basic elements: isolator, air lock, blower, air filter assembly, cage assembly, refuse trays, food cups, waterbottles, and scales. Plastics were used for all components, and air filtration removed particles down to sizes of 0.35 μm^2. An illustration of the trace-element controlled isolator system is given in Figure 27. In order to control the quality and trace element level of diets, complete chemically defined diets were developed based upon aminoacids in place of proteins. As a result of adopting this rigorous approach, the first clear-cut growth effects of tin for the rat were observed in 1968; tin added at physiological levels of 1.5-2.0 ppm to a tin-deficient diet produced an enhanced rate of growth.

Schwarz advances the hypothesis that it is possible that an element such as tin is not required for intrauterine life but may be required later. Within hours after birth, tin is found in tissues

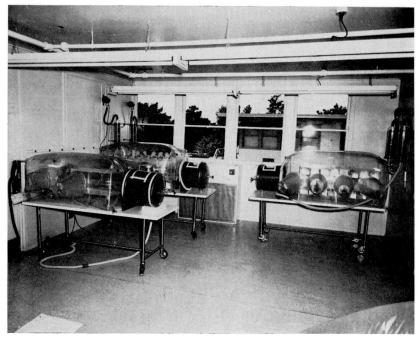

Figure 27: Trace-element controlled isolators. Each isolator holds 32 young rats in a trace-element "sterile" environment. The above three isolators were used by Schwarz (39) in establishing the essentiality of four elements (tin, vanadium, fluorine, and silicon) for growth in the rat. Photograph reproduced with permission from Dr. K. Schwarz.

and originates from the colostrum. It is suggested that tin acts as a signal which stimulates postpartum processes which are indispensable in extrauterine life. Therefore, we have a well-defined area of ignorance for the role of trace elements in the newborn, which is likely to be of even greater importance for the premature infant. The significance of nutrition and biochemical activity of the colostrum is intriguing; if it acts as a trigger for essential metabolic processes then this must concern enzymes in a developing system. Although enzyme reactions tend to be specific for characteristic elements, their efficiency can easily become impaired by the presence of other elements which, because of similarities in chemistry, can participate in substitution reactions detrimental to normal development. The areas of concern, to mention a few, would include gut processes and gut transfer mechanisms, respiration, brain development, and protein synthesis. While it is easy to speculate on such matters, the fact remains that the chemical elements, even when present in trace quantities, can have very dramatic effects on the development and function of biochemical systems. The area of concern must rest with enzyme systems: the tools for carrying out the demanding analytical investigations are available; perhaps what is lacking, overall, is a little imagination and innovative research into this important field of work. Until evidence is forthcoming, speculations are easy to dismiss, but it seems certain that effects are registered in the newborn. Accepting that an individual's response to insults from the chemical elements has an individual character, it seems inevitable that some forms of morbidity in early childhood will accrue, and there remains the question of later effects in life as a result of damage or disturbance to the biochemistry of the body initiated at birth. Until we know more about the essential or nonessential nature of the chemical elements in different chemical forms and variability with geography, far more objective research is required. There are now about fourteen trace elements for which essential functions in warm-blooded animals have been conclusively demonstrated and perhaps some twenty additional elements should be taken into serious consideration as potential pretenders for essential trace-element function. The subject has to be approached from the point of view of multidisciplinary cooperative ventures; it is very complex and the nature of the research is long-term dedication. An objective should be to further our understanding of the interplay of elements in real systems, possibly with little hope of observing large dramatic effects, but rather defining more precisely borderline states of health subject to considerable variability, and for which the addition of a particular element, or changes in concentration, can result in defined patterns of morbidity.

Food and Geography

In the passage of elements from rocks→soils→plants→animals→man various processes of discrimination occur which often reduce the extent to which elements available in the environment enter items of diet. The more technologically advanced a country becomes, the more the natural association with the environment becomes lost. However there are many parts of the world where it is still possible to study man in the surroundings of a natural environment and to observe patterns of morbidity.

In the Soviet Union (East Transbaikal area and Amur area), the Republic of China, and North Korea, provinces with calcium deficiency have been described by Vinogradov (38), Koval'skii (39), and Khobot'ev (40). In these areas a disease, Urov disease, caused by calcium deficiency is endemic and has been described by Damperov (41). Urov disease is manifest in both animals and people early in life and results in an enlargement and stiffening of joints, inhibition of movement, and an effect on the function of epiphyses and cartilage. The majority of human victims are between eleven and fifteen years old. In the past the disease has been attributed to the presence in water of Pb, Cd, Au, Fe, and Ra, also the contamination of grain by Fusarian fungi. In the endemic area strontium is concentrated in the humics of soils and is taken up into plants and animals in place of calcium. Similarities in chemical and physical properties between Ca^{2+} and Sr^{2+}, such as ionic radii, permit Sr to readily enter Ca compounds; the extent to which substitution of Sr for Ca takes place varies considerably. As a result of the substitution of Ca by Sr in bone hydroxyapatite, changes will take place in the crystal structure which will affect both the mechanical properties of bone and bone-forming processes. The possibility of finding areas of calcium depletion and strontium enrichment in the diet of developed countries is remote although still worthy of consideration. Hamilton and Minski (42) report the results of a survey of the strontium content of human bone for the United Kingdom following a survey by Mole (43). The geology of surface deposits and bedrock for the United Kingdom is known in great detail, hence it is possible to identify calcareous areas of rock and compare the distribution of such materials with that of strontium in bone. The results of this study are presented in Figure 28 and illustrate the following:

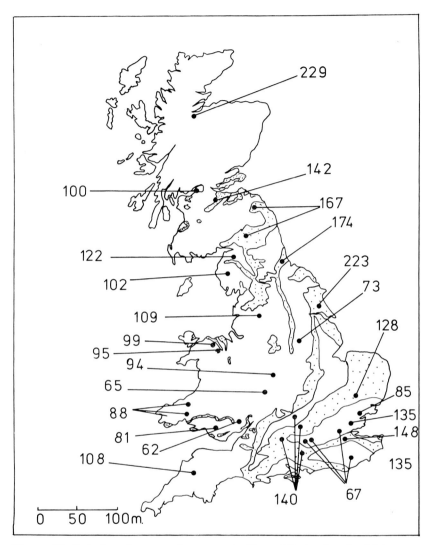

Figure 28: Relations between the concentration of Sr (μg/g of ash) in samples of ashed human bone and the distribution of calcareous rocks (shown by dots) for the United Kingdom. The main type of calcareous rock is either chalk or limestone.

1. higher Sr content of bone for southeast and northwest England is related to the presence of calcareous rocks such as chalk and limestones;

2. two areas of noncalcareous sandstone, one in central England and the other in northwest England, also tend to be associated with high levels of strontium in bone;
3. the level of strontium in bone from individuals residing in noncalcareous regions is low relative to that for calcareous regions.

Mole comments upon the distinctive localised geographical distribution for levels of strontium in bone and observes that—

1. the strontium content of milk is related to the distribution of this element in bone. The strontium content of milk will be directly related to the overall geochemistry of the pastures from which the milk is obtained; however any simple picture is distorted by the distribution of milk throughout the country and in some areas stall feeding of livestock with imported fodder during winter months;
2. with the exception of northwest and southeast England, the level of strontium in bones of children from two to five years of age is not demonstrably different from that of adults;
3. in the newborn, differences in strontium levels in bone may be related to seasons of the year and mode of death, for example congenital malformations.

Mean data for the mass of strontium in human bone ash for various countries indicate a range of values of about a factor three. Using data obtained by Thurber et al. (44) for towns and cities, reported in terms of strontium content per gram of bone ash, we note levels of 104 $\mu g/g$ for Boston, Massachusetts, United States, 160 μ/g for London, England, and 344 μ/g for Recife, Brazil; this data may be a reflection of nutritional standards or levels of strontium in relation to calcium in local natural environments as illustrated for the United Kingdom in Table XXXV. Results for levels of strontium in bone ash from Italy are presented in Table XXXVI and illustrate quite striking regional differences: the ratio of young/adult for Sr is 0.62 for Udine, compared with 0.88 for Rome and 0.98 for Milan; the Rome and Bari areas are associated with calcareous rocks.

TABLE XXXV*

A COMPARISON BETWEEN THE STRONTIUM CONTENT OF HUMAN BONE (μg/g ash) AND WATER QUALITY FOR THE UNITED KINGDOM

Area	ppm Sr (μg Sr/g Ash)	Total Water Hardness†	Category of Hardness†‡
Invernesshire	229	3-160	S-MH
Wigton-Aspatria	166 ± 34 (A)§	12-60	MS
London-S.E. England	146 ± 6	294	SH-VH
Thames Valley	139 ± 6	150-350	MH-VH
Carlisle S. Fringe	124 ± 9 (A)	12-60	MS
Carlisle	119 ± 8	65	MS
Stirlingshire	133 ± 8	30-55	S-MS
Lanarkshire	110 ± 7	25-50	S-MS
Glamorgan	110 ± 8	110-180	S-H
Devonshire	108 ± 9	60-250	S-H
Lancashire	107 ± 10	25-108	S-MH
S. Wales (Cardiff)	106 ± 6	110	S-H
S.E. Lancashire	105 ± 10	19-72	S
Blackpool, Fylde	104 ± 9	24-90	S-MS
Birmingham, Warwickshire	104 ± 5	100	S-H
Worcestershire	104 ± 28 (A)	80-305	S-VH
Westmorland	99 ± 3	18-184	S-MS
Shropshire	94 ± 8 (A)	240	SH-VH
N. Wales	93 ± 4	50	S-SH
Glasgow, Renfrewshire	89 ± 8	9-58	S-MS

* From Hamilton, E. I., and Minski, M. J.: Abundance of the chemical elements in man's diet and possible relations with environmental factors. *Sci Total Environ*, 1:389, 1972/73.
† Values for total hardness refer to data reported for particular major water companies. Categories of hardness provide a general guide for the quality of water in various countries of the United Kingdom.
‡ S = Soft, 0-50 ppm; MS = moderately soft, 50-100 ppm; SH = slightly hard, 100-150 ppm; MH = moderately hard, 150-200 ppm; H = hard, 200-300 ppm; VH = very hard, > 300 ppm.
§ A = Adult bone.

TABLE XXXVI*

THE CONCENTRATION OF Sr (μg/g ash) IN HUMAN BONE (VERTEBRAE) FROM ITALY†

Locality	Age Range	No. of Samples	Sr (μg/g Ash)
Udine‡	< 1 y	30	50 ± 20¶
	> 37 y	6	95 ± 44
Rome§	< 1 y	36	158 ± 39
	> 29 y	3	179 ± 29
Milan‖	< 1 y	28	186 ± 31
	> 44 y	6	189 ± 65
Bari	< 2 m	25	108 ± 17

* From Hamilton, E. I. and Minski, M. J.: Abundance of the chemical elements in man's diet and possible relations with environmental factors. *Sci Total Environ*, 1:382, 1972/73.
† Samples obtained from E. Lanzola. Data obtained by X-ray fluorescence assay.
‡ Sr intake diet 558 μg/day;[12] Ca intake diet 0.62 g/day.
§ Ca intake diet 01.09 μg/day.[46]
‖ Ca intake diet 1.12 g/day.[46]
¶ Standard deviation.

Food—Levels of Elements Altered by Manufacturing Processes

Apart from the deliberate addition of elements to foodstuffs, such as calcium to flour and iodine to salt, most manufacturing processes tend toward a "purified" product and elements not inherently associated with the final product tend to be lost during manufacture. An example of changes in levels of elements in processing crude sugar molasses into refined white sugar is illustrated in Table XXXVII. Tipton and Cook (4) and Schroeder (46) have shown that the mass of chromium in human tissues is significantly lower for United States subjects compared with those from Africa and the Orient. Schroeder and Buckman (47) showed that raw sugars contain more chromium than refined sugars. Chromium exerts a beneficial effect against experimentally induced atherosclerosis and it is tempting to suggest that as North American populations are highly prone to atherosclerosis and myocardial infarction compared with those from Africa and the Orient, this could be related to the consumption of refined sugar in North America. Masironi, Koirtyohann, and Pierce (48) have illustrated that the simple act of polishing rice brings about a loss of essential elements such as Zn and Cu, but not Cd. In countries such as Japan, where rice is a major staple food, high Cd content of rice may cause excessive absorption of this toxic metal. Today evidence is accumulating that marginal deficiency of beneficial trace elements, or excess of harmful ones, may possibly play a role in the etiology of hypertension and of atherosclerotic heart disease.

There is still much to be learned concerning the changes in trace and minor element chemistry brought about by technological production of foods. Possibly one of the most interesting areas is associated with the production of synthetic foods where emphasis is usually placed upon nutritional quality rather than trace element levels. One food of interest is synthetic protein manufactured from the soya bean, which until now has not constituted an essential component of man's diet but has mainly been used as cattle feed. Data obtained by Conner and Shacklette (49) are presented in Table XXXVIII for the levels of

TABLE XXXVII*

SEMIQUANTITATIVE ESTIMATES† FOR THE CONCENTRATION
(μg/g dry weight) FOR VARIOUS ELEMENTS IN DIFFERENT TYPES
OF CANE SUGARS

Element	Barbados Brown Sugar	Demerara Sugar	Refined Sugar	Granulated Sugar
Pb	0.2	0.02	< 0.001	0.002
Ba	0.8	0.3	0.01	0.01
Cs	0.04	0.02		
I	0.03	0.003	< 0.001	< 0.001
Sb	0.08	0.006	< 0.002	< 0.002
Sn	0.1	0.03	< 0.004	0.01
Cd	0.2	0.06	< 0.007	< 0.007
Ag	0.03	0.004	< 0.001	0.002
Nb	0.01	0.003	< 0.007	< 0.007
Zr	0.03	0.004	< 0.001	0.008
Y	0.07	0.006	< 0.005	0.001
Sr	9	2	0.03	0.1
Se	0.2	0.03	0.005	0.007
Ge	0.1	0.03		
Ga	0.2	0.04		
Zn	3	0.5	< 0.02	< 0.02
Cu	3	0.3	0.4	0.08
Fe	49	8	11	0.1
Mn	15	2	0.02	0.01
Cr	3	0.4	0.02	0.08
V	0.4	0.02	0.002	< 0.001
Ti	7	1	0.5	< 0.007
Sc	2	0.1	0.001	0.003
Ca	1,650	81	12	26
K	15,000	1,900	18	123
S	2,040	500	5	3
P	250	25	3	0.2
Si	735	60	2	4
Al	0.7	0.90	0.02	0.007
Mg	1,760	144	2	0.5
F	0.05	0.06	0.02	0.003
B	5	2	0.4	0.3
Be	0.03	0.006	0.002	0.0002
Li	0.1	0.03	0.001	0.0002

* From Hamilton, E. I., and Minski, M. J.: Abundance of the chemical elements in man's diet and possible relations with environmental factors. *Sci Total Environ*, 1:380, 1972/73.
† Assay by spark source mass spectrometry; no relative sensitivity factors applied.

some elements in a high natural protein food, maize, and the soya bean which is subjected to chemical processing to form synthetic meat. A comparison of both sets of data illustrates sig-

TABLE XXVIII
A COMPARISON OF SOME LEVELS OF ELEMENTS IN CORN AND SOYA BEAN*

Element	Corn	Soya Bean
Ba	21 ± 7 ppm	295 ± 101 ppm
B	63 ± 5 ppm	210 ± 29 ppm
Cd	0.51 ± 0.1 ppm	1.1 ± 0.8 ppm
Ca	0.31 ± 0.01%	5.45 ± 0.57%
Co	< 1 ppm	1.8 ± 0.5 ppm
Cu	92.8 ± 33.5 ppm	197 ± 27.5 ppm
F	0.5 ppm	0.48 ± 0.02 ppm
Fe	1.35 ± 0.23%	0.12 ± 0.014%
Pb	< 20 ppm	—
Mg	6.23 ± 0.25%	3.08 ± 0.46%
Mn	290 ± 30 ppm	348 ± 41 ppm
Mo	12 ± 4 ppm	14 ± 5 ppm
Ni	23.8 ± 8.7 ppm	105.3 ± 19.1 ppm
P	1.16 ± 0.01%	1.41 ± 0.04%
K	1.07 ± 0.02%	1.05 ± 0.02%
Si	0.06 ± 0.01%	0.12 ± 0.04%
Sr	15.75 ± 1.5 ppm	305 ± 108 ppm
Ti	< 5 ppm	6.4 ± 3.1 ppm
V	< 5 ppm	—
Zn	1,825 ± 96 ppm	990 ± 127 ppm
Zr	< 20 ppm	—
Mean % ash	1.63 ± 0.13	1.073 ± 0.0189

* Data from Conner, J. J. and Shacklette, H. T.: Background geochemistry of some rocks, soils, plants and vegetables in the conterminous United States. U.S. Geological Survey, Professional Paper 574-F, 1974.

nificant differences in the relative abundance of elements in both materials. As the soya bean is a natural product, element levels will be influenced by those present in the soils in which it is grown; as a result of further chemical and physical processing it can be anticipated that certain changes in the levels of elements will take place which may have some bearing upon some forms of nutritional disorders, particularly if supplied in quantity to people suffering from starvation where any natural element balance is likely to be disturbed.

REFERENCES

1. Bransby, E. R. and Fothergill, J. Reported by J. Longwell: The fluoridation of public water supplies, (d) chemical and technical aspects. R So Health J, 77:361, 1957.

2. Hamilton, E. I.: The chemical elements and human morbidity—water, air and places—a study of natural variability. *Sci Total Environ,* 3: 3, 1974.
3. Shah, K. R., Filby, R. H., and Davies, A. I.: Determination of trace elements in tea and coffee by neutron activation analysis. *Int J Environ Anal Chem,* 1:63, 1971.
4. Anderson, W., Hollins, J. G., and Bond, P. S.: The composition of tea infusions examined in relation to the association between mortality and water hardness. *J Hyg Camb,* 69:1, 1971.
5. Bhat, I. S.: Daily intake of cobalt by the adult population of Tarapur. *Health Phys,* 24:553, 1973.
6. Yamagata, N., Kurioka, W., and Shimuzu, T.: Balance of cobalt in Japanese people and diet. *J Radiat Res,* 4:8, 1963.
7. Schroeder, H. A., Nason, A. P., and Tipton, I. H.: Essential trace metals in man: cobalt. *J Chron Dis,* 20:869, 1967.
8. Tipton, I. H., Stewart, P. L., and Martin, P. G.: Trace elements in diets and excreta. *Health Physics,* 12:1683, 1966.
9. Tipton, I. H., Stewart, P. L., and Dickson, J.: Patterns of elemental excretion in long term balance studies. *Health Phys,* 16:455, 1969.
10. Ministry of Agriculture, Fisheries and Food: *Survey of Lead in Food.* Lond, HMSO, 1972.
11. Thompson, J. A.: Balance between intake and output of lead in normal individuals. *Br J Ind Med,* 28:189, 1971.
12. Lloyd, W. E., Hill, H. T., and Meerdink, G. L.: Observations of a case of molybdenosis-copper deficiency in a South Dakota dairy herd. In Chappell, W. R. and Petersen, K. K.: *Molybdenum in the Environment,* vol. I. New York, Dekker, 1976.
13. Vallee, B. L., Ulmer, D. D., and Wacker, W. E. C.: Arsenic toxicology and biochemistry. *AMA Arch Indust Hlth,* 21:132, 1960.
14. Luh, M-D., Baker, R. A., and Henley, D. E.: Arsenic analysis and toxicity—Review. *Sci Total Environ,* 2:1, 1973.
15. Colbourn, P., Alloway, B. J., and Thornton, I.: Arsenic and heavy metals in soils associated with regional geochemical anomalies in southwest England. *Sci Total Environ,* 4:347, 1975.
16. Porter, E. K. and Peterson, P. J.: Arsenic accumulation, by plants on mine waste (United Kingdom). *Sci Total Environ,* 4:365, 1975.
17. Iwataki, N. and Horiuchi, K.: Arsenic content of blood. *Osaka City Med J,* 5:209, 1959.
18. Walkiw, O. and Douglas, D. E.: Health food supplements prepared from kelp—a source of elevated urinary arsenic. *Clin Toxicol,* 8: 325, 1975.
19. Friberg, L. and Vostal, J. (Eds.): *Mercury in the Environment. An Epidemiological and Toxicological Appraisal.* Cleveland, Ohio, CRC Press, 1972.

20. *Mercury and the Environment.* Paris, OECD Publication, 1974.
21. Bakir, F., Damluji, S. F., Amin-Zaki, L., Murtadha, M., Khalidi, A., Al-Rawi, N. Y., Tikriti, S., Dhamir, H. I., Clarkson, T. N., Smith, J. C., Doherty, R. A.: Methylmercury poisoning in Iraq. *Science, 181:*230, 1973.
22. Hoshino, O., Tanzawa, K., Hasegawa, Y., and Ukiya, T.: Mercury content of hair for Japanese nationals. *Eisei Kagaku, 12:*90, 1966.
23. Suzuki, T., Matsubara-Kahn, J., and Matsuda, A.: Mercury content of hair of Japanese after emigration to Burma or East Pakistan. *Bull Environ Contam 7 Toxicology, 7:*26, 1972.
24. Suzuki, T. and Ohta, U.: Mercury content of hair of Japanese after emigration to Bolivia and Brazil. *Jap J Publ Health, 16:*769, 1969.
25. Amiel, S.: Analytical application of delayed neutron emission in fissionable material. *Israel AEC,* Report IA-621: 58, 1961.
26. Hamilton, E. I.: The concentration of uranium in man and his diet. *Health Physics, 22:*149, 1972.
27. Welford, G. A. and Baird, R.: Uranium levels in human diet and biological materials. *Health Physics, 13:*1321, 1967.
28. Thieme, R., Schramel, P., Klose, B-J., and Waidl, X.: Der Einfluss regionaler Umreitfaktorne auf die spurenelement Zusammensetzung der Plazenta. *Geburtshilfe Frauenheild, 34:*36, 1974.
29. Widdowson, E. M., Chan, H., Harrison, G. E., and Milner, R. D. G.: Accumulation of copper, zinc, manganese, chromium and cobalt in human liver before birth. *Biol Neonate, 20:*360, 1972.
30. Nassi, L., Poggini, G., Nassi, P. A., Vecchi, C., and Galvan, P.: Zinc in natural and artificial milk. *Ann Sclavo, 17:*848, 1975.
31. Rătu, R. and Sporn, A.: Contributii la stabilirea aportului de plumb şi stanniu prin ratia alimentařa. *Igiena, XVI:*533, 1967.
32. Boppel, B.: Lead content of foodstuffs. 4. Lead content of infant foods. *Z Lebensm Unters Forsch, 158:*291, 1975.
33. Murthy, G. K.: Trace elements in milk. *CRC Critical Reviews in Environmental Control.* January 1, 1974.
34. Shearer, T. R., Demetrois, and Hadjimarkos, D. M.: Geographic distribution of selenium in human milk. *Arch. Environ. Health. 30:*230, 1975.
35. Schwarz, K.: Elements newly identified as essential for animals. In *Nuclear Activation Techniques in the Life Sciences.* Vienna, International Atomic Energy Agency, 1972.
36. Schwarz, K.: Tin as an essential growth factor for rats. In Mertz, W. and Cornatzer, W. E.: *Newer Trace Elements in Nutrition.* New York, Dekker, 1971.
37. Schwarz, K. and Smith, J. C.: A controlled environment system for new trace element deficiencies. *J Nutr, 93:*182, 1967.
38. Vinogradov, A. P.: Biogeochemical provinces. *Tr Dokuchaevskoi sessii Akad,* Nauk, USSR, 1948.

39. Koval'skii, V. V.: New goals and problems in the biochemistry of farm animals in connection with the study of biogeochemical provinces. *Biogeokhim labor Akad,* Nauk, USSR, 1957.
40. Khobot'ev, V. G.: Biogeochemical provinces with calcium deficiency. *Geochemistry* (USSR) 8:830, 1960.
41. Damperov, N. I.: The Urov or Kashen-Bek disease. *Medgiz,* 1939.
42. Hamilton, E. I. and Minski, M. J.: Abundance of the chemical elements in man's diet and possible relations with environmental factors. *Sci Total Environ, 1:*375, 1972/1973.
43. Mole, R. H.: Stable strontium in human bone: geographical and age differences in the United Kingdom and their correlation with levels of strontium-90. *Br J Nutr, 19:*13, 1965.
44. Thurber, D. L., Kulp, J. L., Hodges, E., Gast, P. W., and Wampler, J. M.: Common strontium content of the human skeleton. *Science, 128:*256, 1958.
45. Tipton, I. H. and Cook, M. J.: Trace elements in human tissue. Part II —Adult subjects for the United States. *Health Physics, 9:*103, 1963.
46. Schroeder, H. A.: Cadmium, chromium and cardiovascular disease. *Circulation, 35:*570, 1967.
47. Schroeder, H. A. and Buckman, J.: Cadmium hypertension, its reversal in rats by a zinc chelate. *Arch Environ Hlth, 14:*693, 1967.
48. Masironi, R., Koirtyohann, S. R., and Pierce, J. O.: Zinc, copper, cadmium and chromium in polished and unpolished rice. *Sci Total Environ, 7:*27, 1977.
49. Conner, J. J. and Shacklette, H. T.: Background geochemistry of some rocks, soils, plants, and vegetables in the conterminous United States. United States Geological Survey Professional Paper 574-F, 1975.

Chapter 6

METHODS—PREPARATORY TECHNIQUES

WHEN DISCUSSING METHODS of analysis it is common practice to only describe those processes involved in analytical chemistry carried out at the bench, together with details of instrumental procedures and the final calculation of results. My viewpoint of methods concerns all those processes which are required in order to produce numerical or descriptive data in order to solve problems related to the concentration, or distribution, of the chemical elements in defined materials. As an example of an extreme case it is practical to describe in great details the step-by-step procedures which are essential for the laboratory analysis; however, such an approach is usually only acceptable for a defined matrix in which the relative abundances, and sometimes absolute amounts, of various elements do not vary to any significant extent. On the other hand it is not practical to describe all processes for all stages required in an analysis because some are amenable to precise evaluation while others can only be described in rather vague terms. However, the scientific quality of the final data is dependent upon an appreciation of all the stages involved; for instance, it is quite pointless ignoring the specific requirements of sampling and then employing very detailed and precise procedures for the actual analytical determination required by a particular instrumental procedure. It is common practice for the professional analytical chemist to devote all his energy toward the chemical analysis and ignore any matters pertaining to the previous history of the sample, while the person responsible for obtaining a sample of material will often pay proper attention to sampling and assume that once a sample is obtained it can be passed to an analyst in the firm belief that he will always produce the correct answer! The final outcome of such an approach can be disastrous; it is quite simple to produce data of questionable quality, in terms of scientific objectives, and even to produce quite useless data in spite of the fact that the analytical procedures employed in the analysis are perfect. Consider two extreme cases: (1) the selection of a sample of tissue for eventual assay for organic mercury, which is dried after collection and most of the mercury lost and (2) the routine analysis of a sample for one trace element in which the levels of some elements in different samples vary such that the signal selected for assay for the element of concern is interfered with by the other elements. Finally in the field of trace element analysis, serious contamination of samples can occur at all stages and unless the total process can be carried out without contamination then it is questionable whether or not the analysis is worth performing. In discussing methods I will attempt to make comment upon most of the important stages for total analysis of the chemical elements in biological samples, but the reader should be aware that there are no clear-cut answers to many of the problems and that it is quite pointless identifying optimal methods of analysis, for example, the selection of a sophisticated instrumental method which may only be available to a few.

One of the first questions to be asked in seeking analytical data is what is the required quality of the data? Is it desirable to obtain a number for the mass of an element, in say a tissue, which approaches the true mass, that is, high accuracy of assay, or is a number required which is not very accurate but the quality of the data is very reproducible from sample to sample, that is, data associated with high precision of assay. Then comes the important question of the type and quality of data required to tackle a defined scientific objective; in order to consider this matter adequately it is often essential to consult with those conversant with the scientific use of the data in order to ensure that, as far as is reasonable, the required data are obtained. Some of the major points worthy of consideration follow:

1. Is the bulk assay of an element in an aliquot of a sample the required goal? Can the distribution of an element in a sample and differences in chemical form be totally ignored?
2. If an element, identified as a trace element in the bulk sample, is in fact concentrated to quite high levels in a small

discreet volume of the sample, is it feasible to carry out any useful analysis on the bulk sample?
3. Is it realistic to devote all attention to the analysis of a single element when it is known that synergistic and antagonistic associations are involved in biochemical processes?
4. What is the mass of sample required for a defined analytical procedure? What frequency of sampling is required? Is it practical to carry out the optimum procedures in terms of both obtaining the samples and providing sufficient laboratory manpower for their analysis? Quite often rather severe problems of logistics exist particularly when large numbers of samples are involved. It is quite pointless in carrying out a classical epidemiological evaluation dealing with very large numbers of samples when the analytical laboratory can only deal with a throughput of a few samples per day.
5. Can procedures be worked out in order to ensure that the sample can be obtained free from contamination, or that elements are not lost or disturbed after collection? Following death, a sample of tissue can undergo quite significant changes, such as the redistribution of electrolytes by diffusion or cell breakage, and microorganisms can develop, some of which can drastically alter element distributions.
6. Accepting the severe problems of contamination in trace element analysis, can some elements or ratios of elements be used in order to identify the presence of contaminants? Each sample has to be treated on its own merits but it is possible to identify some crude guidelines. For example, a Si/Al ratio of about three if found in human tissues, with the exception of perhaps the lung, will usually indicate contamination from general laboratory dust or soil debris; this ratio is fairly characteristic for most natural mineral debris. Various other ratios such as K/Rb, Al/Sc, Cd/Zn, Ca/Sr, and Se/S can often provide some information of contamination. Contamination from laboratory apparatus and equipment can easily take place, for example Si and B from Pyrex® glass and Fe, Cr, Ni, and Co from various

types of surgical tools; the risk and degree of contamination becomes greater in relation to the total number of operations employed.

Without reference to a well-defined problem and the identification of specific chemical and biological matrixes, it is not possible to provide detailed comments on all the problems likely to be encountered in the analysis of biological materials. However, it is relevant to this monograph to make general comments upon some of the more important parameters.

SAMPLING

It is difficult to identify the most important component, if there is one, in trace element analysis, but undoubtedly one of the most critical is that of sampling, a process which is often ignored or delegated to a minor role. It is quite pointless to attempt to define a rigid set of rules when sampling biological materials, but in the ideal case, a sample may be considered as an aliquot of part of a system to be studied which illustrates the qualities of the mass from which it is taken and should be homogenous in all its intensive properties. When the object of study is large, a series of different samples are required in order to represent the whole. The population of the sampled material should coincide with the population for which the information is required. Failure to recognise the importance of sampling lies at the root of many investigations and can result in false data being obtained or in obscuring the meaning of data which are produced. Proper objective sampling has practical rewards such as the reduction of costs, an enhanced speed in which data are obtained, an increase in the scope of a study, improvement in both precision and accuracy, and also the scientific value of the data obtained. Of the many approaches to sampling, those of probability and nonprobability sampling are most commonly encountered. The former concerns the selection of components of a system in a well-defined manner aimed at obtaining an aliquot truly representative of the system being studied. The latter approach relies upon the subjective judgement of an expert in the selection of samples, but provides no way of evaluating the results since reliability depends upon an individual's judgement which may be good or bad; adoption of this approach cannot be accepted. There is of course a further type of sampling, often practised, which consists of so-called random sampling of a material and for which samples have no relation to the structure of the material or system under examination. The problem of adequate sampling of materials is a very complex matter and has been described by Cochran and Cox (1) and Anders (2). Ingamells, Engels, and Switzer (3) have proposed a laboratory sampling constant which provides an expression for the relative deviation to be expected in a series of determinations using w-gram subsamples. This is important in trace element analysis when the trace element is contained in one particular phase which is randomly distributed throughout the bulk sample. This constant is expressed as $K_s = R^2 w$ where K_s is the sampling constant and w is the weight in grams of a sample which must be removed to ensure 68 percent confidence for a relative sub-sampling error of 1 percent or less; R is the relative standard deviation (%).

As the mass of an element in a sample decreases and becomes associated with small discreet phases, a Gaussian distribution becomes replaced by Poission statistics; as this characteristic becomes more pronounced the Poission distribution often becomes skewed. Liebscher and Smith (4) have described the essential elements as exhibiting a "normal" distribution and nonessential elements as having a "log-normal" distribution. However, this approach must be treated with caution as a normal distribution of an element in a tissue can easily take on the appearance of a log-normal type simply because of contamination or inadequate sampling, which is particularly true when very small amounts of elements are being determined. The application of the correct statistics for sampling and experimental design is very important when it becomes essential to determine very small differences between levels of elements present in a tissue; it is not always correct to assume a true Gaussian distribution of an element, in which case, for small numbers of samples, it is often more appropriate to select the median value rather than the arithmetic mean as a more valid approximation to the mean level of an element. In clinical practice it is often inappropriate to implement the desired requirements for sampling simply because of the lack of material or the lack in knowledge of the true distribution of an element in a tissue. It therefore becomes important to consider the selection of material very carefully in order to overcome sampling problems and enhance the quality of data. For example simple dissection can selectively remove part of a tissue which contains the elements of interest and excludes other areas of no interest but which, if included in a bulk analysis, would only serve to add to matrix problems or dilute the concentration of the element of interest. For instance, in the case of the kidney it may be advantageous to separate the medulla from the cortex, while for bone it becomes essential to separate cortical from trabecula material with particular reference to the marrow content.

Today the tendency is to analyse bulk samples of tissue for the chemical elements but a trend can be observed in which more attention is gradually being paid to the distribution of the elements at the macro and micro levels in order to identify the sites of concentration, and hence to consider those areas where the elements are more directly concerned with biochemical processes.

TISSUE SAMPLING

When working with any human tissue, precautions must be taken to prevent transfer of diseases to the analyst; in the first instance, clean protective gloves must always be worn and discarded to an incinerator immediately after use; the use of face masks is desirable; dissections should be carried out in cabinets supplied with ultra clean air and equipped with ultraviolet light to reduce the transfer of pathogens; all dissecting tools should be sterilised before and after use. There are two reasons for adopting these procedures: one is to prevent the risk of infection to workers, and the other, perhaps of a more psychological nature, is making the analyst acutely aware of the need to work under very controlled and clean conditions. All samples of tissue for any serious study should be obtained by individuals who have received suitable training: contamination at this early stage of processing can invalidate any final analytical data. However, in order to be practical, it is not always possible to work under optimum conditions in which case, for example, when obtaining postmortem material outside a hospital, it is far better to remove a whole organ using normal implements and following transfer to a laboratory to selectively remove the initial cut surfaces in order to discard possible contaminated areas. This final stage of dissection should be carried out with acid-leached quartz knives which can easily be prepared by fracturing a piece of ultra clean quartz crystal, selecting fragments of the appropriate shape, and fixing them to a quartz rod with an adhesive, followed by acid leaching with ultra pure reagents. With experience very complicated dissections can be undertaken with these tools, even microdissection under a microscope. In some circumstances available samples of tissue are known to be contaminated, an extreme case being samples obtained from car accidents, yet they can still be used; one approach described by Maletskos (5) involves placing the total sample in a bath of water coloured with a dye followed by removal and rapid freezing in liquid nitrogen; the frozen sample is then broken with a hammer and only those samples removed which are not stained by the dye as illustrated in Figure 29. In the selection of samples for analysis it is essential to obtain sufficient quantity of material for a variety of measurements which may be required at a later date, for example protein or lipid analysis. It is also desirable to remove a fresh aliquot of a sample for rapid freezing in a mixture of liquid nitrogen and oxygen (50/50), pentane ($-95°C$), or hexane ($-70°C$) in order to provide material suitable for light or electron microscopy. For in-depth and diagnostic studies it is often pointless carrying out detailed and costly analytical determinations for element levels, then to observe high levels in parts of a sample and not be able to relate them to possible histological features. If material suitable for sectioning can be obtained it is also possible to carry out quite simple histochemical staining techniques to identify the sites of deposition or retention of the chemical elements at the cellular level.

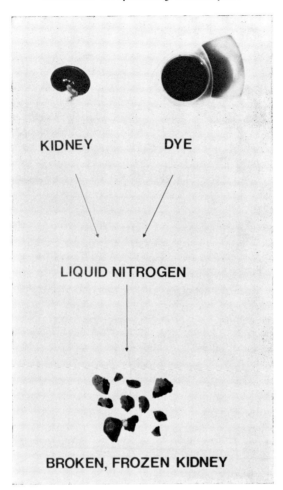

Figure 29: A sample of fresh tissue is placed in a beaker of water containing a coloured dye, removed, and then excess liquid is drained off followed by immersion of the sample in liquid nitrogen until frozen solid. The sample is then placed between two sheets of Teflon and fractured into small pieces by hitting with a hammer; aliquots of the frozen fragments which do not contain the dye are then removed for analysis.

In selecting human tissues for analysis it is essential to consider the biological basis for sampling. Most human tissues are in a state of dynamic flux which becomes "frozen" at the time of sampling; levels of the chemical elements which are ultimately determined may be directly related to influx of elements just prior to sampling, for example as a consequence of recent intake of food. While some tissues do not respond to recent changes in body levels of elements some, such as hair and nail parings, may reflect metabolic events that occurred a few days or months before sampling; Hopps (6) has described the biologic use of hair and nail for analyses of trace elements which serves to illustrate some of the problems and emphasizes the need for more information about biological phenomena responsible for the observed levels and distributions of trace elements in human materials.

SAMPLING LIQUIDS

While the sampling of solid tissues does present considerable problems, they become more severe when sampling liquids such as blood and urine. An ideal technique for sampling blood consists of the insertion of a short, cleaned quartz needle into a vein and collecting a sample by gravity feed. Clearly this procedure is lengthy and uncomfortable for the donor and impractical for the collection of a large number of samples. Often use has to be made of standard methods but lengths of delivery tubing should be kept as short as possible and all materials should be cleaned before use. If suitable leaching techniques are used, contamination from tubing and collection bottles should be minimal but great care has to be directed to contaminants arising from the hypodermic needle which can easily give rise to contamination from elements such as Fe, Cr, and Ni. An example of contamination arising from sampling blood is given in Table XXXIX, while Versieck et al. (7) have described contamination from needles in liver biopsies; for Cr, contamination of ×100 the natural levels can be obtained. It is not possible to generalise on sources of contamination and it becomes essential to carry out adequate blank control experiments. However, as a precaution the first 10 ml of blood should be discarded. Samples of blood must be collected in ultra clean containers which have been tested for loss of leachable elements or surface absorption; such tests should simulate all the features that the sample is likely to be subjected to before removal of an aliquot for analy-

TABLE XXXIX*

RELATIVE CONTAMINATION OF FRESH HUMAN BLOOD FOLLOWING CONTACT WITH VARIOUS MATERIALS TOGETHER WITH RELATIVE ENRICHMENT FACTORS IN NORMAL STEEL HYPODERMIC NEEDLES TO FRESH BLOOD (+ INDICATES SLIGHT ENRICHMENT)

Element	Blood Stored in a PVC Bottle	PVC Delivery Tube Immersed in Blood	Hypodermic Needle Immersed in Blood	Hypodermic Needle
Pb	5	2	5	—
Sr	7	3	2	≈ 133
Cr	3	0	4	$\approx 7 \times 10^6$
Cu	0	0	2	$\approx 1 \times 10^3$
Sn	0	13	0	$\approx 4 \times 10^3$
Mo	0	0	0	$\approx 6 \times 10^4$
Fe	+	0	+	MAJOR
Ni	0	5	3	$\approx 3 \times 10^4$
Sb	0	5	0	$\approx 4 \times 10^3$
V	3	1	0	$\approx 3 \times 10^3$
Cd	7	0	5	$\approx 1 \times 10^3$
Nb	2	2	7	$\approx 1 \times 10^4$
Si	0	0	0	≈ 5
F	10	9	0	≈ 2
B	22	6	0	≈ 220

* From Hamilton, E. I.: Review of the chemical elements and environmental chemistry—strategies and tactics. *Sci Total Environ*, 5:16, 1976.

† Fresh human blood was used as a standard and in the contamination experiments 100 ml of the fresh blood were used for an exposure of 24 h. All samples were analysed by spark source mass spectrometry, but the data are not corrected for effects arising from selective ionization.

sis, for example freezing and time of storage. Unless ultra pure anticoagulants or preservatives can be obtained, they should not be used; hence a separation of serum from red cells has to be carried out directly after delivery; Harrison and Sutton (8) have determined 2,930 ± 300 ppm Ca, 92 ± 715 ppm Sr, and 12 ± 4 ppm Ba in samples of heparin. For blood it is important to obtain as much information concerning the donor as possible with particular reference to sex, age, address, duration at that address, and type of work. While samples of blood can be readily obtained for hospitalised individuals, anomalies in element levels can be observed as illustrated in Table XL. These features may be related to contamination, altered metabolic activity as a result

TABLE XL*

THE CONCENTRATION (μg/ml) OF SOME ELEMENTS IN BLOOD OBTAINED FROM HOSPITALS COMPARED WITH A RANDOM SELECTION OF BLOOD FROM HEALTHY INDIVIDUALS AND MEAN U.K. BLOOD DATA

Element	Hospital Samples Mean (n = 30)	Range	U.K. Normal Samples Mean (n = 30)	Range	U.K. Mean (n = 2,500)
Fe	431 ± 93	93 – 615	491 ± 26	448 – 544	493 ± 4
Cu	1.7 ± 0.8	1.1 – 5.4	1.2 ± 0.2	1.0 – 1.5	1.2 ± 0.002
Zn	8.0 ± 0.8	5.3 – 16.0	6.7 ± 0.8	4.7 – 8.4	7.0 ± 0.2
Cl	3,150 ± 750	2,540 – 6,290	2,990 ± 230	2,570 – 3,440	3,020 ± 77
Br	8.2 ± 9.3	3.7 – 38.5	4.7 ± 0.7	3.7 – 6.6	4.8 ± 0.1
Ca	73.2 ± 25.7	54.5 – 162.5	61.5 ± 5.9	54.2 – 74.7	60.4 ± 0.3
K	2,200 ± 300	1,740 – 2,950	1,840 ± 90	1,173 – 2,050	1,900 ± 20
Rb	3.2 ± 0.9	2.1 – 5.8	2.7 ± 0.6	1.8 – 4.7	2.8 ± 0.1
P	414 ± 107	316 – 789	323 ± 17	294 – 350	335 ± 10
S	1,800 ± 140	1,550 – 2,320	1,770 ± 50	1,680 – 1,880	1,770 ± 14

* From Hamilton, E. I.: The chemical elements and human morbidity—water, air and places—a study of natural variability. *Sci Total Environ*, 3:9, 1974.

of being confined to bed, a change in diet, the use of drugs, or disease.

Samples of urine can be very important for diagnosis of some forms of disease but can prove to be very difficult to obtain free from contamination arising from articles of clothing; this becomes an important factor when sampling industrially exposed workers. There are many problems associated with sampling of urine; there is the controversy of the significance of a so-called twenty-four hour sample, the use of creatinine for normalisation of twenty-four hour samples, and variability in content of suspended solids which differ greatly from individual to individual. Faeces can provide valuable material for metabolic balance studies but can be difficult to obtain, and unless special laboratory conditions are available, objections are often raised to such analyses. Because both faeces and urine are bulky materials, rather large containers are required although several proprietary products are available; data are given in Table XLI indicating the levels of some elements in two types of collection containers which are often used. If large numbers of analyses are required

TABLE XLI*

THE RELATIVE CONCENTRATION (ppm dry weight) FOR SOME ELEMENTS PRESENT IN CONTAINERS USED FOR THE COLLECTION OF EXCRETA†

Element	Faeces Container	Cellulose Bag for Urine
Bi	0.03	0.2
Pb	13	38
Ce	0.2	0.3
La	0.2	1.0
Ba	29	18
Cs	0.02	0.1
Sn	3	0.5
Cd	0.01	0.8
Ag	0.3	0.4
Mo	0.2	0.4
Zr	1.0	0.09
Y	0.008	0.08
Sr	8	8
As	11	14
Zn	702	108
Cu	32	57
Ni	13	41
Co	4	3
Fe	54	281
Mn	5	5
Cr	4	11
V	0.5	2
Ti	193	179
Ca	920	2,650
S	>680	>350
P	1,110	343
Si	537	276
Al	0.2	0.03
Li	0.5	2

* From Hamilton, E. I.: Review of the chemical elements and environmental chemistry—strategies and tactics. *Sci Total Environ,* 5:15, 1976.

†It is standard practice to ash excreta in sampling containers.

for faeces and urine, then it is advisable to restrict the work to one special room equipped with an appropriate air handling system and sterile working conditions. During drying phases and ashing, objectional odours are present; if these can be extracted by one venting system, a gas burner or heated metal gauge can be situated at the outlet and total combustion of the organic vapours achieved, thus releasing only carbon dioxide to the external air.

TABLE XLII*

DATA FOR THE WATER AND ASH CONTENT OF SOME HUMAN TISSUES

Tissue	No. of Samples	% Water	% Ash†
Blood	104	79.9 ± 0.2	1.2 ± 0.02
Brain	10	78.2 ± 0.3	1.7 ± 0.02
Whole Kidney	18	74.9 ± 1.5	1.3 ± 0.1
Kidney (cortex)	8	74.9 ± 1.5	1.8 ± 0.07
Kidney (medulla)	8	74.9 ± 1.5	1.1 ± 0.06
Liver	11	70.6 ± 1.6	1.5 ± 0.09
Lung	10	79.2 ± 0.7	1.3 ± 0.06
Lymph Nodes	6	70.1 ± 3.1	1.3 ± 0.1
Muscle	6	72.6 ± 1.1	1.3 ± 0.03
Ovary	6	80.1 ± 0.9	1.4 ± 0.10
Testis	5	80.7 ± 1.6	1.4 ± 0.02

* From Hamilton, E. I., Minski, M. J. and Cleary, J. J.: The concentration and distribution of some stable elements in healthy human tissues for the United Kingdom. *Sci Total Environ*, 1:349, 1973.

† Low temperature ashing.

SAMPLE PREPARATION

One of the first measurements on tissues or biological liquids is usually that of weighing the "fresh weight"; this is not always possible because of loss of water or liquid phases during transit, in which case it seems permissible to dry the sample to constant weight and then estimate the wet weight using data provided in Table XLII. However, prior to any treatment of the sample it is advisable to check that sufficient untreated material has been set aside for any histological examination, also that a sufficient mass of the sample is available for all the analytical requirements, but retaining some for storage and future use. If analyses are to be made for very volatile species, then a separate aliquot must be set aside before it is subjected to any processing.

Sample Homogenisation

It is assumed that by this stage samples for special treatment have been identified and set aside, in which case the first problem is to reduce the sample to a homogeneous state suitable for sampling and analysis. Sample homogenisation may be achieved by wet or dry techniques or a combination of both. The use of simple systems, such as the Stomacher (illustrated in Fig. 30) or a simple polytetrafluoroethylene (Teflon®) press used under vacuum, can serve to separate cell contents from membrane debris. If such separations are used, any subsequent filtration must be carried out under very carefully controlled conditions together with adequate blank assays.

Vacuum freeze drying provides the simplest technique for removing water and can be carried out using industrial systems,

TABLE XLIII*

EXAMPLES OF CONTAMINATION ARISING FROM GRINDING GRAPHITE IN VIALS CONSTRUCTED OF DIFFERENT MATERIALS

Element	Polystyrene Pot ($\mu g/g$)	Ceramic ($\mu g/g$)	Agate ($\mu g/g$)	Tungsten Carbide ($\mu g/g$)
W	<0.1	<0.01	<0.1	>10
Ba	<0.2	>10	<0.2	>0.02
Sn	<0.4	<0.08	0.3	0.2
Cd	<0.4	<0.04	0.3	0.2
Sr	<0.1	<0.02	<0.1	>0.01
Rb	<0.2	<0.02	<0.2	>0.02
Br	<0.1	<0.02	<0.1	0.1
Zn	<0.2	2.0	<0.2	4.0
Cu	<0.2	0.4	<0.2	3
Ni	<0.1	0.8	<0.1	3
Co	0.05	0.2	<0.1	>10
Fe	<2	>10	>10	>10
Mn	0.04	1.5	0.6	1.0
Cr	0.2	1.5	0.4	1.0
V	<0.1	5	2	0.8
Ti	5	>10	7	>10
Sc	0.04	0.08	<0.04	<0.04
Ca	<2	>10	<2	7
K	<0.5	>10	<2	7
Cl	<0.04	>10	>10	<0.04
S	5	<0.06	>10	<0.06
P	<0.02	>10	1	0.8
Si	<0.5	>10	>10	>10
Al	<0.02	>10	7	0.7
Mg	<2	>10	<2	<0.2
Na	<0.02	>10	2	5
F	<0.02	0.2	<0.02	<0.02

The data are normalised to 10 cc vials and represent overall contamination, i.e. grinding, mass spectrometric electrode preparation.

* From Hamilton, E. I., Minski, M. J., and Cleary, J. J.: Problems concerning multi-element assay in biological materials. *Sci Total Environ*, 1:10, 1972.

provided they are free from contamination (note that when the vacuum is released only clean filtered air is allowed to enter the system); alternatively quite simple pure quartz systems can easily be fabricated. A related technique consists of freezing the sample at liquid nitrogen temperatures and carrying out the grinding stage at these temperatures when the sample is extremely brittle and most tissues can be reduced to less than 300 mesh size in about ten seconds. Commercial instruments are available such as illustrated in Figure 31 or, alternatively, small vessels can be fabricated from Teflon or boron nitride and used on conventional grinding equipment. Most of these techniques require the addition of metal grinding balls which have to be encased in a

Figure 30 (Left)

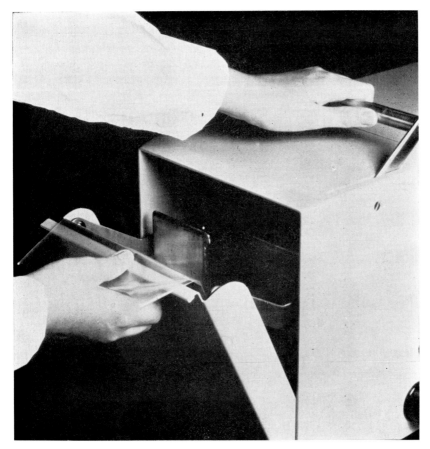

Figure 30 *(Right)*

Figure 30: Homogenisation of tissue samples in a Colworth Stomacher Laboratory Blender 80 (Seward Lab, Suffolk, United Kingdom). *Left*, a sample is placed in a leached polyethylene bag together, if required, with a suitable solvent phase. *Right*, the bag is lowered between the two reciprocating paddles before the front is closed and the homogenisation process started. Illustrations kindly supplied by Seward Laboratories, United Kingdom.

noncontaminating material. In order to reduce the dry solids or powders to provide a homogeneous mix, some form of grinding is required and this process, if used, cannot be achieved without some degree of contamination being introduced. Contamination

Figure 31: SPEX (Spex Industries, Inc., Metuchen, New Jersey, United States) Freezer Mill. In the illustration the lid is open and the partly exposed block houses the grinding vial which can be seen on the top of the unit; it consists of a plastic tube, a plastic covered metal rod for impaction-grinding, and a top plug for sealing the unit.

of ultra pure graphite powder in various types of grinding vials is illustrated in Table XLIII.

Polystyrene introduces the minimum quantity of contamination but is soft and can only be used for gentle mixing. Most of the contaminants which are added to the graphite after grinding can be related to the chemical composition of the grinding vials;

in the case of tungsten carbide, which is very hard, contamination from cobalt arises from the use of an alloy used to bind the tungsten carbide liner to the stainless steel outer case. Although not as hard as tungsten carbide, which tends to be very brittle, agate is preferred for grinding, especially the pale grey types free from complex banded structures; coloured types should be avoided. Following use, grinding vials should be adequately cleaned and periodically the internal surfaces should be repolished with very fine diamond paste in order to reduce the removal of pitted surface material into the samples. Adequate cleaning of grinding vials does present a problem; ultrasonics can be used for some materials but agate vials are prone to fracture. The most important aspect of this stage is to determine the contamination characteristics of each vial at regular intervals and to select materials most appropriate for particular analyses; clearly, if silicon is to be determined agate should not be used. Examples of some commercially available grinding vials are illustrated in Figure 32. When grinding large samples the speed

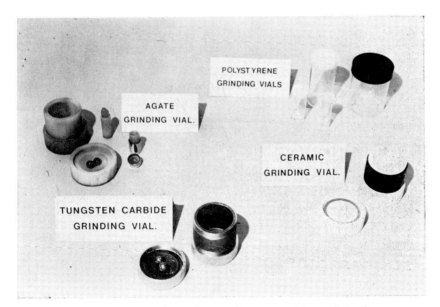

Figure 32: Various types of grinding and mixing vials used in the preparation of tissue powders for trace element analysis.

of the operation can be enhanced by frequent sieving in order to remove the very fine phases which tend to become wrapped around larger fragments and reduce efficiency of grinding. Plastic sieves can be used but contamination from zinc is likely, while great care must be taken to ensure that during sieving and transfer of powder the fine fraction is not partially lost as airborne dust. The final powder should be eventually transferred into some form of cleaned plastic container for storage. The surface of some plastics can become highly charged with static electricity and, when opened, materials can be lost or nonrepresentative aliquots removed.

Dry Ashing

Apart from the removal of organic phases, dry ashing provides a second phase of concentrating elements, usually reducing all materials to simple inorganic forms which can then be dissolved in acids to provide true solutions of samples. Conventional muffle furnaces have limited application, contamination from ceramic liners can be overcome by using silica linings, but the major problems arise because of draughts and thermal gradients across most ovens or furnaces as described by Hamilton, Minski, and Cleary (9) and illustrated in Figure 33; the loss of many elements through thermal volatilisation can also lead to severe cross contamination of samples. Even if the internal furnace temperatures can be controlled, exothermic reactions taking place in samples, at particular sites, can result in localised heating; this problem is commonly encountered during the ashing of bone.

An alternative is to employ low temperature ashing in an atmosphere of nascent oxygen. A stream of pure oxygen, excited by a radio-frequency discharge, can effectively decompose organic substances at a temperature of about $100°C$; in this process, metastable oxygen species are formed by the discharge and have sufficient energy to rupture all C-C and C-H bonds. The decomposition process is essentially a surface phenomenon, and therefore, the rate of ashing is enhanced if some form of agitation can be used, thus continuously exposing fresh surfaces to oxida-

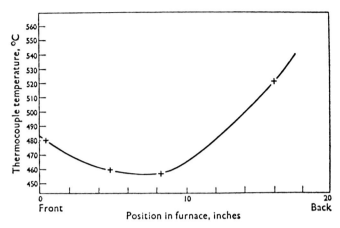

Figure 33: Relationship between temperature and position of samples in a laboratory furnace as determined with standard melting point sentinels.

tion. Sample weights are usually restricted to about 1 g of material which is thinly spread on the floor of a shallow silica tray. Even with this procedure some elements are lost by volatilisation but, in theory at least, if single chamber oxidation units are used the volatilised species can be trapped in a cooled finger as illustrated in Figure 34. In order to study the loss of elements, Hamilton, Minski, and Cleary (10) introduced various radiotracers, by intravenous injection, into individual rats which were killed four hours later. This approach permits the various nuclides to become incorporated to some degree into tissues, thus partly overcoming some of the problems of the selective loss of an element by simply adding a tracer, which may have a different chemical form to that of an element present in a sample. The dry samples were then ashed in quartz trays in two separate, commercially available, low temperature ashers for twenty hours at $100 \pm 10°C$, results of this experiment are given in Table XLIV. Following ashing, the total γ-activity of the dish was determined and also any activity remaining after the ash was removed as some elements can be avidly retained on the walls of materials used for ashing. Low temperature ashing is a useful method, and in spite of the loss of some elements it appears to give fair-

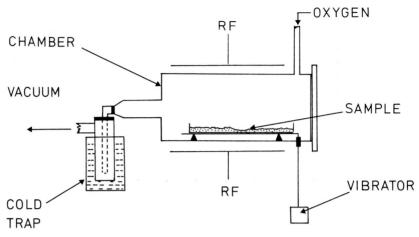

Figure 34: A cold trap system used to retain volatiles lost during low temperature ashing together with a vibrator to expose fresh surfaces of a sample to oxidation.

ly reproducible losses for uniform methods of operation and permits adequate ashing under clean, controlled, and reproducible conditions. The great advantage of dry ashing is that no reagents are required; however it is worth mentioning two other dry techniques which I have found useful.

1. The deliberate heating of samples to ≈ 1,000°C under vacuum and trapping all volatile species on cold fingers inserted into apparatus as illustrated in Figure 35. This technique has been used for the quantitative removal of Se, As, Zn, Sb, Pb, Tl, and Hg from large samples of dried cortical bone.
2. The Schöniger oxygen flask technique can be used for samples weighing up to about 250 mg; an external source of heat can be used, such as a focussed heat lamp, directed onto the sample wrapped in a piece of paper free of the elements of interest; problems can arise because of retention of some elements on the platinum sample holder, but the technique can sometimes be of use especially when sample size is small; ultra pure quartz combustion flasks are re-

TABLE XLIV*

LOSS OF SELECTED ELEMENTS FROM RAT TISSUES AFTER LOW TEMPERATURE ASHING IN TWO TYPES OF COMMERCIALLY AVAILABLE INSTRUMENTS

Element	Radionuclide Used	Tissue	% Loss by Volatilisation†		% Loss by Retention†	
			A	B	A	B
Sodium	^{22}Na	Blood	3	0	tr	0
		Kidney	7	0	tr	0
		Femur	1	0	6	0
Arsenic	^{24}As	Blood	2	11	tr	tr
		Kidney	33	11	2	tr
		Femur	0	0	0	5
Zinc	^{65}Zn	Blood	1	0	0	0
		Kidney	4	0	0	0
		Femur	0	0	0	0
Strontium	^{85}Sr	Blood	2	0	0	0
		Kidney	0	tr	0	0
		Femur	8	0	0	0
Silver	^{100m}Ag	Blood	3	13	tr	6
		Kidney	18	23	1	5
		Femur	26	22	0	25
Caesium	^{137}Cs	Blood	tr	0	tr	0
		Kidney	0	0	0	0
		Femur	0	0	0	0
Iodine	^{131}I	Blood	23	66	0	0
		Kidney	nd	80	nd	0
		Femur	nd	14	nd	5
Bromine	^{82}Br	Blood	81	91	0	0
		Kidney	nd	93	nd	tr
		Femur	nd	34	nd	1
Gold	^{198}Au	Blood	nd	0	nd	tr
		Kidney	nd	0	nd	tr
		Femur	nd	0	nd	13

* From Hamilton, E. I., Minski, M. J., and Cleary, J. J.: Problems concerning multi-element assay in biological materials. *Sci Total Environ, 1:*7, 1972.

† A = LTA 600 instrument (Tracerab, Weybridge, Surrey, England)
B = IPC instrument (International Plasma Corporation, Hayward, California, United States)
tr = trace
nd = not determined

quired preferably with some form of intermittent flow of clean oxygen.

Wet Ashing

With the exception of radioactivation techniques (see page 267) the main problems in wet ashing arise in the preparation

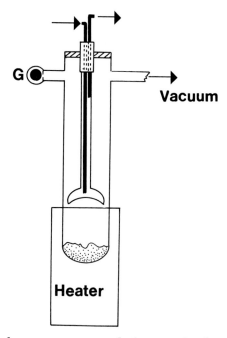

Figure 35: A simple system constructed of quartz, for the pyrochemical separation of elements from acid digest residues or untreated bone. With this type of system it is possible to ash up to about 1 g of dried tissue, but if larger amounts are used, problems arise because of the condensation of organic materials. The tube containing the sample is pumped down to about 10^{-3} torr determined by the vacuum gauge G; the central tube, terminating in a concave structure, contains a flow of cold water, and is placed in a heating block and the whole unit located within a protective screen in case of breakage. A sample, such as bone, is heated to about 900°C for one hour and volatile elements are condensed on the surface of the concave collector. If radio-frequency heating is used, extraction time of only a few minutes are required.

of reagents in a very pure form. This problem has been solved, for most practical purposes, by the use of subboiling distillation stills as described by Kuehner et al. (11) and Mitchell (12); it is now possible to prepare pure reagents containing less than 10 ppb of most elements while many are below the 1 ppb level; an example of this type of still is illustrated in Figure 36. Recently a further type of subdistillation still constructed of polypropylene has been described by Dabeka et al. (13) for preparing high purity hydrochloric and hydrofluoric acids; the authors investigated Teflon for hydrofluoric acid and showed that it provided a persistent source of iron, nickle, chromium, and manganese, while Vycor® was a source of silicon and zirconium and linear polyethylene for titanium, aluminium and zinc. However, the problem of obtaining ultra pure reagents is not concerned simply with the stills; very special and ultra clean working booths are required to prepare clean reagents, and suitable containers are required for their storage; further, the quality of the reagents has to be monitored regularly, which can become very time consuming. Thiers (14) and Zief and Mitchell (15) have described methods for the control of contamination in trace element analysis, while problems of surface absorption and leaching of elements from plastics and glass have been described by Karin, Buono, and Fasching (16), Struempler (17), and West, West, and Iddings (18).

Open bench dissolutions can be carried out if systems are adequately protected as illustrated in Figure 36, but more commonly closed systems such as Teflon bombs are to be preferred; sev-

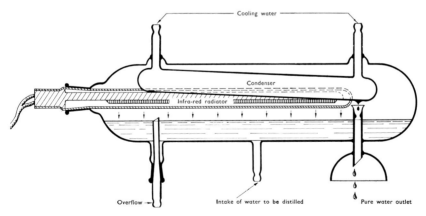

Figure 36: Subdistillation still manufactured by Quartz & Silice, France. Units are made of transparent silica glass and the heating element is an infrared heater encased in quartz. Depending upon available models, output of ultra pure water is between 200 and 1500 cm³/hour.

eral very useful systems for digesting samples in Teflon bombs and other devices have been described by Lautenschläger (19), illustrated in Figure 38, and Tölg (20). When using Teflon encased in metal, the possibility exists of the gradual migration of metal into the Teflon, hence the need to monitor for contamination regularly.

Sansoni, Kracke, and Winkler (21) have described a technique, based on the use of nitric acid and hydrogen peroxide, for the ashing of very large samples of tissues, while Sansoni and Kracke (22) have described an automated version which can be carried out at temperatures below 100°C, in sealed systems, hence volatiles are not lost. The technique was developed for use in the assay of radioactive nuclides where reagent purity is not

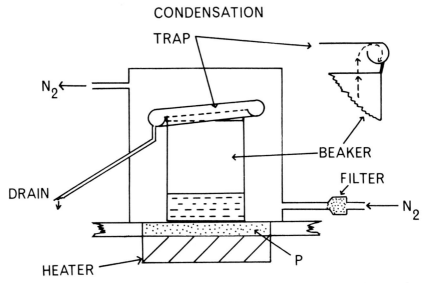

Figure 37: A system for reducing the entry of contaminants during the evaporation of liquids. The condensation trap is made from the base of a glass beaker and about 2 cm of wall has been curled inwards to form a channel in which condensed phases can drain. The system is purged with pure nitrogen and particulate material arising from the gas cylinder is removed by a filter. The heater base consists of a piece of Corning Pyroceram™ set into the top of a laboratory bench, and heating is carried out from below thus overcoming problems which arise from corrosion of metal faced heaters in laboratories.

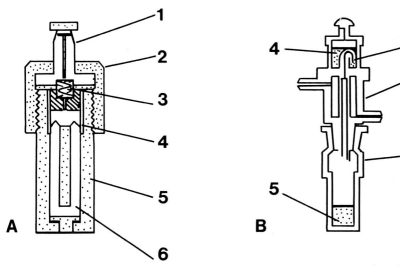

Figure 38: Two types of commercially available digestion systems described by Lautenschläger (19).

A: 1 = Lifting and sealing piece.
   2 = Screw cap.
   3 = Tension spring.
   4 = PTFE cover.
   5 = Pressure bomb.
   6 = PTFE reaction vessel.

B: 1 = Cooling and absorption chamber.
   2 = Reaction vessel.
   3 = PTFE frit in the absorption chamber.
   4 = Absorption liquid.
   5 = Reagent solution.

important, but a distinct advantage of the technique is that large samples can be ashed.

The determination of trace levels of elements in liquid reagents is possibly more difficult than assay for trace elements in solids. One of the problems in determining reagent blanks often arises because of the practice of adding, to a separate container, aliquots of reagents which are then evaporated to dryness when hopefully no residue will remain. The behaviour of an element in such a sample system can be different to that when a real sample is used and some elements may be lost by volatilisation or through absorption on container walls. For biological materials it is often essential to destroy organic matter, and techniques have been described by Gorsuch (23).

The Laboratory Environment

The importance of the quality of the internal environment of a laboratory engaged in the assay for minor and trace levels of the elements is often underrated; failure to consider such matters in sufficient depth and detail can have dire consequences. Often very large sums of money are spent on the purchase of sophisticated instrumentation in the belief that their acquisition will automatically guarantee the quality of data; unless suitable laboratory environments are available it seems inevitable that the quality of data will be poor. Of the many types of laboratory engaged in environmental analysis, consider three:

The General Purpose Laboratory Concerned with Duties Which Do Not Normally Consider Minor or Trace Element Analyses

When analyses are required for trace constituents, it is often left for individuals to improve environmental quality; hence any precautions taken will depend upon the experience of individuals, while the quality of the data which is produced will reflect their past experiences. As a result it is inevitable that the quality of the data will tend to be variable.

Laboratories Concerned with Special Tasks and Responsibilities

An example of such a laboratory is one concerned with determining whether or not the level of a constituent in a material meets certain statutory requirements. Usually such laboratories have to deal with a very high work load and often the tendency is to direct effort to those methods which have adequate detection limits and which are amenable to routine procedures for defined matrixes. Often little interest is expressed for levels of elements which fall below some acceptable value. Apart from errors inherent in the selected methods (fundamental to a particular technique or limitations imposed by the gross chemical matrix of samples), there is a risk that such data can enter the literature without qualification with regard to quality.

Special Laboratories

Special laboratories are now becoming more common and are responsible for producing data of high quality on a very large number of matrixes. A subsidiary purpose is often to establish standard methods of analysis and improve both methods and techniques. The emphasis of such laboratories lies in the field of research and development rather than providing routine services. Such laboratories are usually well equipped with analytical equipment and expertise, both of which are essential in order that flexibility of analytical approach can be implemented and cross-standardisation techniques deployed. By concentrating expertise in one place, further flexibility is achieved by providing the means for undertaking difficult types of analysis in an economic manner.

In relation to the three types of laboratories which have been considered, each serves a useful purpose for defined fields of work, but it does not follow that each laboratory can, in practice, be called on to undertake analyses outside its sphere of experience without the staff undergoing some form of training. Many demands for trace element analysis in environmental materials are often presented to laboratories which have no experience in the specific type of analysis required; it is assumed by many that experience in one type of analysis automatically means adequate experience in all types of analysis. This is an illusion and is well exposed in programmes for the analysis of Reference Materials, which will be described later, and has lead to the identification of some laboratories of a superior type and others as inferior, yet the competence of analysts in both types of laboratories may be similar. In order to tackle current problems, with particular reference to medical studies, it is essential that laboratories all over the world produce data of similar quality; if this is not achieved then clinical findings concerning levels of the elements may be related to analytical uncertainties rather than morbidity. In modern minor and trace element studies, very clean working conditions are required; many laboratories even fail to meet minimum requirements of cleanliness demanded by the problems, and there are few guidelines to indicate acceptable conditions. Reasons for failure are many, ranging from ignorance to awareness but lack of funds to remedy matters. In my experience it is not uncommon, at least during the initial phase of establishing a laboratory with the required degree of cleanliness, for such duties to consume at least 20 percent of the total work load.

Some of the problems and difficulties in establishing laboratories suitable for trace element chemistry have been identified, but there is a further problem which is particularly associated with new buildings, namely one of ensuring that designs agreed upon by those who have to use the laboratories are in fact implemented by the architects. Often architects and builders have limited experience in the construction of modern chemical laboratories and are more concerned with external appearances; for example, a series of separate extract vents from fume cupboards is often more desirable than closeting them all together, often with the result that the extract fumes soon find their way into the air input systems! Hamilton (24) has discussed some of the requirements in the design of radiochemical laboratories, while McDermott (25) has provided a very valuable handbook for ventilation and contamination control of laboratories. Each laboratory seeks its own solution but it is worth recording some personal experiences in establishing a comprehensive multielement laboratory, described by Hamilton, Minski, and Cleary (10), which required a very high degree of controlled cleanliness.

The nature of many environmental studies requires the segregation of activities, isolation, and restricted entry into laboratories. This was achieved by erecting a prefabricated wood structure as illustrated in Figure 39A-D: an air lock at the entrance to the laboratory separates internal from external air. The only drawback encountered with this structure was the lack of insulation, with excessive heat loss in the winter and heat gain in the summer; it was during the latter season that the number of electronic faults could be directly related to ambient room temperature. Characteristics of local climatic features need to be seriously considered when establishing trace element laboratories; also the availability of such fundamental requirements as a stabilised and reliable supply of electricity can become an important factor in a developing country.

Figure 39: Stable Element Laboratory of the late Radiological Protection Service, Sutton Surry, England. A: General external view showing the extended entrance block which also served as an air lock.

Figure 39B: Sample receiving room; at the rear is an automatic freeze dryer, to the left is the sample storage cupboard supplied with its own clean air input and extract system.

Figure 39C: Sample preparation room; to the rear is a lamina flow enclosure used for mixing dusty samples; to the left are air extract hoods through which air from the building was vented; to the right are various items used for grinding.

Within the laboratory, adequate space was allocated for benches, free floor space for standing instruments, and wall space for cupboards. All internal surfaces were painted with white polyurethane paint, dust traps were reduced to a minimum, and no metal fittings were used unless sealed or ducted in an inert material. All sources of volatile materials were reduced to a minimum and all sources likely to produce organic vapours, such as rotary pumps, were fitted with separate sealed traps; through pipes, volatiles were vented to the external atmosphere downwind of the prevailing winds as illustrated in Figure 39E; the extract piping should have an internal diameter of at least 6 cm in order to prevent the development of back pressure in the system. It can be argued that laboratories will contain several tons of analytical equipment composed of metal, but they rarely act as sources of metal contamination, although some attention should be paid to sources of finely divided metal, such as copper windings and brush contacts of electric motors. In wet and humid environments, particularly marine environments, special attention has to be paid to metal corrosion, for example aluminium, galvanised iron, and nickle coated materials; provided that surfaces are degreased they can be coated with protective lacquers. The laboratory was supplied with a simple system of air conditioning as illustrated in Figure 40. In spite of a very heavy external dirt load, the system performed satisfactorily as illustrated in Table XLV.

More sophisticated systems employing cyclonic air scrubbing

Figure 39D: Early view of the mass spectrograph room containing a spark source mass spectrograph (central); to the left is a photodensitometer subsequently attached to a PDP 8E, 8K core computer for automated processing; to the right (background), are small experimental mass spectrometers and sample conditioning units, foreground thermal ionization unit. At a later date, the spectrograph was fitted with a lamina flow cell and a direct electrical detection system and Hall probe for electrostatic and magnetic scanning.

Figure 39E: Illustration of an oil trap attached to a vacuum pump and associated extract system to remove oil vapour from laboratory environments.

TABLE XLV*

A COMPARISON BETWEEN THE CONCENTRATION OF ELEMENTS IN LABORATORY AIR AND EXTERNAL AIR BORNE PARTICULATE MATTER

Element	Maximum Concentration in Cleanest Laboratory Air (μg/g of Dust)	Mean Concentration (12 Samples) in External Air (μg/g of Dust)
U	$< 1 \times 10^{-2}$	0.1 ± 0.1†
Bi	$< 2 \times 10^{-2}$	1.5 ± 0.4
Pb	$< 4 \times 10^{-2}$	$2{,}150 \pm 620$
Ce	$< 2 \times 10^{-2}$	3.9 ± 1.1
La	$< 1 \times 10^{-2}$	1.9 ± 0.6
Ba	0.3	35.9 ± 9.6
Cs	$< 1 \times 10^{-2}$	1.3 ± 0.5
I	$< 1 \times 10^{-2}$	2.9 ± 1.4
Sb	$< 3 \times 10^{-2}$	14.8 ± 5.2
Sn	$< 5 \times 10^{-2}$	9.6 ± 6.7
Cd	0.1	2.8 ± 0.9
Ru	$< 2 \times 10^{-2}$	1.7 ± 0.5
Zr	0.5×10^{-2}	4.2 ± 1.2
Y	$< 9 \times 10^{-2}$	1.6 ± 0.6
Sr	$< 1 \times 10^{-2}$	13.5 ± 4.9
Rb	$< 1 \times 10^{-2}$	22.6 ± 8.3
Br	$< 2 \times 10^{-2}$	23.3 ± 8.3
Se	$< 2 \times 10^{-2}$	0.6 ± 0.2
As	$< 8 \times 10^{-2}$	53.3 ± 15.9
Ge	$< 2 \times 10^{-2}$	17.9 ± 7.0
Ga	$< 1 \times 10^{-2}$	30.2 ± 12.8
Zn	2	$1{,}640 \pm 510$
Cu	$< 2 \times 10^{-2}$	213 ± 66
Ni	0.5	69.7 ± 22.1
Co	0.1	9.0 ± 3.7
Fe	6	$3{,}230 \pm 650$
Mn	$< 6 \times 10^{-3}$	116 ± 46
Cr	$< 6 \times 10^{-3}$	39.2 ± 12.2
V	$< 5 \times 10^{-3}$	259 ± 72
Ti	3	258 ± 93
Ca	4	$2{,}690 \pm 880$
K	—	$7{,}920 \pm 3{,}930$
Cl	—	$1{,}470 \pm 1{,}100$
S	—	$20{,}000 \pm 5{,}800$
P	1.5	$1{,}150 \pm 670$
Al	6	$3{,}000 \pm 1{,}080$
Mg	18	$2{,}390 \pm 950$
Na	134	$2{,}950 \pm 1{,}240$
F	0.1	1.0 ± 0.4
Be	$< 1 \times 10^{-3}$	3.2 ± 1.4
Li	$< 8 \times 10^{-4}$	14.7 ± 8.7

* From Hamilton, E. I., Minski, M. J., and Cleary, J. J.: Problems concerning multi-element assay in biological materials. *Sci Total Environ*, 1:5, 1973.
† Standard error.

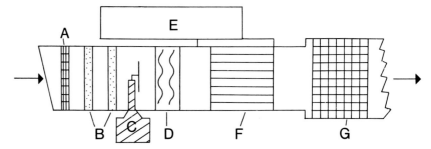

Figure 40: Air cleaning system to the Stable Element Laboratory.

A. Coarse polypropylene insect screen.
B. Coarse fibre glass filters impregnated with a tacky substance.
C. Air input fan.
D. Silica sheathed heater coils.
E. Air cooling coils and humidity control.
F. 95 sq ft of filtering media, flow capacity 2,000 cfm, 90% efficiency, A.F.I. test.
G. Ultra-HEPA filter, 0.3 μm, 99.97% efficient on DOP test.

At appropriate positions, visible manometers were fitted in order to indicate when the filters required changing. Total size of system 12 × 2 × 2 feet.

and electrostatic dust precipitators are not always an advantage as they require regular cleaning and have to be backed up by duplicate items in case of failure and during maintenance.

Within the laboratory the clean conditioned air was ducted at both high and low levels in polypropylene trunking such that the input air first entered the cleanest rooms and then passed to rooms with less stringent requirements and was finally evacuated through exhaust fume hoods. Audible warning systems were used to indicate failure of the air conditioning system, while the efficiency of the filters was regularly monitored through visual manometers. The air in the laboratory was maintained at positive pressure relative to the external air, except in the room containing the extract hoods which was at a slightly negative pressure. The entrance to the laboratory was fitted with an air lock and a shoe-change barrier. Within the laboratory, commercially available sticky mats were placed at the entrance of each room to remove dirt adhering to the soles of overshoes. Laboratory garments were made of untreated cotton and washed by the laboratory staff using liquid detergents followed by several washes in deionized water; normal procedures for laundering garments were not used because the treatment involves the use of phosphates and starch which can lead to contamination from phosphates, boron, alkaline earths, and various other elements and their compounds. As a result of using this system, the main contaminants encountered in the laboratory were attributable to glass debris, cotton and paper fibres, female cosmetics, human skin and hair debris; data are provided in Table XLVI indicating some possible contaminants. For some studies it was neces-

TABLE XLVI*

SOME SOURCES OF ELEMENTS FROM DIFFERENT MATERIALS WHICH CAN GIVE RISE TO SAMPLE CONTAMINATION

Material	Most Abundant Elements Present
Dust	Na, Mg, Al, Si, P, S, Cl, K, Ca, Ti, Cr, Mn, Fe, Co, Zn, As, Se, Rb, Ba, Re, Pb
Paint	Na, Al, Si, S, Cl, K, Ca, Ti, Zn, Cr, Ba, Pb
Clothing debris	Na, Al, Si, P, S, Cl, K, Ca
Sweat	Na, K, Cl, Zn, S
Tobacco ash	Cl, K, Ca, Si, Br, Sb
Cosmetics	B, Na, Al, Si, P, S, Cl, K, Ca, Ti, Fe, Zn, Zr, Ba
Polyvinylchloride (PVC)	Si, Mg, Al, Ca, Cl, Zn, Cd, Ni, Na, Pb
Cellulose type adhesive tape	Na, Ca, Ba, Mg, Zn, Ti, Sr, Cr

Element	Face Powder 1 (Solid) % Dry Wt	Face Powder 2 (Loose Powder) % Dry Wt
Ba	0.004	0.034
Zn	0.38	3.50
Fe	0.23	0.11
Ti	0.90	0.63
Ca	0.10	6.0
K	0.013	0.025
Cl	0.015	0.063
S	0.03	0.04
P	0.04	0.14
Si	19.3	9.0
Al	0.30	0.14

Other elements present in face powder: Pb, Rare Earths, I, Ag, Sb, Cd, Nb, Sr, Ge, Cu, Ni, V, Mg, Na, F, B, Li. Elements present in lipsticks: Br (0.05%), (0.4%), Bi (4.2%), Fe (0.16%), Ti (6.2%), Ba, Zr, Zn, Ca, Cl, S, P, Si, Al.

* From Hamilton, E. I., Minski, M. J., and Cleary, J. J.: Problems concerning multi-element assay in biological materials. *Sci Total Environ*, 1:3-4, 1973.

sary to improve the quality of the air and this was achieved by the use of local laminar flow ultra-filtered air stations which do not interfere with the general flow of air in the laboratory. A class of air cleanliness approximating United States Federal Class 10000 standard was achieved for general air; the main air input provided air to Class 100 (less than 100 particles per cubic foot with a mean diameter of five microns). However, instruments, analytical procedures, and human traffic generated considerable numbers of dust particles. With normal movements the human body is capable of generating about one million particles of < 0.3 microns diameter per minute; the rate of deposition of such debris is low and it tends to be carried in the air flow to the extract hood. Through the adoption of these procedures the air quality of the laboratory, over a period of three years when the laboratory was operational, was sufficient for most analytical purposes. It is of course possible to use more refined systems of air control such as laminar flow (ceiling to floor air movement), but they are very expensive to purchase and can be difficult to maintain.

All laboratory consumables were purchased in bulk after being screened for possible contaminants. This approach proved to be valuable in identifying plastics which contained significant levels of zinc, cadmium, and lead; it also showed that common items such as disposable tips of micropipettes can contain leachable quantities of zinc and cadmium, while one commonly used paper towel, selected because it did not produce finely divided paper debris, contained up to 150 ppm of leachable zinc.

The first major problem encountered by the laboratory was to obtain an adequate supply of clean water. Failure to obtain water of sufficient purity was instrumental, during the early period of the laboratory, in adoption of nonaqueous methods of sample preparation. The main problems encountered were as follows:

1. Very large variations with time were encountered in the abundance and mass of elements and organic debris in the raw water supply.

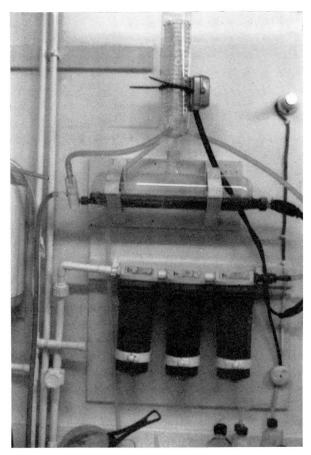

Figure 41: Preparation of clean water. A general view of a quartz still and three prefilters. Left, coarse fibre filters; middle, carbon filter; and right, 0.45 micron fine filter.

2. Water for general washing and non-ultra trace analysis was prepared using the system illustrated in Figure 41, and a more comprehensive system is illustrated in Figure 42. However, a major problem arose because of very great difficulties in storing the water in vessels prior to use. For ultra-trace element work, water was prepared from subboiling distillation and the water was used directly from the system. Small quantities of clean reagents were prepared by

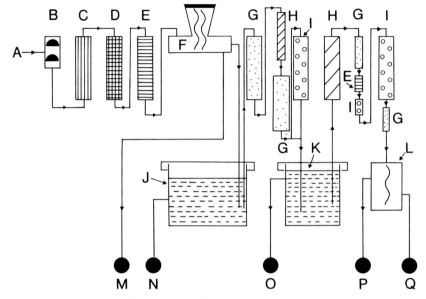

Figure 42: Advanced total clean water system.

A. Raw water input.
B. Coarse Teflon filter.
C. Fibre filter.
D. Carbon filter.
E. 0.45 micron pleated paper filter.
F. Quartz still.
G. Mixed ion-exchange beds.
H. Teflon filters.
I. Carbon filters.
J. First stage purified water for general washing and for further filtration; water stored in a 80 litre leached polythene tank.
K. Container for second stage of purification; water used for final rinsing of apparatus and chemical uses.
L. Third stage of purification in a sub-distillation still.
M. Hot water overflow from still used for general washing.
N. Ordinary distilled water outlet.
O. Outlet for deionized water.
P. Outlet for hot water from sub-distillation still.
Q. Outlet for highest quality water.

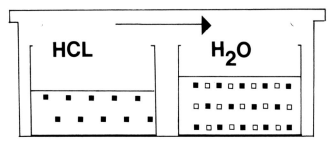

Figure 43: Isopiestic distillation of hydrochloric acid. Initially two containers, one containing analytical grade concentrated hydrochloric acid and the other ultra pure water, are placed in a sealed polythene tank. Following a period of about twenty-four hours the acid will have partly distilled over into the tank containing the water to provide a solution of dilute HCl.

vacuum distillation in ultra clean quartz systems, hydrofluoric acid in Teflon stills, and for special work, acids were prepared from subboiling stills. Large volumes of dilute hydrochloric acid and ammonia solutions were also readily prepared by isopiestic distillation, illustrated in Figure 43.

A further major problem in laboratories concerned with trace element analysis is the need to ensure that glass and plastic containers are adequately cleaned; a suitable system for cleaning using the combined action of condensed nitric acid vapour and steam leaching is illustrated in Figure 44; this procedure provides an adequate reduction of contaminants. Excellent trace element analysis can be performed in normal laboratories, but in describing practical requirements for a clean laboratory I have in mind the need to control environments for the total process of trace element investigations and not solely the end analytical stage; further, there is need to achieve conditions suitable for the analysis of any element and the most extreme requirements are required in multielement analysis when levels of common elements such as calcium or sodium may be just as important as those for rare elements.

Stardards and Harmonisation of Methods

If we assume that all reasonable precautions have been taken

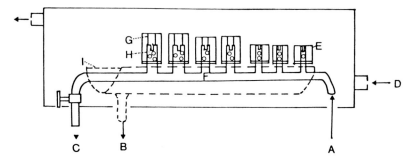

Figure 44: Acid and steam leaching apparatus for cleaning plastics and glassware.
A. Inlet for steam or acid vapour.
B. Exit for condensed waste phases.
C. Control valve made of Teflon to regulate flow of acid vapour and steam.
D. Inlet for ultra clean nitrogen purified by filtration and acid scrubbing.
E. Inverted beakers to be cleaned.
F. Main manifold made of quartz.
G. Quartz support rods.
H. Perforated quartz rods of main manifold for acid vapour and steam; the diameter of the uprights increases with distance from the source in order to insure an even flow of leaching vapours.
I. Quartz trough for collecting condensed phases.

prior to the final assay for an element, what can be done to ensure that this final and critical step is performed correctly? Ideally, any method should be free from problems of matrix, should be selective, and (for economic reasons) is better the simpler and cheaper it is. It is quite irrelevant what method is used and it is pointless engaging in discussions on the merits of manual versus instrumental and destructive versus nondestructive techniques. Further, I see no point in identifying a specific method or instrument as they may only be available to a few individuals. All methods require some form of internal laboratory standardisation but this in itself cannot guarantee that acceptable data will be obtained, this can only be accomplished in a satisfactory manner by participation in programmes of interlaboratory standardisation procedures and use of reference materials for defined matrixes. Briefly, the following types of standards now constitute an important part of analytical chemistry:

1. Reference Materials—these mainly consist of solids or powders which are issued by various bodies with certified levels of the chemical elements which are homogenously distributed throughout the materials. Provided that the Reference Material has a composition very similar to the samples of interest, an ability to obtain the correct value for the standard provides some confidence for the quality of analysis on samples. However, it does not follow that if the standard can be analysed correctly the overall quality of environmental data is also acceptable, simply because the problems of sampling, sample preparation, and contamination, or loss of elements are not covered by the standards. The range of biological reference materials is quite large and further materials are still required. For example, liquid samples such as whole blood, serum, or urine are not available, and possibly will never be available, simply because of the problems of storage; in their place dried products can be prepared of which possibly serum is urgently required. With increasing attention being paid to the chemical form of the elements and organic-inorganic associations, there is also a need for standards with certified levels of such compounds. There is a need also for standards containing very low levels of elements in different chemical forms with particular reference to blood, muscle, milk, and urine. The potential list of materials is large but, bearing in mind the high cost of preparation, the number made available should be kept as small as possible, at the same time providing general cover for most of the matrixes of interest.

2. The next type of standard is the intercomparison material. Usually these consist of various materials in which the elements are homogenously distributed and they are then passed to various laboratories for analysis; the results are passed to a central body where they are tabulated and the relevant statistics applied to produce mean values for the concentration of the elements. As more and more laboratories analyse these materials confidence grows in the quality of the mean data and gradually they can serve as Reference Materials; a good example of this type of

standard is Bowen's Kale (26), data for which are presented in Table XLVII. The first returns of interlaboratory comparison exercises are most illuminating and usually illustrate a very wide range of values, often spreading over several orders of magnitude. This may be accounted for by a real lack of expertise or technique for the analysis of some elements, but also because the experiences of the laboratories participating in the exercise are very different. If samples are sent to laboratories with suffi-

TABLE XLVII*

ESTIMATES FOR THE CONCENTRATION OF SOME ELEMENTS IN HUMAN BLOOD (μg/g)

Element	Blood	Element	Blood
Ag	0.0108 (0.003-0.2)	Mn	0.031 (0.0008-0.6)
Al	0.30 (0.1-6)	Mo	0.001 (0.001-0.16)
As	0.0077 (0.002-0.8)	N	
Au	0.00035	Na	2,018 (1,360-2,700)
B	0.13 (0.04-0.25)	Ni	0.047 (0.02-0.4)
Ba	0.055 (0.04-0.07)	O	
Bi	0.009?	P	369 (330-440)
Br	3.33 (1.3-10)	Pb	0.25 (0.09-0.6)
Ca	62.1 (53-71)	Pd	
Cd	0.0065 (0.005-0.007)	Ra	$6.6 \cdot 10^{-12}$
Ce		Rb	2.33 (1.2-6)
Cl	2,919 (2,600-3,300)	Re	
Co	0.058? (0.0002-0.2)	Ru	
Cr	0.054 (0.006-0.33)	S	1,928?
Cs	0.0032	Sb	0.0047 (0.001-0.008)
Cu	1.11 (0.53-2)	Sc	
Dy		Se	0.193 (0.07-0.32)
Eu		Si	4.03 (1.2-100)
F	0.55 (0.02-1.6)	Sm	
Fe	464 (350-525)	Sn	0.17? (0.015-0.3)
Ga	0.00052?	Sr	0.039 (0.016-0.24)
Ge	0.44	Ta	
Hf		Te	0.0055
Hg	0.0083 (0.001-0.013)	Th	0.0005?
I	0.060 (0.015-0.13)	Ti	(0.025-0.1)
In		Tl	0.00048?
Ir		U	0.00072 (0.0001-0.08)
K	1,737 (1,450-2,770)	V	0.0079? (0.008-1.5)
La		W	0.001?
Li	0.0088 (0.003-0.3)	Zn	6.53 (3.4-13.7)
Mg	37 (28-46)	Zr	0.0063?

* From Bowen, H. J. M.: Problems in the elementary analysis of standard biological materials. *J Radioanal Chem*, 19:215, 1974.

cient experience, it would be expected that the scatter of results would be less and an acceptable level for the abundance of an element in a sample could be more easily obtained. However, these exercises serve another purpose in providing some informa-

TABLE XLVIII*

EXAMPLES OF SOME INTERNATIONAL STANDARDS APPLICABLE TO ENVIRONMENTAL CHEMISTRY

A. TYPES OF MATERIALS

United States Geological Survey Rock Standards, Washington, D. C.

G1† Granite (exhausted)	AGV-1† Andesite
W1† Diabase (exhausted)	PCC-1† Peridotite
G-2† Granite (replacement for G1)	DTS-1† Dunite
GSP-1† Granodiorite	BCR-1† Basalt

National Bureau of Standards, U. S. Dept. of Commerce, Washington, D. C. 20234

1571‡ Orchard leaves	97a‡ Clay (flint)
1573‡ Tomato leaves	81a‡ Glass (sand)
1577‡ Liver, bovine	606‡ Calcium carbonate
1630 Coal (Hg)	607-618† Multi-element series 500 ppm, 50 ppm, 1 ppm, 0.02 ppm glass mixes
1631 Coal (S)	
1633† Fly ash	1677-1681 Carbon dioxide in nitrogen
27c‡ Iron ore	1673-1675 Carbon dioxide in nitrogen
120b‡ Phosphate rock	1683-1687 Nitric oxide in nitrogen
69a‡ Bauxite	16049-1609 Oxygen in nitrogen
1b‡ Limestone (Argillaceous)	1610-1613 Hydrocarbons in air
88a‡ Limestone (Dolomitic)	1636-1638 Lead in reference fuel
607‡ Feldspar	1641-1642 Mercury in water
70a‡ Feldspar (potash)	1634 Fuel oil
99a‡ Feldspar (sodic)	

International Atomic Energy Agency, Vienna, Austria

W3‡ Fresh water	H6‡ Human blood serum (total iodine and bound iodine)
Air 3‡	
A2‡ Animal blood	MA‡ MI dried oysters
A3/1‡ Calcined bone	Soil 5‡
A6‡ Fish solubles	SL1‡ Lake sediment
A7‡ Milk powder (Na, K, Ca, Sr)	V1 Corn flour (Hg)
A7‡/1 Milk powder	V2 Wheat flour (Hg)
H1 Urine man (U, Th, Pu)	V2/1 Wheat flour
H1‡/1 Urine man	V5‡ Wheat flour (iodine)
H4‡ Muscle	V4‡ Potatoes
H5‡ Bone	

Powdered Kale†

* From Hamilton, E. I.: Review of the chemical elements and environmental chemistry—strategies and tactics. *Sci Total Environ*, 5:19, 1976.
† Data available for most elements of the Periodic Table.
‡ Limited data available for elements.

B. EXAMPLES OF DATA OBTAINED FOR THE NBS BOVINE LIVER (NBS-SRM 1577)§ (ppm dry wt)

Element	Range of Results NBS	Range of Results Other	Recommended Interim Value	NBS Certified Value
Antimony		< 0.1-.014 (2)	0.014	
Arsenic	0.050-0.059 (1)		0.055	
Cadmium	0.24-0.32 (3)	< 0.1-0.35 (6)		0.27 ± 0.04
Calcium	117-125 (1)		123	
Cesium		0.013 (1)	0.013	
Chlorine	2,533-2,656 (1)		2,610	
Chromium		0.18 (1)	0.2	
Cobalt	0.17-0.19	0.15-0.29 (3)	0.18	
Copper	181-204 (3)	120-202		193 ± 10
Iron	259-299 (2)	232-280 (6)		270 ± 20
Lead	0.26-0.41 (2)	0.33-2.5 (3)		0.34 ± 0.08
Magnesium	600-611 (1)		605	
Manganese	9.3-11.3 (2)	10.4-27.9 (7)		10.3 ± 1.0
Mercury	0.014-0.017 (2)	0.006-0.070 (7)		0.016 ± 0.002
Molybdenum	3.08-3.34 (1)		3.2	
Nitrogen	10.25-11.03% (2)			10.6 ± 0.4%
Potassium	0.93-1.01% (2)	0.756-0.99% (3)		0.97 ± 0.06%
Rubidium	17.5-19.2 (2)	16 (1)		18.3 ± 1.0
Selenium	1.05-1.17 **(2)**	0.95 (1)		1.1 ± 0.1
Silver	0.054-0.081 (1)	< 0.55-0.11 (2)	0.06	
Sodium	0.223-0.267% (2)	0.20-0.267% (3)		0.243 ± 0.011%
Strontium	0.136-0.142 (1)		0.14	
Uranium	0.00055-0.00090 (1)		0.0008	
Zinc	120-134 (3)	78.5-242 (10)		130 ± 10

C. EXAMPLES OF DATA OBTAINED FOR THE NBS ORCHARD LEAVES (NBS-SRM 1571) § (ppm dry wt)

Element	NBS	Other	Interim	Certified
Aluminium	341-428 (2)	99-420 (14)	400	
Antimony		2.7-3.5 (3)	2.9	
Arsenic	11.3-15.9 (2)	10-18.3 (7)		14 ± 2
Barium		40-52 (5)	45	
Bismuth	0.10-0.14 (1)		0.11	
Boron	31-34 (1)	23.7-38 (13)		33 ± 3
Bromine	9.4-10.5 (1)	8.9-9.5 (2)	10	
Cadmium	0.09-0.12 (2)	< 0.1-.45 (9)		0.11 ± 0.02
Calcium	2.052-2.125% (2)	1.63-2.41% (15)		2.09 ± 0.03%
Chlorine	680-780 (1)	790 (1)	750	
Chromium	2.34-2.38 (1)	1.8-2.5 (2)	2.3	
Cobalt	0.18-0.23 (1)	0.1-0.18 (5)	0.2	
Copper	10.8-13.7 (3)	9.9-20 (25)		12 ± 1
Iron	282-319 (3)	151-367 (21)		300 ± 20
Lanthanum		1.2-2.1 (2)		
Lead	41.4-47.9 (3)	37-53 (12)		45 ± 3

Element	Range of Results NBS	Range of Results Other	Recommended Interim Value	NBS Certified Value
Lithium	14 (1)			
Magnesium	0.6136-0.6370% (1)	0.40-0.71% (14)	14	0.62 ± 0.02%
Manganese	88.6-95.0 (2)	52-144 (2)		91 ± 4
Mercury	0.149-0.170 (3)	0.10-0.18 (15)		0.155 ± 0.015
Nickel	1.2-1.5 (2)			1.3 ± 0.2
Nitrogen	2.723-2.784% (2)	2.50-2.86% (8)		2.76 ± 0.05%
Phosphorous	0.206-0.211% (1)	0.14-0.31% (13)		0.21 ± 0.01%
Potassium	1.444-1.506% (2)	1.11-1.62% (18)		1.47 ± 0.03%
Rubidium	11.0-12.5 (2)	10.3-12 (3)		12 ± 1
Scandium		0.04-0.2 (3)	0.04	
Selenium	0.076-0.091 (2)	0.08-0.21 (5)		0.08 ± 0.01
Sodium	71-90 (3)	40-524 (12)		82 ± 6
Strontium		23-45 (6)	37	
Uranium	0.026-0.031			0.029 ± 0.005
Zinc	22.8-28.2 (3)	18-81 (25)		25 ± 3

§ Data given in terms of µg/g dry weight. Data are presented for individual values reported for NBS results and mean values reported by different laboratories. Values in parentheses are for the total number of groups reporting values. The NBS-certified values are considered to be the accepted levels present, while the recommended interim values represent best estimates obtained so far. Both materials can be obtained from the United States National Bureau of Standards, Washington, D.C.

tion concerning the quality of laboratories which perhaps have little experience in minor or trace element analysis but, nevertheless, are called upon to produce data used in morbidity and environmental studies.

Examination of first returns for laboratories situated in many countries clearly illustrate the need for considerable improvement in the quality of analytical data but, once more, it must be acknowledged that the use of standards only constitutes one part of the overall problem of determining levels of elements in environmental and biological materials. A final warning concerns human nature; it is inevitable that a laboratory undertaking analysis of an intercomparison or reference material will pay particular care to the analysis, and it does not follow that the quality of analysis obtained on a standard reflects the day-to-day ability of a laboratory; this point should be borne in mind when considering the spread of data from first returns. Examples of some available reference and intercomparison materials are given in Tables XLIII and XLIX and Figure 45.

The whole question of environmental standards for protection of the public revolve around measurements of different quantities as illustrated in Table L. Of all the chemical elements, lead has possibly received most attention but on a routine basis it is still a difficult element to determine. The results

TABLE XLIX

RESULTS FOR SOME INTERLABORATORY INTERCOMPARISON EXERCISES ILLUSTRATING THE NEED FOR IMPROVEMENTS IN THE QUALITY OF DATA FOR A VARIETY OF MATRIXES

A. INTERNATIONAL ATOMIC ENERGY AGENCY RESULTS FOR VARIOUS RADIONUCLIDES IN DIFFERENT MATRIXES

Material	Radionuclide	No. of Institutes	No. of Results	Mean	Standard Deviation
Vegetation	Pu-239	8	35	8.7 pCi/l	3.6
Water	Rn-222	13	59	121.5 pCi/l	9.9
Soil	Ra-226	11	62	0.56 pCi/g	0.32
Soil	Ra-228	7	42	0.76 pCi/g	0.21
Feldspar	K (via ^{40}K)	32	151	126.0 g/kg	4.7
Water	H-3 (Tritium)	10	42	1331.0 pCi/l	282.0
Animal blood	Cs-137	48	217	86.3 pCi/kg	24.6
Human urine	U	17	17	31.7 µg/l	—
	Th	8	5	11.9 µg/l	—
	^{239}Pu	11	8	0.73 pCi/l	0.24

B. EUROPEAN ECONOMIC COMMUNITY—INTERCOMPARISON PROGRAMME AND HARMONISATION OF TECHNIQUES; RESULTS FOR Pb, Hg, AND Cd IN BIOLOGICAL FLUIDS AND WATER*

Element	Material	No. of Analyses	Mean	Median	Range	Coefficient of Variation Interlaboratory
Pb	Human blood†	55	23.3	18.1	1-115	77.5
Pb	Human urine‡	33	62.6	62.3	5.3-159	58.8
Pb	Water‡	51	209.0	196.6	24.7-669.8	54.9
Hg	Human blood†	18	2.6	2.4	0.2-9.0	80.5
Hg	Human urine‡	29	7.5	4.6	1-87.5	206
Cd	Human blood†	17	1.9	0.96	0-11.0	143
Cd	Water‡	22	38.2	35.0	18-116.7	51.2

C. SUMMARY OF IAEA INTERCOMPARISONS OF TRACE-ELEMENT ANALYSIS 1966-1969[§]

Matrix	Element	Mean (dry wt)	Standard Deviation ±1	Error (95% Confidence Limits of Overall Mean) ±1	Number of Institutes	Number of Results
Flour (Hg-treated)	Hg	4.59 ppm	1.37	0.70	20	101
Flour (untreated)	Hg	42.5 ppb	14.7	9.0	20	86
Flour (fumigated)	Br	8.03 ppm	1.17	0.70	12	70
Flour (unfumigated)	Br	56.2 ppb	12.4	19.8	7	23
Dried animal blood	Cu	2.07 ppm	0.66	0.45	11	61
Dried animal blood	Mn	0.19 ppm	0.11	0.10	7	29
Dried animal blood	Mo	—	—	—	7	17
Dried animal blood	Fe	2,826 ppm	246	225	9	46
Dried animal blood	Zn	17.6 ppm	2.9	2.2	11	56
Dried animal blood	Cr	0.22 ppm	0.26	0.33	8	22
Animal bone	Cu	5.7 ppm	1.3	1.4	7	32
Animal bone	Mn	20.0 ppm	3.2	3.3	6	28
Animal bone	Mo	0.5 ppm	—	—	4	8
Animal bone	Fe	1,057 ppm	428	534	5	23
Animal bone	Zn	183.3 ppm	24.6	30.3	6	30
Animal bone	Cr	2.0 ppm	0.70	0.74	6	15
Mussel shells	Cu	1.63 ppm	1.03	1.04	9	45
Mussel shells	Mn	71.4 ppm	9.9	8.3	9	44
Mussel shells	Mo	0.48 ppm	0.19	0.23	7	28
Mussel shells	Fe	1.054 ppm	345	286	8	39
Mussel shells	Zn	7.6 ppm	3.5	1.7	9	43

° From Hamilton, E. I.: Review of the chemical elements and environmental chemistry. *Sci Total Environ,* 5:20, 1976.
† μg per 100 ml.
‡ μg per litre.
§ From *IAEA Report,* 30th Vienna 1970. Vienna, IAEA, 1970.

of a European Intercomparison programme for assay of lead in whole human blood are given in Figure 46. Because of small differences between the accepted and toxic levels for this element, only slight errors in analysis can alter a picture from one of no concern to one for which remedial action is required. In general, signs of lead poisoning are not usually observed for blood lead levels of less than 30 μg/100 mls, while it is rare to observe levels of less than 10 μg Pb/100 mls. In fact even when presumed

178 The Chemical Elements and Man

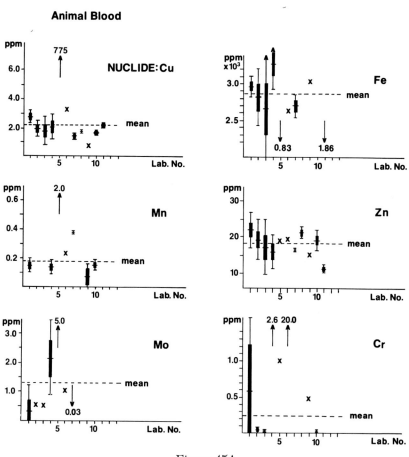

Figure 45A

Methods—Preparatory Techniques

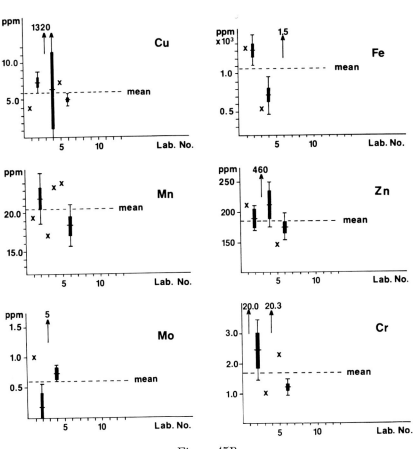

Figure 45B

180   The Chemical Elements and Man

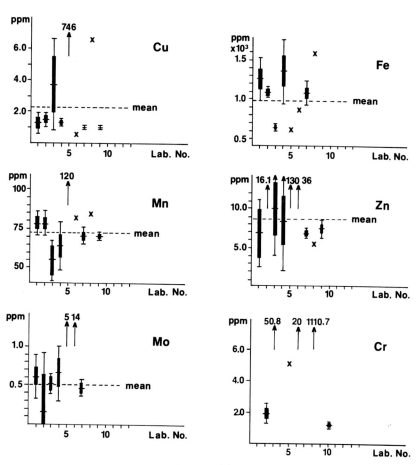

Figure 45C

Figure 45: Some results of the International Atomic Energy Agency intercomparison exercises for animal blood, animal bone, and mussel shells. From Tugsavul, A., Merten, D., and Suschny, O.: The reliability of low-level radiochemical analysis; results of intercomparisons organised by the Agency during the period 1966-1969. Vienna, International Atomic Energy Agency, March 1970.

TABLE L*
THE ROLE OF ANALYTICAL DATA IN ESTABLISHING ENVIRONMENTAL STANDARDS

Environmental standards; definition and terms used by the United Nations Conference on the Human Environment and by the World Health Organization (The items in italics are related to the measurement of pollutants. Therefore since the results of analytical procedures form the basis upon which economic, administrative, medical or legal decisions are taken it is essential that the levels determined should be reliable.)

Exposure: the AMOUNT of a particular physical or chemical agent that reaches the target.

Target (or receptor): the organism, population or resource to be protected from specified risks.

Risk: the expected frequency of undesirable effects arising from a GIVEN EXPOSURE to a pollutant.

Criteria (or exposure-effect relationships): the quantitative relations between the exposure to a pollutant and the risk or magnitude of an undesirable effect under specified circumstances defined by environmental variables.

Primary protection standard: an accepted MAXIMUM LEVEL of a pollutant (or its indicator) in the target or some part thereof, or an accepted MAXIMUM INTAKE of a pollutant or nuisance into the target under specified circumstances.

Derived working levels (or limits): MAXIMUM ACCEPTABLE LEVELS of pollutants in specified media other than the target designed to insure that under specified circumstances a primary protection standard is not exceeded (derived working levels are known by a variety of names, including environmental or ambient QUALITY STANDARDS, MAXIMUM PERMISSIBLE LIMITS and MAXIMUM ALLOWABLE CONCENTRATIONS. When derived working levels apply to products such as food or detergent, they may be known as product standards).

The maximum acceptable release of a pollutant from a given source to a specified medium under specified circumstances may be termed a discharge (or effluent or EMISSION STANDARD or a RELEASE LIMIT). Effluent charges levied on the release of pollutants and materials taxes or price adjustments levied on materials which may become pollutants may also be used to limit the release of pollutants (in order to meet discharge standards or release limits, it may be necessary to set various types of TECHNOLOGICAL STANDARDS or codes of practice concerned with the performance and design of those technologies or operations leading to the release of pollutants).

Derived working levels and the various means used to meet them are collectively termed derived standards and other controls.

Action level: the LEVEL OF A POLLUTANT at which specified counter-measures, such as the seizure and destruction of contaminated materials, evacuation of the local population or closing down the sources of pollution, are to be taken.

* From Hamilton, E. I.: Review of the chemical elements and environmental chemistry—strategies and tactics. *Sci Total Environ,* 5:18, 1976.

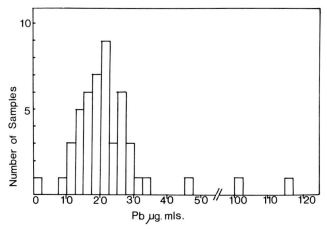

Figure 46: Interlaboratory comparison for the determination of lead in human blood. From Berlin, A. and Smeets, J.: *European Intercomparison Programmes and Harmonisation Techniques with Special Reference to Heavy Metals and Biological Fluids.* Luxemburg, Commission of the European Communities, Report V/F/1760/74e, 1974.

sensitive techniques of clinical examination are used, such as neurological response, it is rare to identify real toxic effects even for many elevated environmental levels. Berlin and Smeets (27) have concluded that present methods of analysis are not adequate when it is necessary to determine small differences in blood lead arising from different environmental exposures. Unfortunately the same comments apply to the assay of ALA in urine as an index of exposure to lead. Assay for another indicator of lead in the body, namely ALA-D in blood, was shown to be reliable, relatively free from analytical problems and operator techniques, and has been described by Berlin, Castilho, and Smeets (28). Since 1973, considerable improvements in the quality of analyses have taken place but problems still remain; it is likely to be many years before reliable data becomes common, rather than the exception, throughout the world. In terms of improving the health of man, and identifying the role of the chemical elements and disease, matters do not rest solely with analyses; dose-response relationships have to be determined for large populations and individuals and considerably more information is required in order to describe the participation of the chemical elements in defined biochemical and metabolic pathways.

So far my comments have been restricted to total assay for elements; when investigating biochemical phenomena it often becomes essential to separate biochemical compounds prior to element analysis. The problems presented by such tasks are indeed formidable, although a vast number of techniques are available for carrying out the separation of such compounds. It is beyond the scope of this monograph to describe the required methods and techniques. Suffice it to say that special methods, such as activation analysis, are often required, but even in this case ionizing radiation can lead to structural damage of compounds.

Finally, in many types of analysis it is common practice to prepare standard solutions of the elements; often the analyst is wary of using glass containers for storage because of the possibility of loss of an element to the glass walls or the entry of elements into the solution through leaching. In many laboratories various types of plastic containers are often used but two types of problems can arise:

(1) Although plastics are made of organic materials they can contain considerable amounts of metals which can be removed by the various techniques which have been described. Once cleaned properly they can be used as reliable containers for standards. Some of the most severe problems of contamination are associated with new articles made of Teflon, especially if they have been formed by turning on a metal lathe; common contaminants which can be removed by leaching are Pb, Zn, Cu, Fe, Cr, Ca, K, Mg, and Al.

(2) It is very difficult to seal screw top bottles so that they are airtight, and because of the nature of the surface properties of plastics, slight changes in room temperature are sufficient to loosen the lids thus allowing the gradual loss of liquids by evaporation and an increase in concentration of elements in solution. Hamilton (29) has discussed the problem of storage of standard solutions in plastic bottles and showed that, while fairly full bottles can experience an increase in salt concentration of about

2 percent over a period of one year, as the volume of standard is reduced the rate of concentration of salt can be quite high. For example in the case of a 1 liter bottle containing 80 ml of solution there was an increase in concentration of the solution of 48 percent over a period of one year.

PRECONCENTRATION OF THE ELEMENTS

The mass of an element which can be determined in samples by any method has limitations in terms of the presence of other large or very small amounts of the elements; the former can usually be overcome by simple dilution techniques in order to bring concentrations into the practical range of the measuring techniques; the latter presents far more formidable problems but, in principle, any element can be subjected to some form of concentration technique, which will vary for different matrixes. Difficulties arise in carrying out such processing without the introduction of contaminants, without the loss of an element, and finally a requirement to obtain an element in a fairly pure form if the full benefit of preconcentration is to be realised. Skogerboe and Morrison (30) have discussed various aspects of the technique; if the limits of detection of a particular method are known, then it is possible to select the optimum sample size and identify an appropriate method of separation. It is of course possible to perform various modifications to the final instrumental phase of assay in order to improve limits of detection, but often they have very limited use and eventually it becomes impossible to separate the signal from the element from that arising from background noise of a detector. Concentration techniques offer the distinct advantage of removing problems associated with the matrix of the sample, and the technique allows a direct comparison of the signal from the sample with that of a standard containing the element of interest in a fairly simple and convenient chemical form.

The literature on analytical chemistry is replete with methods and techniques and they will not be dealt with in detail in this monograph. Precipitation or coprecipitation techniques rarely produce a pure product but the simplicity of the technique often has much to recommend. The purity of the final precipitate can often be improved by repeated reprecipitations, but each stage increases the likelihood of contamination or loss. Solvent extraction, a technique involving the partition of an element in solution between two immiscible phases, can often provide adequate separations by a single stage of extraction and is also amenable to batch and automated processing. By altering the pH of the medium, or by the addition of complexing or chelating compounds, quite complex sequential separations for many elements can be achieved. Chromatographic techniques employing paper, ion-exchange resins, and gaseous separations have marked a significant advance in separation and concentration technology; for example, one type of ion-exchange resin, Chelex 100, containing iminodiacetate functional groups, provides chelation groups which are particularly useful in the separation of the transitional group of elements; analytical applications of ion-exchangers have been described by Rieman and Walton (31). In radioactivation analysis (see page 267), contaminating elements present in the materials used for concentration and separation procedures are of no concern; very elegant and comprehensive schemes of separation have been devised based upon ion-exchange resins, including automated systems, some of which are available commercially. However, for nonradioactive techniques considerable attention has to be paid to the presence of contaminants which are invariably present in most materials, some of which can be removed by quite simple techniques while others present considerable difficulties, some of which have been reviewed by Tölg (20). Apart from the actual materials employed, many techniques require the use of large volumes of reagents which have to be purified prior to use. By the use of special chemical techniques some elements can be concentrated from large samples by very elegant methods, for example the separation of trace impurities of arsenic from solutions by the in situ generation of arsine gas followed by absorption in a solution in which a colour reaction takes place and the end measurement is made by spectrophotometry (Haywood and Riley, 32). Tavlaridis and Neeb (33) have described the use of gas chromatography for elements extracted by diethyldithiocarbamates from aqueous solutions followed by sequential combustion using a flame ionization detector; cations such as Zn, Cd, Cu, Ni, and Pb can be determined in the concentration range of 0.6-0.06 ppm. Radziuk and Loon (34) have described a very simple gas chromatograph for the separation of selenium alkanes using an atomic absorption spectrophotometer as a detector. With increased interest in the chemical form of the elements in biological tissues, chromatographic separation and concentration techniques have become very important tools of investigation.

TABLE LI

TECHNIQUES FOR THE SEPARATION OF ELEMENTS AND COMPOUNDS PRIOR TO CHEMICAL ASSAY

Freeze drying Concentration of elements from clear solutions
Precipitation e.g. Hydroxides, sulphides, ammonium pyrrolidine dithiocarbamate, Chitosan
Distillation e.g. As, Cl, Ge, Hg, Re, Ru, Sb, Se, Sn (usually as chlorides or hydrides)
Electrodeposition e.g. Cu, Zn, Cd, Pb, Hg, Au
Solvent Extraction e.g. Separation of an element present in an aqueous solution by extraction into an immiscible solvent phase at controlled pH. Common solvents: carbon tetrachloride, chloroform, ethyl acetate, methyl-isobutyl-ketone (MIBK) and alcohols.
  a) Chelates—Dithizone (Diphenylthiocarbazone)
     Ammonium pyrrolidine dithiocarbamate (APDC)
     Oxine
     Cupferron
     Elements extracted: Mn, Fe, Co, Cu, Zn, Cd, Sn, Pb, and several others.
  b) Inorganic Chlorides—organic, e.g. Fe, Mo, Ge.
     Fluorides—ethers, ketones, e.g. Nb, Ta.
     Nitrates—ethers, ketones, e.g. U, Pu, Np, Am.
Ion-Exchange Resins, Papers
  Organic—Acid, Basic, Bifunctional, Chelation types.
  Inorganic e.g. Zirconium phosphate (alkaline metals from alkaline earths, Cs)
    Zirconium molybdate (Separation Mg—Ca—Sr—Ba)
    Zirconium antimonate (Separation K—Rb—Na)
    Alumina—U
Organic media e.g. Celluloses—Fe, Bi, Hg, U
  Alginic Acid—Zn, Ag, In, Pb
  Chitosan—Fe, Co, Ni, Cu, Zn
Modified Substrates e.g. Controlled pore size glass beads—Reagent immobilised by silylation (e.g. upon silica gel) for anion separation of arsenate, selenate, vanadate, phosphate
Physical e.g. Molecular filtration (paper, hollow fibres)
  Dialysis
  Foam, bubble formation

If controlled environment laboratories are available, then the preconcentration of the elements and their compounds can be accomplished in a routine manner. Before they are used it is essential to determine whether or not the materials used in the total analytical process are likely to result in the addition of contaminants and at what level in relation to those elements present in the samples; data are provided in Table LI indicating some commonly used separation and concentration techniques.

REFERENCES

1. Cochran, W. G. and Cox, G. M.: *Experimental Design*. New York, Wiley, 1957.
2. Anders, O. U.: Representative sampling and proper use of reference materials. *Anal Chem,* 49:33A, 1977.
3. Ingamells, C. O., Engels, J. C., and Switzer, P.: Effect of laboratory sampling error in geochemistry and geochronology. *24th International Geochemical Cong, Section 10:*405, 1972.
4. Liebscher, K. and Smith, H.: Essential and nonessential trace elements. *Arch Environ Health,* 17:881, 1968.
5. Maletskos, K.: Private communication, 1967.
6. Hopps, H. C.: The biologic bases for using hair and nail for analyses of trace elements. *Sci Total Environ,* 7:71, 1977.
7. Versieck, J., Speecke, A., Hoste, J., and Barbier, F.: Trace-element contamination in biopsies of liver. *Clin Chem,* 19:472, 1973.
8. Harrison, G. E. and Sutton, A.: Alkaline earths in heparin. *Nature (London),* 197:809, 1963.
9. Hamilton, E. I., Minski, M. J., and Cleary, J. J.: The loss of elements during the decomposition of biological materials with special reference to arsenic, sodium, strontium and zinc. *Analyst,* 92:257, 1967.
10. Hamilton, E. I., Minski, M. J., and Cleary, J. J.: Problems concerning multi-element assay in biological materials. *Sci Total Environ,* 1:1, 1972.
11. Kuehner, E. C., Alvarez, R., Paulsen, P. J., and Murphy, T. J.: Production and analysis of special high-purity acids purified by sub-boiling distillation. *Anal Chem,* 44:2050, 1972.
12. Mitchell, J. W.: Ultrapurity in trace analysis. *Anal Chem,* 45:492A, 1973.
13. Dabeka, R. W., Mykytiuk, A., Berman, S. S., and Russell, D. S.: Polypropylene for the sub-boiling distillation and storage of high purity acids and water. *Anal Chem,* 48:1203, 1976.
14. Thiers, R. E.: *Contamination in Trace Elements*. New York, Wiley, 1975.

15. Zief, M. and Mitchell, J. W.: *Contamination Control in Trace Element Analysis.* New York, Wiley, 1976.
16. Karin, R. W., Buono, J. A., and Fasching, J. L.: Removal of trace elemental impurities from polyethylene by nitric acid. *Anal Chem, 47:* 2296, 1975.
17. Struempler, A. W.: Adsorption characteristics of silver, lead, cadmium, zinc, and nickel on borosilicate glass, polyethylene and polypropylene container surfaces. *Anal Chem, 45:*2251, 1973.
18. West, F. K., West, P. W., and Iddings, F. A.: Adsorption of traces of silver on container surfaces. *Anal Chem, 38:*1566, 1966.
19. Lautenschäger, W.: Sample preparation for atomic absorption spectroscopy (AAS). Beckman Information 2-76, 3, 1976. [Obtainable from Beckman Instruments International. S.A., Geneva, Switzerland]
20. Tölg, G.: Extreme trace analysis of the elements—I. Methods and problems of sample treatment separation and enrichment. *Talanta, 19:* 1489, 1972.
21. Sansoni, B., Kracke, W., and Winkler, R.: Rapid assay of environmental radioactive contamination with special reference to a new method of wet ashing. In *Environmental Contamination of Radioactive Materials,* Proceedings of a Seminar, Vienna 24-28th March, 1969. FAO, IAEA, WHO. Vienna, International Atomic Energy Agency, 1969.
22. Sansoni, B. and Kracke, W.: Rapid determination of low-level alpha- and beta-activities in biological materials using wet ashing by OH radicals. In *Environmental Contamination of Radioactive Materials,* Proceedings of a Seminar, Vienna 24-28th March, 1969. RAO, IAEA, WHO. Vienna, International Atomic Energy Agency, 1969.
23. Gorsuch, T. T.: *The Destruction of Organic Matter.* New York, Pergamon, 1970.
24. Hamilton, E. I.: The design of radiochemical laboratories. Legal aspects of radiochemistry. *Proc Soc Analyt Chem, 8:*203, October 1971.
25. McDermott, H. J.: *Handbook of Ventilation for Contamination Control.* Michigan, Ann Arbor Sciences Pub Inc, 1976.
26. Bowen, H. J. M.: Comparative elemental analyses of a standard plant material. *Analyst, 92:*124, 1967.
27. Berlin, A. and Smeets, J.: *European Intercomparison Programmes and Harmonisation of Techniques with Special Reference to Heavy Metals in Biological Materials.* Luxembourg, Commission of the European Communities. Health Protection Directorate. V/F/1760/74; 1974.
28. Berlin, A., del Castilho, P., and Smeets, J.: European Intercomparison Programmes. In, *Proc Int Symp Environmental Health Aspects of Lead,* Amsterdam, 2-6 Oct, 1972. Luxembourg, Comm European Communities, 1972.

29. Hamilton, E. I.: Storage of standard solutions in polythene bottles. *Nature, 193*:200, 1962.
30. Skogerboe, R. K. and Morrison, G. H.: Trace analysis: Essential concepts. In Kolthoff, I. M., Elving, P. J., and Sandell, E. B. (Eds.): *Treatise on Analytical Chemistry,* pt I, vol 9. New York, Wiley, 1971.
31. Rieman, W. and Walton, H. F.: Ion exchange in analytical chemistry. In *Int Ser Mon Anal Chem,* vol 38. Oxford, Pergamon, 1970.
32. Haywood, M. G. and Riley, J. P.: The spectrophotometric determination of arsenic in sea water, potable water and effluents. *Anal Chem Acta, 85*:219, 1976.
33. Tavlaridis, A. and Neeb, R.: Gas-chromatographic multielement analysis using simple and fluorinated diethyldithiocarbamates. *Z Anal Chem, 282*:17, 1976.
34. Radziuk, B. and Loon, J.: Atomic absorption spectroscopy as a detector for the gas chromatographic study of volatile selenium alkanes from Astragalus racemosus. *Sci Total Environ, 6*:251, 1976.

Chapter 7

INSTRUMENTAL METHODS

IN USING THE TERM instrumental methods, I refer to those modern instruments which are commercially available; instruments are not new to analytical chemistry but they have become more sophisticated and have been developed with the ultimate objective of permitting quantitative and rapid analysis for the chemical elements present in a wide variety of material with the minimum of difficulty. Modern instruments are essential for trace element analysis but a Utopian state in which utter reliance can be placed upon the quality of the data they produce is nowhere in sight; the problems of analysis are great and often some instruments only add to the problems. The earlier parts of this monograph have identified some of the many problems which confront the analysis of the elements prior to any final assay for mass or concentration by instrumental methods. It is quite pointless to attempt a discussion of all the problems, as each requires an appreciation of the level of the element or elements to be determined, the matrix in which they are present, chemical form, and problems associated with a particular instrument. An objective of most types of analysis is to determine a signal characteristic of an element such that it can be clearly differentiated from other signals registered by an instrument; also that the signal will alter in strength in direct proportion to changes in concentration; further, the change in strength of the signal with concentration should be the same as that recorded for the element in a simple matrix, such as a dilute acid. When dealing with real samples other elements, invariably present in the material, alter the characteristic signal of the element to be determined; the higher the resolving power of the instrument the better the chance of separating the required signal from interferences but this can be difficult to achieve and often results in the loss of sensitivity. One series of interferences can be directly related to the overall matrix of the sample in which other elements provide signals which overlap those of the element to be determined; the presence of other elements can enhance or quench the sample signal and for very low concentrations of an element this type of interference can totally obscure that from the sample.

The method of excitation of elements can be strongly influenced by the presence of other elements, often related to chemical form; for example, in thermal techniques some elements tend to volatilise as compounds and do not produce characteristic element signals while others may be so inert that hardly any signal is produced. Each instrument has its own set of characteristic requirements for optimum assay and they will vary with different matrixes. A typical response curve for different concentrations of an element in a sample is given in Figure 47. The range over which optimum response takes place can vary considerably; in the case of flameless atomic absorption (see page 238),

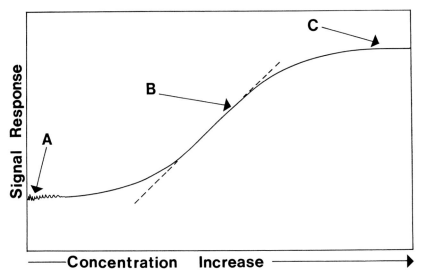

Figure 47: Ideal type of response curve. A = background signal, no response from element signal detected above background noise. B = linear response, proportional increase in signal strength with increase in concentration. C = area of saturation for which no increase in signal strength occurs with increase in concentration, for example because the detector has become saturated.

the useable part of the response curve may only cover a change in concentration of a few ppm while in the case of inductive coupled plasma emission optical spectrometry (see page 258) a linear response may be valid from the ppm to approaching 100 percent of an element.

The number of elements determined in a sample should be defined by scientific objectives but when dealing with human tissues it is advantageous, at an early stage at least, to determine as many elements as practical. Today attention has been placed upon assay for a few elements arising from pollution and they tend to be treated in a manner divorced from the element matrix in which they exist, hence important synergistic effects may be lost. Finally, before commenting upon individual methods, it does not matter what type of technique or instrument is used in the analysis of elements, all that counts is the quality of the data for defined objectives. Many newcomers to the analysis of the elements have limited or no experience in analytical chemistry and tend to rely upon the instrument as an infallible tool often with sad consequences. It is very unfortunate to observe those who ignore the problems of sampling and pass material for analysis to chemists without any discussion on the type or reason for which the data are required. Similarly chemists often exclude themselves from sampling and eventual use of data, solely concern themselves with the sample once it enters the laboratory, and then pass results on with no concern for the manner in which they are used.

Colorimetry and Spectrophotometry

The literature of analytical chemistry is replete with methods and techniques describing analysis for trace elements by colorimetry and spectrophotometry which testifies to the importance of these methods. Some of the earliest measurements for levels of trace elements were made by visual techniques in which a color, specific for a particular element, is developed in a solution and compared with the naked eye against a standard containing a known amount of the element. Of the many excellent text books on the subject, that by Sandell (1) provides basic details and applications; this volume was first produced in 1944 marking a major step in evaluating the problems of minor and trace element analysis. Precision, accuracy, and sensitivity of detection can be high for a large number of elements but in many cases an analysis requires a separation of an element in a pure form and the use of rather large amounts of reagents. The methods tend to be manpower intensive and require considerable technique which today is often not acceptable.

Colour Methods

The use of the human eye to detect colour is limited to the wavelength band between about 400 and 800 mμ, with an optimum at about 560 mμ, although an individual's eye-response in distinguishing between colours and depth of colour varies considerably. Nevertheless, quite a large number of analyses for various trace elements can be made simply by the use of reagents which produce specific colours for elements and then comparing the developed colours with a series of standard which may be liquids, coloured glass or plastics (see Hawkins, Canney, and Ward, 2), or paper strips appropriately coloured with crayons or paints; the background illuminating light passes through, or is incident upon, both standards and samples, and therefore natural or artificial illumination can be used.

Provided that interfering compounds can be removed, the development of some colours with specific reagents is often simple, hence the technique can be used in the field. Some commercially available kits consist of a series of graded colour discs together with the reagents packaged in plastic pillows which are simply opened in the field and added to a known volume of sample, the colour developed and compared to the standard discs; examples of some of these systems are given in Figure 48, together with some detection limits for various elements and anions in Table LII. There are of course limitations in the use of such techniques for the assay of biological materials, usually because of lack of sensitivity, but they have a part to play in many programmes of environmental research and monitoring. For example, it is possible to prepare the colorimetric reagent in a liquid phase which can then be sprayed onto the surface of a sample, the colour developed in situ, and distribution of elements examined; an example of this technique is given in Figure 49 for detecting the presence of reactable zinc on foliage, a useful technique when determining the distribution in the field of fallout zinc around a smelter, thus providing a basis for establishing sampling procedures for more refined methods of assay.

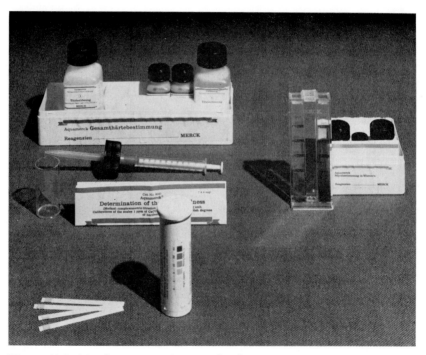

Figure 48A: Merck systems. *Rear*—A kit for carrying out titrations in the field. *Right*—Colour system. The unit consists of three compartments; the coloured solution containing the element of interest is placed in the central tube which is flanked on both sides by cells which are coloured; the intensity of colour corresponds to a known concentration of the element. Altogether there are 10 increments of depth of colour in each cell. *Front*—Dip sticks (Merkoquant) made of plastic on the end of which an element or compound test paper is heat sealed. The stick is dipped into a sample solution and a colour is produced which is compared with a fixed series of colours representing known concentrations of an element or compound which is supplied with the kit. Illustration supplied by British Drug Houses Ltd, Poole, Dorset, U.K., on behalf of E. Merck, Darmstadt, FRG.

Instrumental Methods

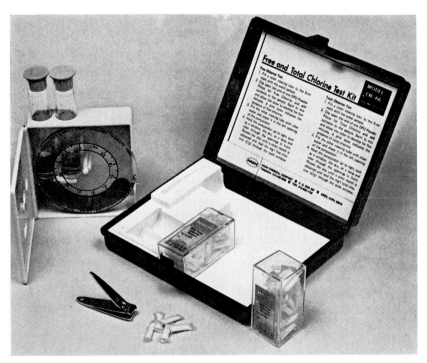

Figure 48B: Hach system. A simple system consisting of a graded colour disc, a comparator box, and chemicals supplied in plastic pillows, to produce specific colours with different elements and compounds; a small clipper is supplied to open the pillows. In the comparator box the solution blank is placed in the left (behind which the colour disc is positioned) and the sample in the right hand hole; the disc is then rotated until the depth of colour in both tubes (sample and blank) is the same, concentration is then read from the disc. Illustration furnished by Hack Chemical Company, Ames, Iowa, United States.

When dealing with samples which have the same colour as that produced by the element specific reagent, the sample can be sprayed with the reagent and, after production of the colour, can be pressed against a piece of white fine-grained filter paper to provide a print of the element distribution; an example of this technique has been described by Hamilton (3) and illustrated in Figure 50 for the distribution of zinc (production of a red colour by reaction with dithizone) in a sample of human

TABLE LII

EXAMPLES OF LEVELS AT WHICH ELEMENTS CAN BE DETECTED USING RAPID VISUAL COLOUR METHODS

	Hach Chemical Company mg/l	E. Merck Limited mg/l		Tintometer Company ppm Lovibond
		Aquamerck	Merckoquant	
Aluminium	—	—	250	4
Ammonia	3	—	—	2
Bismuth	—	—	—	2
Bromine	—	—	—	1,750
Calcium	—	—	250	400
Chlorine	3	1.5	—	4
Chromate	1.5	—	250	0.2
Cobalt	—	—	1,000	—
Copper	5	—	500	5
Cyanide	0.2	—	—	—
Fluoride	—	—	—	1.6
Hydrogen sulphide	3.0	—	—	—
Iodine	5	—	—	36
Iron	10	50	500	28
Potassium	—	—	2,000	40
Lead	—	—	—	2
Manganese	3	—	500	2
Nickle	—	—	500	10
Nitrate	50	—	500	6
Nitrite	0.2	20	50	—
Phosphate	5	10	—	80
Silica	40	3	—	20
Sulphide	—	—	500	—
Sulphate	—	—	—	80
Tin	—	—	—	12
Zinc	—	—	250	10

liver. The use of an element-selective colour reagent can also be applied to histological preparations. Using light microscopy it is possible to identify the subcellular distribution of several elements with high resolution. If quantitative data are required, the coloured areas can be scanned with a microphotometer; if the optical resolution is sufficient, several elements can be determined at once. Before any of these techniques are used, it is first essential to determine that any observed distribution of an element is not disturbed as a result of histological techniques. Standard techniques for Fe, Mg, Al, Ca, V, As, Co, Ni, Cu,

Zn, Si, Sn, Ba, Bi, Hg, and Pb have been described by Pearse (4), Lillie (5), and Bancroft (6); the technique is not new and the use of the Prussian blue reaction for the determination of iron was described by Perls in 1867. For the imaginative reader, I suggest a reading of Feigl (7) and Feigl and Anger (8), which provides abundant data for the colour technique for a wide range of organic and inorganic substances. Quite clearly, such techniques are best suited for high concentrations of elements

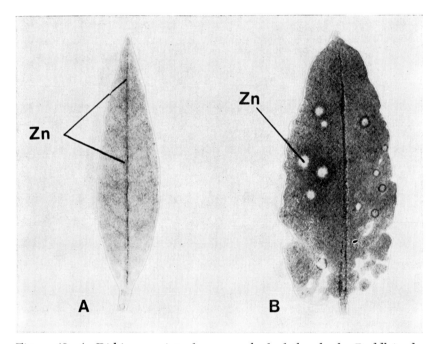

Figure 49: *A:* Dithizone print of a young leaf of the shrub, *Buddleia davidii*, on filter-paper showing the presence of zinc in the vascular system. Mag. ×1.5. *B:* Dithizone print of the same shrub but for an old leaf. The vascular system was not broken during pressing of the print and no zinc patterns are obvious as illustrated in *A*. However, numerous "zinc spots" were observed, some are undoubtedly related to dust particles trapped on leaf hairs while others appear to be related to an exudate from the leaf. The dark area along the median line of the leaf is composed of dust particles trapped along the depression of the central vein. This method may prove to be of value in investigating atmospheric particulate matter trapped on leaves. Mag. ×1.5. From Hamilton, E. I.: *Sci Total Environ.*

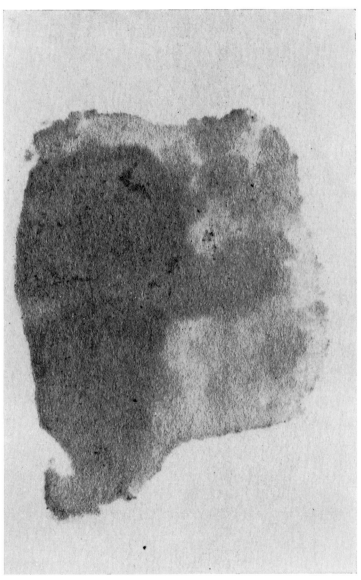

Figure 50: The distribution of zinc in a section of freeze-dried human liver. Dithizone print recorded on fine-grained filter paper. In this example a very dilute solution of dithizone was used; differences in intensity of colour represent small differences in zinc levels. Artifacts can be introduced when using this technique depending upon the direction in which the cells are cut and the migration of leachable zinc through the tissue. Mag. ×2.

Spectrophotometry

Spectrophotometry is concerned with the measurement of the relative capacity of chemical systems to absorb incident radiant energy at specific wavelengths. There are two fundamental laws describing the absorption of monochromatic light in clear and homogeneous media. The Bouguer-Lambert law is an expression of the relationship between the absorption of light in passing through different thicknesses of an absorber, while Beer's law states that the fraction of monochromatic light absorbed in passing through a solution is directly proportional to the concentration of the absorber. Combining both laws results in the following expression:

$$\log I_0/I = abc = A \text{ (absorbance)}$$

where I_0 = intensity of the incident beam of light
 I = intensity of the transmitted beam of light
 a = a constant of absorptivity whose numerical value depends upon the choice of concentration units
 b = the optical path length
 c = the concentration of the coloured solution

When b is known, a can be calculated for a solution of known concentration at a specific wavelength. By knowing A, a, and b for an unknown solution, the concentration, c, can be calculated provided that Beer's law is obeyed.

A spectrophotometer consists of a source of stabilized light, of continuous radiant power, containing sufficient intensity for the region of the spectrum in which the characteristic absorption bands of the element complex in the solution are found. Radiant energy of the required wavelength is obtained through a monochromator (a device for converting a heterogenous beam of radiation—electromagnetic or particulate—to form a homogenous beam by absorption or refraction of unwanted components) by which a relatively narrow spectral band width is obtained through a system of narrow slits and a dispersion media such as a prism or grating. The resolution of an instrument depends upon the degree to which the adjacent wavelengths can be separated. Light of the selected wavelength then passes through a transparent cell containing the solution to be analysed followed by a detector (such as a photovoltaic or barrier layer cell), photoemissive cells, or an electron multiplier phototube. Instruments using filters (filter photometer) or monochromators (spectrophotometer) can be used in association with single or double beam optical systems; in the former the beam of light, after passing through the filter system, traverses a reference cell and the output, adjusted to 100 percent transmittance or zero absorbance, is then passed through the sample solution; in double beam instruments, a beam splitter is used so that one part of the light beam passes through the reference cell and the other through the sample solution, and the ratio of the two intensities are compared. This technique allows for changes in light intensity with time while the measurement is being made and also corrects for other variables such as those arising from electronics of the system.

A very high degree of cleanliness is required when undertaking spectrophotometric measurements; for most measurements the solutions should be perfectly clear and free from tissue debris (commonly used to clean the cell), fingerprints should not be present on the external walls of the cells which must be dry. When undertaking several measurements, matched cells should be used; if not available, differences in absorption of light for each cell should be noted and appropriate values used to correct final data. It is also essential to ensure that the internal surfaces of the cells do not become etched with time and thus give rise to light-scattering phenomena. In using reagents for developing colours for specific elements, other elements or constituents of the sample solution which can also influence the colour should be absent. If contamination is suspected, interfering ions can sometimes be removed, or their effects reduced, by altering pH, using oxidation-reduction stages, by forming suitable complexes, or by determining optical densities at several wavelength settings.

Measurements of absorbance should be made in the central part of the absorption curve, very dense solutions should be diluted while solutions with weak absorbance, for example less than 0.2 absorbance, should be measured in cells with a longer path length. With experience it is possible to select an optimum sample size in order to obtain the most suitable absorption value. Photometric measurements are usually made at the wavelength of maximum absorbance, although Beer's law is often obeyed at other wavelengths and the selection of a less sensitive position can be advantageous when dealing with dense solutions. Turbid solutions (Turbidity and Nephelometry) can also be analysed and have special applications, such as the formation of coloured lakes and in the analysis of aluminium and SO_4^{-2} which is estimated in the form of barium sulphate suspension.

There are a great variety of reagents for developing specific colours such as diphenylthiocarbazone (dithizone) for Ag, Bi, Cd, Cu, Pb, Sn, Tl, and Zn, by extraction into carbon tetrachloride or chloroform at a specific pH; curcumin for beryllium; bathophenanthroline or 1,10-phenanthroline for iron; dimethylglyoxime for nickel; and tiron for titanium. Numerous excellent books have been written on the subject, for example Sandell (1), Snell and Snell (9), Charlot (10), Boltz and Schenck (11), and Thomas and Chamberlin (12). In clinical chemistry, especially when a large number of routine measurements have to be made, photometric and spectrophotometric techniques have become very important tools of investigation when incorporated into various commercially available autoanalyser systems (Du Cros and Salpeter, 13). Spectrophotometric methods are also suitable for assay in the field and such work benefits considerably from the use of portable battery-operated spectrophotometers, such as the Bausch and Lomb Mini-Spec 20™ as illustrated in Figure 51; this instrument has a 20 nm bandwidth, solid state electronics, with a precision grating covering the range 400-700 nm. In recent years spectrophotometry has tended to become overshadowed by other methods, such as atomic absorption spectrometry, but the number of new instru-

Figure 51: Bausch & Lomb Spectronic Mini 20 precision spectrophotometer. This battery-operated portable instrument contains a precision, diffraction grating monochromator that provides a continuous 400-800 nm wavelength range and a narrow 20 nm spectral slit width. Built-in stray light and second order filters are inserted automatically. The effective battery charge allows a minimum of 400 readings; wavelength accuracy is better than 3 nm at 546 nm, wavelength readibility 1 nm. Size: 15.2 cm (L) × 8.9 cm (W) × 4.8 (H), weight 0.45 kg. Photograph supplied by V. A. Howe & Co, Ltd, United Kingdom.

ments and new models appearing regularly on the market attest to the widespread popularity and utility of this method of analysis.

In hospitals, routine requirements for chemical and biochemical analysis are well defined and hence are amenable to automated methods of which the Technicon® and similar automated methods are good examples. A further example is the GEMSAEC centrifugal fast analyser system described by Anderson (14); a miniaturised version taking up a space of only 1 ft^3 has been described by Scott and Burtis (15). The system only requires a few microlitres of physiological fluid and is based upon an automated colourimetric analysis ideally suited to clinical laboratories. In principle any colourimetric method can be used; hence it is suitable for a large number of elements and compounds as well as for initial rate, fixed time, and variable time enzyme kinetic analyses. The basic system consists of a multiple cuvette array in the rotor arm of a centrifuge run between 500 and 5000 rpm. Each cuvette containing the coloured solution passes in front of a stationary photometric detector. The transfer of solution, mixing, and addition of reagents is achieved by use of centrifugal fields. Using a 17 place rotor at 600 rpm data are produced at a rate of 10,200 data points each minute; therefore there is a need to interface the output with a computer and associated data display systems. A quartz iodine-tungsten lamp provides a light source, there are six interference filters, and measurements can be made over the range 340-620 nm. Apart from measurement of enzyme reaction rates at high speed, chemiluminescence can also be determined; most studies to date have concerned the following cations and anions: SO_2, O_3, Ca, Cr, Cu, Fe, Mg, Se(IV), Zn, Cl^-, I^-, NO_2^-, PO_4^{-3}, S^{-2}, and SO_4^{-2} which can be determined with at least as good precision and accuracy as more conventional methods.

REFERENCES

1. Sandell, E. B.: *Colorimetric Determination of Traces of Metals*. New York, Interscience, 1959.
2. Hawkins, D. B., Canney, F. C., and Ward, F. N.: Plastic standards for geochemical prospecting. *Econ Geol*, 54:738, 1959.
3. Hamilton, E. I.: Review of the chemical elements and environmental chemistry—Strategies and tactics. *Sci Total Environ*, 5:1, 1976.
4. Pearse, A. G. E.: *Histochemistry*, vol 2, ed 3. London, Livingstone, 1972.
5. Lillie, R. D.: *Histopathologic Technique and Practical Histochemistry*. New York, McGraw, 1965.
6. Bancroft, J. D.: *An Introduction to Histochemical Technique*. London, Butterworths, 1967.

7. Feigl, F.: *Spot Tests in Organic Analysis.* London, Elsevier, 1966.
8. Feigl, F. and Anger, V.: *Spot Tests in Inorganic Analysis.* New York, Elsevier, 1972.
9. Snell, F. D. and Snell, C. T.: Colorimetric Methods of Analyses, ed 3, volumes II, IIA, IVA, IVAA. Cincinnati, Ohio, Van Nos Reinhold, 1959, 1967, 1970.
10. Charlot, G.: *Colorimetric Determination of Elements.* London, Elsevier, 1964.
11. Boltz, D. F. and Schenck, G. H.: *Handbook of Analytical Chemistry.* New York. McGraw, 1963.
12. Thomas, L. C. and Chamberlain, G. J.: *Colorimetric Chemical Methods,* ed 8. Salisbury, UK, The Tintometer Co. Ltd., 1974.
13. Du Cros, M. J. F. and Salpeter, J.: Automated methods for assessing water quality come of age. *Environ Sci Tech, 9:*929, 1975.
14. Anderson, N. G.: New computor-interfaced fast analysers. *Science, 166:* 317, 1969.
15. Scott, C. D. and Burtis, C. A.: A miniature fast analyzer system. *Anal Chem, 45:*327A, 1973.

Electrochemical Methods

One of the more important aspects of analysis for the chemical elements in biomedical applications is the determination of the chemically active constituents, that is the nature of the chemical form of an element which participates in defined biochemical processes. Many analyses are only concerned with assay for the total mass of an element present in a sample, although it is possible to carry out preliminary chemical separations in order to isolate defined chemical species. Such processes involve chemical separations without the introduction of contaminants and assume, in the first instance at least, that the compound of interest can be uniquely separated and does not undergo chemical exchanges as a result of separation procedures; clearly the extent to which such processes can take place depends upon the stability of the compound to be separated. Following bulk analysis of an element, attempts are then made to reconstruct natural associations which represent the natural dynamic state of biochemical systems. It is therefore of some interest to consider those methods which can provide some information for the real levels of elements which will respond when present in particular chemical

Instrumental Methods

species. Although a complete answer to the problem is not available, the use of electrochemical techniques opens up many new lines of approach; many of the methods are not new and some have remained dormant for several decades awaiting refinements in electronic systems through which very exacting measurements can now be made.

In relation to analysis for the chemical elements, electrochemistry is concerned with the Nernst equation and Faraday's laws.

Nernst Equation

This equation forms the basis for one type of analysis in which a probe is inserted into a solution which will respond to various ions in a selective manner; such probes are called ion-selective electrodes. The Nernst equation provides a means of interpreting the potentiometric output of an ion-selective electrode in terms of ion activities:

$$E = E_o - (0.19841 \ T/n) \log a_i$$

where E = Emf of an electrode expressed in Mv
E_o = Standard emf of an electrode expressed in Mv
T = Absolute temperature in °K
n = An integer corresponding to the electron change in an electrode reaction
a_i = Activity of an ion "i"

In order to explain the performance of ion selective electrodes, the following relationship is commonly used:

$$E = E^o - (0.19841 \ T/n) \log (a_i + K_{ij} \ a_j^m + \ldots)$$

where K_{ij} is the selectivity coefficient of the interfering ion "j" in relation to the ion to be determined "i," and "m" is an integer of fraction. The value of K_{ij} is used to determine whether or not an ion-selective electrode can be used in the presence of known impurities.

Faraday's Laws

These laws may be summarised as follows:

1. The mass of substance dissolved or liberated as a result of the electrolysis of solution is proportional to the quantity

of electricity which passes through the electrolyte.
2. The masses of different products released by passing a given quantity of electricity through different electrolytes are proportional to the chemical equivalents of the substance concerned.
3. The products of electrolysis appear only at the electrodes.

All three laws are obeyed over a wide range of concentrations, temperature, and electric current.

In electrochemical analysis it is possible (a) to observe changes in the magnitude of the current passed through an electrolyte (coulometry), (b) to determine products deposited on an electrode surface (electrogravimetry), or (c) following deposition onto an electrode, to observe characteristics of current associated with post deposition on polarisable electrodes (this technique is suitable for element analysis, especially at trace levels, and is referred to as polarography and voltammetry).

Ion-Selective Electrodes

A schematic diagram illustrating the basic instrumentation for ion-selective electrode measurements is given in Figure 52 and various types of electrodes in Figure 53. The most critical part of an ion-selective electrode is the membrane which senses and selects, allowing only those ions of interest to pass from a sample solution on one side to an internal solution on the other side, which contains a fixed activity of an ion to which the membrane is selective. When an ion-selective electrode is placed in a solution of a sample, there is a flux of ions across the membrane in the direction of the solution which contains a lower activity of the mobile ion. Ions carry an electrical charge, and hence, an electrical potential is established which opposes further ion migration. The resulting equilibrium which is reached reflects the potential across the membrane required to prevent further net movement of ions. Measurement of the potential developed between the two solutions is made with a high impedance voltmeter with reference to some form of standard electrode (Covington, 1), for example an Ag/AgCl pair based upon the standard hydrogen reference electrode. In practice a standard curve is pre-

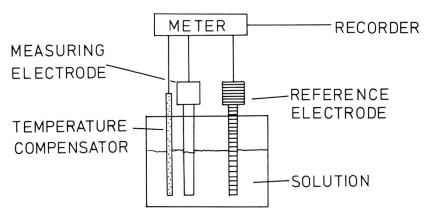

Figure 52A: Diagram illustrating a simple ion-selective electrode system.

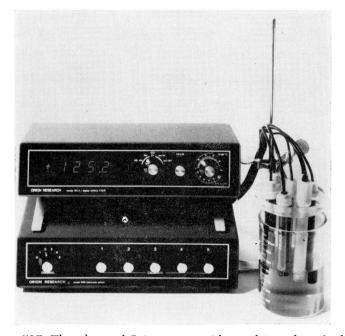

Figure 52B: The advanced Orion system with a multiswitching facility.

pared relating electrode potential versus log A of the standard solution. Most modern instruments consist of digital pH/mv meters covering a range of +1000 mv to −1000 mv with a precision of measurement of 0.1 mv; thus direct activities can be obtained from the meter scale.

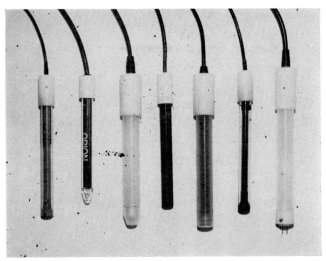

Figure 53A: Various types of ion-selective electrodes. Left to right: Reference, pH, Liquid Membrane, Solid State, Gas Sensing, Combination, Flow-thru.

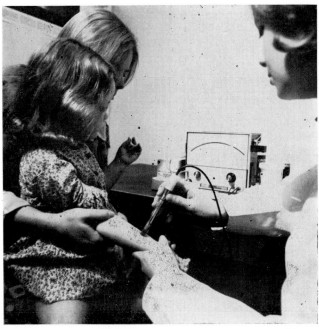

Figure 53B: Ion-selective electrode for assay of sodium or chlorine in skin secretions.

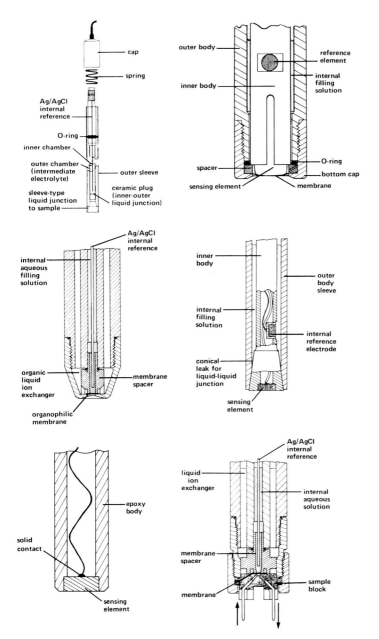

Figure 53C: Diagrams illustrating various sensing systems used in ion-sensitive electrodes.
Figures supplied by Orion Research Inc., Cambridge, Massachusetts, United States.

Ion-selective electrodes have an advantage over many other methods in speed of response to ion activity; the method can be partly classed as nondestructive, within the requirements of pH adjustment and the addition of reagents essential to adjust ionic strength of the sample in order to minimise matrix effects. Very small samples are required, often no sample pretreatment is necessary, while some measurements of clinical interest can be made in situ. Ion-selective electrodes provide numerical data for the activity of selected ions which can be related to concentration of ions. In any real system an element can occur in a wide variety of chemical associations, such as complexes and ion pairs, and by judicious selection of complexing and decomplexing agents, valuable information can be obtained for the chemical form of some elements in solutions or slurries. However, very often a major problem arises because of lack of knowledge of the chemical form of the elements in real solutions which presents problems in standardisation. The preparation of ideal ionic standards presents no serious difficulties but rarely will real solutions consist solely of free ions. Considerable developments have taken place in approaches to standardisation; for example, electrodes used for standards and samples are treated in an identical manner in order to reduce nonideal responses caused by the elements in different chemical forms. The method of standard addition analysis is also used, taking care that the increments of standards added closely bracket the sample signal. The method can be used for continuous assay but then it becomes essential to ensure that slimes or deposits do not form on the ion-selective membrane.

At present a wide variety of ion probes are commercially available, for example, assay for ammonia, calcium, chloride, cyanide, fluoride, iodide, nitrate, potassium, silver, sodium, sulfide, and water hardness. The sensing membrane can consist of a wide variety of materials such as glass, solid state devices, liquids, ion-exchange resins, sparingly soluble metals and metal chelates, impregnated graphite, and enzyme-reactive materials. Ion-selective electrodes have provided an answer to the difficult analysis for fluoride ions in aqueous media, which can be determined at subnanogram levels in volumes as small as ten microliters. In fluoride ion assay a Total Ion Strength Adjustment Buffer (TISAB) is used containing sodium chloride, acetic acid, sodium acetate, and sodium citrate at a fixed pH; the use of citrate ions complexes iron and aluminum thus displacing into solution free fluoride ions. A further application of ion-selective electrodes is in potentiometric titrations which offer enhanced accuracy at the expense of longer times for analysis, but the technique is still dependent upon using the response curve for which the addition of standard increments follows a linear response. The technique can be based upon a variety of reactions such as acid-base neutralisation, redox reactions, and complexation; the titration reaction must be stoichiometric, take place rapidly, and go to completion.

Apart from commercially available electrodes many types of electrodes can be fabricated in the laboratory and offer the possibility for studying membrane transport phenomena on biological simulated membranes. Moore (2) has described the use of an ion-exchange calcium electrode in biological fluids to study calcium fractions in relation to malignancy and hypercalcemia; in normal subjects, variation in total calcium was accounted for by a corresponding variation in protein-bound calcium, 80 percent bound to albumin and 20 percent to globulins. In patients with malignancy and hypercalcemia, an increase in total serum calcium was accompanied by a progressive rise in ionized total calcium at a constant fractional rate corresponding to 45 percent of the total. Khuri (3) has described the application of a variety of ion-selective electrodes in biomedical research with particular reference to micro-electrodes capable of in situ, extra- and intracellular analyses; these analyses are at present restricted to major elements such as sodium, potassium, calcium, chloride, and fluoride activities which have a very important bearing upon physiological functions.

Ion-selective electrodes mark a renaissance of analytical potentiometry, initiated by the work of Cremer (4) and Haber and Klemensiewicz (5) on the H^+ selective glass electrode followed by the work of Eisenman and his co-workers (6); more recent descriptions of the techniques have been presented by Pungor (7), Koryta (8), Durst (9), Lakshiminarayanaiah (10), and Berman and Herbert (11).

Polarography and Voltammetry

Polarographic analysis can be used for the determination of any substance, molecular, organic or inorganic, which can be reduced or oxidised, at a mercury electrode in any solvent whose dielectric constant is sufficiently high to permit a reasonable ionic strength to be maintained by the addition of some indifferent electrolyte, see Murray and Reilley (12), Milner (13), Browning (14), Brainina (15), Meites (16), Neeb (17), and Purdy (18).

The technique was originally described by Heyrovsky and Kuta (19) but it is only quite recently that the technique, because of improvements in electronics, has emerged as one of the most powerful tools available for the analysis of trace levels of elements when present in different chemical forms. The classical electrode consists of a drop of mercury which appears at the orifice of a capillary tube, falls through an unstirred solution of a sample, and is replaced by another drop from a mercury reservoir thus presenting a new electrode surface with each drop. As the drop of mercury falls through the sample solution, current-voltage time relationships taking place on the surface of the very small electrode form the basis of polarography. Mercury has the most negative potential utility of any electrode material and is only limited by anodic dissolution; when this occurs other electrodes such as Au, Pt, or graphite can often be used. Upon potential scanning of the dropping mercury electrode (DME) very little current flows until a potential region is reached at which the element of interest in the sample solution can be reduced; at this point the current increases abruptly followed by a levelling off and becomes essentially parallel to the background current level. A schematic diagram of a DME cell is illustrated in Figure 54 and a DME polarogram consisting of three major steps is illustrated in Figure 55; the potential on the polarographic wave, midway between the two current levels, is called

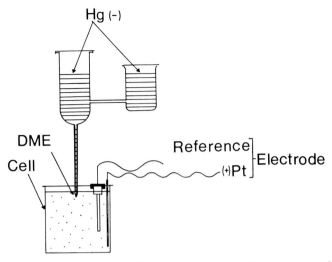

Figure 54: Schematic diagram illustrating the main features of a dropping mercury electrode (DME) cell as used in polarography. Alternative electrodes are the stationary electrode (HMDE) or a rotating glassy carbon electrode coated with a very thin layer of mercury.

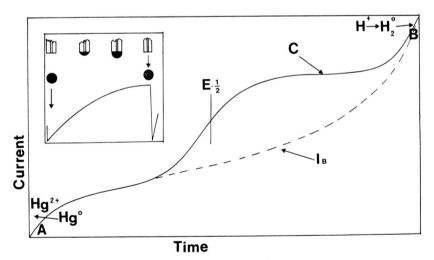

Figure 55: DME Polarogram.
  A. Positive potential limiting reaction.
  B. Negative potential limiting reaction.
  C. Diffusion current plateau.
  $E_{\frac{1}{2}}$. Half-wave potential.
  I_B. Background current.
Insert. Illustrates current time relationships with fall of the mercury drop.

the half-wave potential. In d.c. polarography, the area of interest is the diffusion plateau since the only means of ion transport for an unstirred solution is that of diffusion, hence in this region sample ions are being reduced as rapidly as they diffuse up to the electrode. The diffusion (concentration) gradient established with the passage of each drop is destroyed when the drop falls, hence each new drop of mercury enters a solution which is essentially unchanged.

A polarograph consists of a potentiostat to control the cell potential, a timer to synchronise data acquisition with the periodic fall of each drop, and a gating delay system to permit current sampling at the same time of drop fall; finally there is a system for scanning the desired potential range.

Recent advances in polarography include the application of single potential pulses of increasing height applied to consecutive drops toward the end of each drop life; the Faradiac current is sampled in the remaining time portion of the pulse interval. As the pulse is only applied to each drop for a period of about fifty milliseconds, a diffusion gradient is not developed and hence more Faradiac current is available for measurement. In derivative pulse polarography two sampling times are used, one just before pulse amplification and one at the end of the pulse; signals are stored in a memory capacitor where only the difference in current is displayed and is a measure of the change which has occurred in the region of the half-wave potential where the current is changing rapidly with potential. The recorded polarogram is a peak-shaped incremental derivative of the conventional polarogram and sensitivity is enhanced from 10^{-5}-10^{-6}M for d.c. polarography to 10^{-8}-10^{-9}M for derivative pulse polarography. Sensitivity can be improved by various techniques, such as the displacement of an element bound in a particular form by the addition of an inert substance which results in the release of an equivalent amount of the element to be determined.

In recent years increased attention has been paid to the technique of inverse voltammetry (anodic stripping voltammetry) in which the element of interest is first electrolytically enriched at an electrode from a stirred nitrogen purged solution, followed by a polarographic determination. A commonly used material for the pre-enrichment is mercury with which reducible elements form an amalgam; a typical anodic stripping voltammetric cell is illustrated in Figure 56. By recording the current voltage curve in the anodic direction with a rapidly changing direct voltage, anodic current-maxima are recorded which are directly dependent on the concentration of the metal deposited in the mercury amalgam. The whole process of an initial stage of enrichment provides preconcentration of elements from solutions without the need for chemical processing and hence a reduction in the entry of contaminants. However the stage of deposition can be lengthy (typical times of 0.5 h are not uncom-

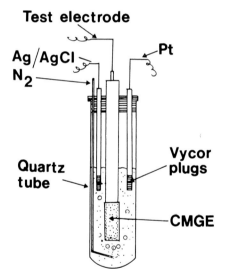

Figure 56: Illustration of a typical cell for use in anodic stripping voltammetry. The composite mercury graphite electrode (CMGE) is prepared by electrodeposition of a very thin layer of mercury onto a wax-impregnated graphite rod. [A commercially available system has been described by Matson, W. R. and Roe, D. K.: Trace metal analysis in natural media by anodic stripping voltammetry. In *Analytical Instrumentation,* Vol. 7, New York, Plenum Press, 1967.] The sample solution is stirred by passage of a fine stream of nitrogen bubbles through the cell. Overall length of cell three inches.

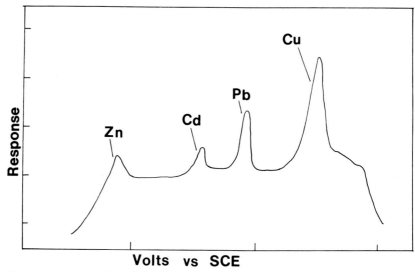

Figure 57: Typical anodic stripping curve for Zn, Cd, Pb, and Cu response versus a saturated calomel electrode (SCE). The Pb peak corresponds to about 100 ng of the metal.

mon), and the solution has to be stirred continuously in order to present fresh material in contact with the amalgam; apart from chemical requirements for the plating stage, temperatures need to be controlled and it is essential to prevent the build-up of gas bubbles on the surface of the amalgam. The method is simple and involves one variable, namely the scan rate for stripping which is the potential at which the electrode is scanned; an example of data output is presented in Figure 57.

The sensitivity of assay by anodic stripping voltammetry is high and by using derivative techniques 10^{-11}M solutions can be analysed. Using mercury (metal or mercury-impregnated materials) the following elements can be determined: Ga, In, Sn, Pb, Sb, Bi, Cu, Ag, Zn, Cd, Fe, Co, and Ni. Matson, Griffin, and Schreiber (20) have discussed the rapid sub-nanogram simultaneous analysis for Zn, Cd, Pb, Cu, Bi, and Tl in biological samples. The electrolytic concentration stage is carried out at about 200-400 mv above the half wave potential in d.c. polarography, or above the peak potential in voltammetry. The percentage of the total mass of an element deposited depends upon the time allowed for deposition and degree of stirring; difficulties can be experienced if intermetallic compounds form between the deposited metals and the electrode. Compared with conventional polarography, the composition of the plating solution is very important; by selection of appropriate plating solutions improved separations and enhanced sensitivity can be obtained, which can also be achieved by altering the composition of the solution after the deposition stage. Multielement analysis can be achieved for some elements in defined matrixes provided that sufficient differences exist in half-wave or plating potentials, while for some types of analysis automated techniques can also be used.

The cheapness of anodic stripping equipment has attracted much attention but in some of the earlier work very poor data were obtained, not because of basic faults in the technique but usually because of contamination of reagents or systems and the use of unsuitable plating solutions. Except for a few simple mixtures most samples have to be processed prior to instrumental analysis but speed of assay can be enhanced by batch processing. Some of the problems of reagent purity can be overcome by making up the appropriate mixtures and then electroplating contaminants onto a pool of mercury; a commonly used system, illustrated in Figure 58, takes about fourteen days to clean several litres of solution.

Apart from the analysis for bulk levels of elements in solution, when sharp, well-defined signals are obtained, if the element of interest is present in various chemical forms the response slope of the signal becomes altered reflecting the presence of different chemical forms of an element. By careful examination of the shape of such complex curves, together with the selection of different electrolytes, the identification of different species can be obtained, but for quantitative analysis reference standards of known composition are required. Electrochemical methods have an important role to play in investigating diseases. Probing the brain with microelectrodes has been discussed by Adam (21) with particular reference to investigating results of chemical neurotransmission for which action potentials are

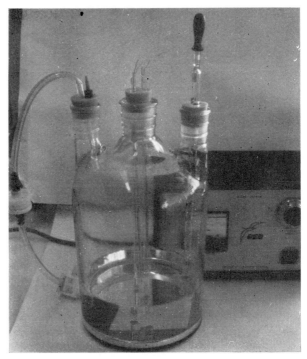

Figure 58: The Model 2014P Reagent Cleaning System [Environmental Sciences Associates, Burlington, Massachusetts, United States]. The four l flask contains a mercury pool, counter and reference electrodes, nitrogen bubbling system, and dispensing pipette. The system is basically of the potentiometric type capable of extracting metals from a solution down to about 5×10^{-10} M after about seven days of operation. If it is possible to maintain a clean system, then extraction down to 5×10^{-12} M can be obtained after thirty days plating at a pH of between 4 and 7. The following solutions can be purified by this method: dilute hydrochloric, nitric, perchloric, and sulphuric acids (about 1 M); ethylene diamine; water; sodium acetate; sodium chloride; sodium citrate; sodium EDTA; sodium nitrate; and any reagents made of nonreducible salts. Extractable elements: Ag, Au, Ba, Bi, Ce, Cd, Cr, Co, Cu, Fe, Ga, In, Ir, Mn, Ni, Os, Pb, Pd, Pt, Rh, Ru, Si, Sn, Te, U, V, and Zn.

about 100 mv in amplitude and last between 1 and 2 ms. Lead can be routinely analysed by anodic stripping voltammetry on body fluids, such as blood and urine, with a sensitivity of about 0.1×10^{-9} g on 0.1-0.5 cc samples. However, although this method has the advantage of high sensitivity it can be plagued with problems of contamination.

REFERENCES

1. Covington, A. K.: Reference electrodes. In Durst, R. A. (Ed.): *Ion-Selective Electrodes*. Washington, D.C., National Bureau of Standards, Special Publication 314, 1969.
2. Moore, E. W.: Studies with ion-exchange calcium electrodes in biological fluids: Serum applications in biomedical research and clinical medicine. In Durst, R. A. (Ed.): *Ion-Selective Electrodes*. Washington, D.C., National Bureau of Standards, Special Publication 314, 1969.
3. Khuri, R. N.: Ion-selective electrodes in biomedical research. In Durst, R. A. (Ed.): *Ion-Selective Electrodes*. Washington, D.C., National Bureau of Standards, Special Publication 314, 1969.
4. Cremer, M. Z.: Uber die Ursache der elektromotorischen Eigenschaften der Gewebe, zugleich ein Beitrag zur Lehre von der Polyphasisch in Electrolytketten. *Z Biol*, 47:562, 1906.
5. Haber, F. and Klemensiewicz, Z.: Uber elektrizsche phasengrenzkräfte. *Z Physik Chem (Leipzig)*, 67:385, 1909.
6. Eisenman, G. (Ed.): *Glass Electrodes for Hydrogen and Other Cations: Principles and Practice*. New York, Dekker, 1967.
7. Pungor, E. (Ed.): *Ion-Selective Electrodes*. Budapest, Akademiai Kiado, 1973.
8. Koryta, J.: Theory and applications of ion-selective electrodes, Part II. *Anal Chim Acta*, 91:1, 1971.
9. Durst, R. A. (Ed.): *Ion-Selective Electrodes*. Washington, D.C., National Bureau of Standards, Special Publication 314, 1969.
10. Lakshiminarayanaiah, N.: *Membrane Electrodes*. New York, Acad Pr, 1976.
11. Berman, H. J. and Hebert, N. C. (Eds.): *Ion-Selective Microelectrodes*. New York, Plenum Pr, 1974.
12. Murray, R. W. and Reilley, C. N.: *Electroanalytical Principles*. New York, Wiley, 1963.
13. Milner, G. W. C.: *The Principles and Applications of Polarography and Other Electrochemical Processes*. London, Longmans, Green and Co, 1958.
14. Browning, D. R.: *Electrometric Methods*. New York, McGraw, 1969.
15. Brainina, Kh. Z.: *Stripping Voltammetry in Chemical Analysis*. New York, Wiley, 1974.
16. Meites, L. (Ed.): *Handbook of Analytical Chemistry*. New York, McGraw, 1963.
17. Neeb, R.: *Inverse Polarographie und Voltammetrie*. Berlin, Verlag Chemie Weinheim Bergstr, 1969.

18. Purdy, W. C.: *Electroanalytical Methods in Biochemistry*. New York, McGraw, 1965.
19. Heyrovsky, J. and Kuta, J.: *Principles of Polarography*. Prague, Chez Acad Sci, 1965.
20. Matson, W. R., Griffin, R. M., and Schreiber, G. B.: Rapid sub-nanogram simultaneous analysis of Zn, Cd, Pb, Bi and Tl. In Hemphill, D. (Ed.): *Trace Substances in Environmental Health—IV*. Columbia, Missouri, U Missouri, 1971.
21. Adams, R. N.: Probing brain chemistry with electroanalytical techniques. *Anal Chem*, 48:1129, 1976.

Atomic Emission Spectroscopy

Three types of emission spectroscopy will be considered, namely optical, absorption, and fluorescence spectroscopy, all of which are dependent upon the production of atomic spectra through thermal processes by atomisation of a sample and involves an input of energy into a population of atoms which is converted to light energy in the form of a spectrum consisting of radiation of a number of discrete wavelengths. In an electrically neutral atom, the atomic number of an element is given by the number of protons in the nucleus, which is equal to the number of electrons. Each electron is identified by a unique set of quantum numbers which define its energy and also limit the transitions of energy changes which are possible. When an electron moves from one energy state to another the energy (E) of the photon, which is emitted or absorbed, is given by the following equation:

$$E = E_1 - E_2 = h\nu$$

where E_1 and E_2 are the energies of the electron in the upper and lower states respectively

h is Planck's constant (6.623×10^{-27} ergs sec)

ν is the frequency of the radiation

The discrete wavelength associated with any particular electronic transition may therefore be described as follows:

$$E = \frac{hc}{\lambda}$$

where c is the velocity of light (3×10^{10} cm sec^{-1})

λ is the wavelength of the radiation, i.e. $\nu = c$

When an atom is in the ground state, its electrons are at their lowest energy levels; when energy is transferred to a population of such atoms, for instance by means of thermal or electrical excitation, energy transfer takes place as a result of collision processes. In emission spectroscopy the intensities of the spectral lines are determined by the number of atoms in the excited state, the duration of time in which they remain excited, and the probability of the electrons returning to the ground state by a transition corresponding to the particular spectral line. The number of excited atoms present in a source are given by the following expression:

$$N_1 = \left(\frac{N_o S_i e}{S_o}\right)^{\frac{-(E_1 - E_o)}{kT}}$$

where
- N_o = the number of atoms in the ground state of energy E_o
- N_1 = the number of atoms in the excited state of energy E_1
- S_i and S_o = the statistical weights of the excited and ground states respectively
- k = the Boltzman constant
- T = the effective excitation temperature

It therefore follows from the above equation that the degree of excitation is dependent upon temperature; an increase in temperature increases the number of excited atoms and hence, provided that appreciable ionization does not occur, the sensitivity of emission analysis is increased. Some elements such as the alkali metals have low excitation energies and can be readily excited in low temperature flames while others with high excitation energies require very hot environments for appreciable excitation. It is therefore a feature of all thermally excited types of emission that the sensitivity of detection will be dependent upon the excitation energies of individual elements or their compounds; easily excited materials will give rise to very intense emissions which through secondary collision and optical processes can provide very broad spectral emissions and interfere with adjacent spectral lines of weaker intensities. Further, for a discrete sample in which spectral information is collected over a period of time, the easily excited elements will be rapidly emitted followed by those of high excitation energies.

Unlike optical emission spectroscopy, which is concerned with excited atoms, atomic absorption spectroscopy is concerned with atoms in the ground state, which are more abundant than those in the excited state. A few lines in the emission spectrum are reversal lines (resonance lines) and if the atom is subjected to radiation of a particular wavelength, a photon can be absorbed and an electron raised to the appropriate energy level; because the number of reversal lines available are small compared to those for optical emission, the spectrum is much simpler and many of the problems of complex emission are reduced. Finally, atoms which are excited by absorption of resonance radiation also re-emit the absorbed energy in the form of fluorescence radiation, which is utilised in fluorescence spectroscopy (which is also associated with very simple spectra). Quite clearly, in any thermal excitation process all three types of excitation occur, but they can be observed independently by using appropriate types of instrumentation.

Optical Emission Spectroscopy

In emission spectroscopy, the sample is mixed with a conducting matrix such as ultra pure graphite, formed into or placed in an electrode, electrically heated, and the light emitted is passed through an optical system (usually an air path but vacuum systems are also used) where discrete wavelengths are separated and then impinged onto some form of photodetector where the signal is recorded; the basic features of the instrument are illustrated in Figure 59. The dispersion of most optical systems is such that all elements can be recorded simultaneously, hence the method is admirable for multielement analysis.

The most commonly used method of excitation is the d.c. arc which is possibly one of the most sensitive general sources available for excitation for most of the elements. The precision of assay by this technique of excitation is usually about ±30 percent, simply because the arc tends to dwell on more conducting

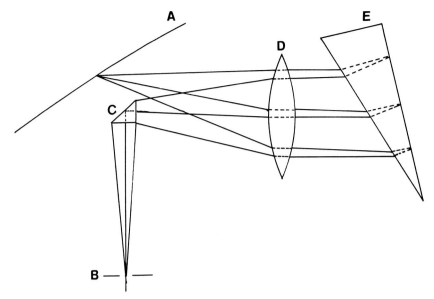

Figure 59: Diagram of a Littrow prism spectrometer.
        A. Photoplate detector.
        B. Slit in front of source.
        C. Reflecting prism.
        D. Lens.
        E. Silvered-back quartz prism.

The prism can be rotated to select the appropriate wavelength range and also can be positioned laterally along the lens to allow optimum focussing as the wavelength changes.

areas of the sample resulting in preferential emission of some elements contained within small volumes. Some of the problems can be overcome by using internal standards consisting of an element, or its compound, which is intimately mixed with the sample, thus providing a monitor for emission of the element of interest. A limitation of this technique concerns the ability to prepare homogenous sources with reference to the constituents of the sample. The standard must be in a chemical form which is similar to that of the element to be determined, otherwise volatilisation of the standard, relative to the sample, may be enhanced or reduced. Carbon (graphite) is a suitable material for the preparation of sources because it can be obtained in a very pure form, is electrically conducting, and has an ionization potential equal to or better than most elements, hence carbon vapours will not depress the excitation characteristics of most other elements. Low currents are used to excite the easily volatilised constituents of a sample and higher currents for the most refractory constituents; hence, by selective volatilisation elements may be recorded according to volatility. Stability of source excitation can be improved by the addition of spectroscopic buffers, such as the alkaline earths or metals, which reduce excitation temperatures and interference from ionized atoms. It is also possible to add various carriers to improve the rate of release of some elements and compounds, while some advantages accrue in the use of controlled atmospheres in the source region. Source excitation can also be accomplished by use of a spark which tends to improve precision of assay; each spark represents a discrete input of energy into the sample and as the process is of a random nature the discharge does not tend to dwell upon one area of the sample. Compared with the d.c. arc source, very little material is consumed; hence, sensitivity of detection is reduced but the source remains cool, thus time-dependent selective volatilisation phenomena are not so significant compared with the d.c. arc. Apart from any restrictions related to the need to prepare a sample which is a good conductor, both solids and liquids can be analysed although the method is mainly used for dry powders. For the analysis of liquids various techniques are used, such as the gradual seepage of a liquid through the floor of a porous carbon cup which is then sparked, or by the use of a rotating disc which is allowed to dip into a liquid and is then sparked against a counter electrode.

Finally apart from the d.c. arc and spark mode of excitation, laser probes have also been used and are now commercially available. An advantage of the laser as an excitation source lies in the ability to analyse microsize sources of materials, whether they are electrically conducting or not. The laser can be used in the form of a microprobe in which the laser first produces an energetic vapour, which is then raised to energy levels for spectral emission. The point of impact of the laser can be identified by the formation of small craters, typically some fifty microns in diameter, and consuming about 0.1 microgram of a biological matrix. The technique is of interest when determining the general distribution of the chemical elements but quantitative analysis can prove to be difficult because of the need to prepare adequate standards.

Following excitation of the elements in the source, the light beam is then passed through a system of defining slits and into an optical dispersion system which usually consists of a prism or diffraction grating. Following dispersion of the incoming light into discrete wavelengths, they are then recorded on a photographic plate (optical spectrograph) or by single or a series of photomultipliers (optical spectrometer).

Details of the technique of emission spectroscopy have been described by Ahrens and Taylor (1), Scribner and Margoshes (2), Dawson and Heaton (3), and DeKalb, Kniseley, and Fassel (4). In recent years improvements have been made in the detection systems with particular attention to semiconductor photoelectric arrays, as described by Horlick, Codding, and Leung (5). When used with a photographic plate detector it is possible, with appropriate sequences of exposure conditions, to simultaneously record all elements present in a sample provided that they can be detected; a list of commonly obtained detection limits are given in Table LIII. Interpretation of photoplates can now be accomplished using automated photodensitometers and associated computer systems. In optical spectrometers, it is common practice to set the detectors for a restricted number of elements and such instruments, termed quantometers, can be used for simultaneous assay for up to about eighty elements and are widely used for on-line quality control analysis in industry. The initial cost of optical spectrographs and spectrometers is high but they are dependable instruments and capable of high quality analysis, provided that adequate attention is paid to the complex problem of matrix effects. Although specialist instruments, once they have been developed for defined matrixes they can be used by normal laboratory technical staff. In clinical studies they

TABLE LIII*

DETECTION LIMITS (ppm) FOR SEMIQUANTITATIVE ANALYSIS BY EMISSION SPECTROMETRY

Ag	0.1	Nb	3
Al	10	Na	100 (3)†
As	100	Nd	100
Au	10	Ni	30
B	30	Os	30
Ba	3	P	1,000
Be	1	Pb	1
Bi	3	Pd	1
Ca	1	Pr	100
Cd	10	Pt	3
Ce	100	Rb	100,000 (10)†
Co	1	Re	30
Cr	3	Rh	3
Cs	3,000 (30)†	Ru	30
Cu	0.3	Sb	100
Dy	30	Se	1
Eu	10	Si	3
Er	10	Sm	100
F	300‡	Sn	3
Fe	30	Sr	3
Ga	1	Ta	100
Gd	30	Tb	100
Ge	3	Te	300
Hf	20	Th	100
Hg	3,000	Ti	2
Ho	10	Te	10
In	1	Tm	10
Ir	30	U	300
K	2,000 (30)†	V	1
La	30	W	30
Li	30 (1)†	Y	3
Lu	10	Yb	1
Mg	0.3	Zn	100
Mo	3	Zr	3
Mn	1		

* Reproduced from Spectrochemical analysis, 2nd ed., 1961, written by S. L. H. Ahrens and S. R. Taylor with permission of publishers, Addison-Wesley/W. A. Benjamin Inc., Advanced Book Program, Reading, Massachusetts, U.S.A.

† A second exposure is necessary.

‡ A separate exposure is required.

can form the major tool for screening analyses and are capable of a high throughput of samples. Special laboratories are required for these types of instruments for which a very high degree of cleanliness can be maintained over long periods of time. They are fairly rugged and have been widely used in mobile laboratories for field investigations. Direct emission spectrographic methods for trace elements in biological materials has been described by Bedrosian, Skogerboe, and Morrison (6). A recent description of trace analysis by spectroscopic methods has been given by Winefordner (7).

REFERENCES

1. Ahrens, L. H. and Taylor, S. R.: *Spectrochemical Analysis*, 2nd ed. Reading, Mass, Addison-Wesley, 1961.
2. Scribner, B. F. and Margoshes, M.: Emission spectroscopy. In Kolthoff, I. M., Elving, P. J. and Sandell, E. B. (Eds.): *Treatise on Analytical Chemistry*, part I, vol 6. New York, Wiley, 1965.
3. Dawson, J. B. and Heaton, F. W.: *Spectrochemical Analysis of Clinical Material*. Springfield, Thomas, 1967.
4. DeKalb, E. L., Knisely, R. N., and Fassel, V. A.: Optical emission spectroscopy as an analytical tool. *Ann NY Acad Sci, 137*:235, 1966.
5. Horlick, G., Codding, E. G., and Leung, S. T.: Automated direct current arc time studies using a computer-coupled photodiode array spectrometer. *Applied Spectrometry, 29:*48, 1975.
6. Bedrosian, A. J., Skogerboe, R. K., and Morrison, G. H.: Direct emission spectrographic method for trace elements in biological materials. *Anal Chem, 40*:854, 1968.
7. Winefordner, J. D.: *Trace Analysis—Spectroscopic Methods for Elements*, vol 46, Chemical Analysis Series, New York, Wiley-Interscience, 1976.

Flame Photometry

Some two hundred years ago it was demonstrated that the colours developed in flames (which are more stable than arc or spark sources), when various compounds of the elements were combusted, were related to emissions from the chemical elements. About one hundred years ago, this phenomenon was first used for purposes of element analysis and forms the basis of modern flame spectroscopy. Early work in the modern era of analytical chemistry was mainly directed towards assay for readily ionizable compounds, such as those of the alkali metals and alkaline earths, and marked a period of significant advances in clinical studies related to assay for levels of electrolytes in human serum. Early instruments used a coal-gas-air flame into which, through a capillary, sample solutions were injected by aspiration; selection of wavelength was accomplished using simple glass or gelatin filters and the selected wavelength detected by means of a photocell. The requirements of a suitable filter are high transmittance at the desired wavelength, with a narrow wavelength band in the high transmittance region, and low transmittance at other wavelengths. Many simple filters provide a spectral band width of between 350 Å and 400 Å, transmission being measured at one half of the maximum; other filters such as the interference type provide better spectral isolation, for example 110 Å at half wavelength maxima. These simple and inexpensive systems were, and still are, suitable for analysis of elements such as Na and K, and in some matrixes Ca, but the emission from this element tends to be somewhat erratic.

Modern flame instruments use laminar flow, premix gas, high temperature flames with very careful control of gas mixtures and the rates of injection of sample solutions into the flame; wavelength selection is achieved by use of a spectrometer or a monochromator, maximum dispersion being obtained with grating monochromators. The selected energy corresponding to the wavelength of the element to be determined is finally detected using photoconductive, photovoltaic, or photomultiplier system. As flames have only limited energy available for volatilisation, only those lines of low excitation potential are readily observed. The total energy, E, of the radiation emitted by a source is given by the product of the number of atoms which undergo transitions, n, and the energy of photo emission, hv, that is $E = n \cdot hv$. Apart from problems concerned with the complex chemistry of combustion processes in a flame, characteristic wavelengths of the various elements are subject to considerable interference: the presence of band spectra reflecting molecular processes which are generated in the hot core of a flame, continuous radiation reflecting processes of dissociation, ionization and association spectral overlap, cation-anion reactions (for example $Ca + PO_4$ produces a stable product and depresses Ca emission), production of stable oxides which reduce levels of elements in the flame—all of which interact to various degrees depending upon the gross chemical composition of the sample matrix. Standardisation is usually accomplished by reference to separately determined standard curves established on solutions which contain the same species of cations and anions and at similar levels to those found in the sample. Another technique utilises the addition of an internal standard consisting of an element, not to be measured in the sample, which must be chemically pure and have the same physical and chemical characteristics and excitation potential as the element to be analysed, for example the use of Li when analysing for Na or K.

A further technique, namely multiple standard addition, is also widely used in flame photometry. The principle is simple; first a signal is obtained from the sample solution, then increments of a standard containing a known mass of the element to be determined are added to a known volume of the sample solution and the signal consisting of the original sample signal plus the increment from the standard is remeasured. As the enhanced signal, following the addition of aliquots of standard, can be related to a known mass of added element it is possible, by simple proportion, to calculate the mass of an element present in the untreated sample solution. The technique operates on the assumption that any changes, apart from those related to blank levels, related to sample matrix will affect both sample and standard in the same manner, hence matrix effects are cancelled. Too often this technique is incorrectly used because increments of standard are added which do not conform to the linear response part of the standard curve. It is essential to establish whether or not a measurement is influenced by sample matrix and compare the response of incremental additions of standard to a simple aqueous solution with that of a sample. With the linearity of response established, the increments of standard should be added such that only small increases in the element signal are obtained in order to approximate the general relation between the mass of sample element to the mass of all other elements and compounds present in the sample. While some matrix effects may result in enhancement of the signal from the sample, in general depression of the sample signal occurs with a lower

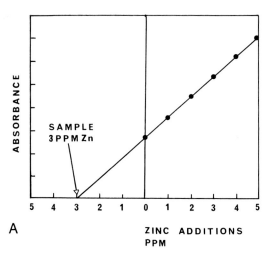
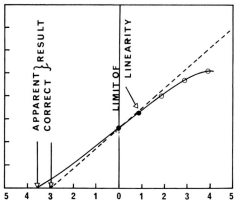

Figure 60: Standard addition analysis. *A*. The use of incremental additions of known amounts of zinc to a sample to determine the concentration of zinc in the sample. *B*. Errors arising because of using a nonlinear concentration response. *C*. An illustration of the validity of the method for different matrixes. A is an aqueous calibration curve, B and C are for samples having different matrixes.

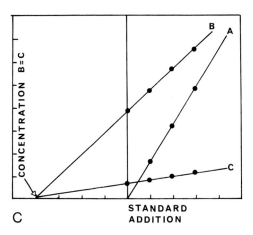

slope to the element response curve compared with that obtained for a pure aqueous calibration. The technique of standard addition analysis is illustrated in Figure 60.

Flame emission spectroscopy has tended to become overshadowed by atomic absorption spectroscopy, but it is a fairly versatile method and often advantageous in terms of lower levels of detection for Na, K, Rb, Ca, Sr, Ba, some rare earths, and Tl. The various methods of atomic spectroscopy have been described by Deans (1), Herrmann and Alkemade (2), Mavrodmeanu (3), and Schrenk (4).

REFERENCES

1. Deans, J. A.: *Flame Photometry.* New York, McGraw, 1960.
2. Hermann, R. and Alkemade, C. Th. J.: *Flammenphotometrie.* Berlin, Springer-Verlag, 1960.
3. Mavrodineanu, R.: *Bibliography on Flame Spectroscopy. Analytical Applications.* Washington, D.C., National Bureau of Standards, Miscellaneous Publication 281, 1967.
4. Schrenk, W. G.: *Analytical Atomic Spectroscopy.* New York, Plenum Pr, 1975.

Atomic Absorption Spectroscopy

In 1814 Kirchoff and Bunsen utilised the principles of atomic absorption to determine the presence of elements in atmospheres of stars; Kirchoff's Law states "Matter absorbs light at the same wavelengths at which it emits light." In 1935, Woodson (1) described an analytical application based on this law for the assay of mercury, but it was not until 1955 that Walsh (2) laid the foundations of atomic absorption spectroscopy as a very versatile analytical technique for the analysis of a large number of metals, particularly when present at trace levels. Although the cost of instruments varies considerably, depending upon the degree of sophistication, many excellent instruments are inexpensive and as a result have been avidly used often by individuals with little experience in analytical chemistry. Interest in these instruments has been enhanced by manufacturers' claims of high sensitivity, ease of operation, and the suggestion, at least through the manner of advertising, that a "black box" for element analysis was now available. In practice this is far from the truth and although instruments can be cheap and reliable, easy to use, and provide very high sensitivity, they are associated with some of the problems inherent to any flame emission system; failure to recognise matrix effects related to simultaneous emissions of spectra from elements and molecules present in samples results in poor analyses. In conventional systems analysis is restricted to liquids; techniques for the analysis of solids are available but suffer from matrix problems and usually only one element at a time is analysed, although assay for several elements simultaneously is possible as described by Meret and Henkin (3) and Ganjei et al. (4).

As we have seen in atomic emission spectroscopy, the process concerns the release of energy when atoms elevated to excited states in flames return to lower energy levels; sensitivity can be increased by increasing the temperature of the flame. In conventional flames with temperatures between 2 and 4000°K, only about 1 percent of the atoms are excited, the remainder being found in the ground state but capable of absorbing radiation at discrete energy (wavelength) proportional to particular excited states which can be provided by light energy.

The principle of the method of atomic absorption spectroscopy is illustrated in Figure 61 and involves the formation of an atomic cloud of elements present in a sample by atomisation in a flame illuminated by a lamp (light source) which emits the spectra (emission) of the element to be determined; light of a wavelength corresponding to a resonance emission line of an element is absorbed by the atomic vapour of that element. We are concerned with the degree of absorption of the emitted radiation generated by the lamp by the same element in the atom vapour cloud. Unlike emission spectroscopy, we are not concerned with having to resolve all components of the spectrum; therefore, the requirements of the optical dispersion system are not stringent, in fact all a monochromator has to do is resolve the line of interest from any others present in the source lamp; a photodetector functions to determine the intensity of the signal

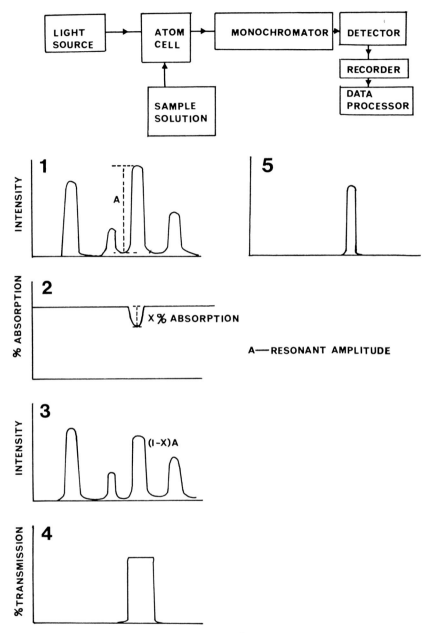

Figure 61: Principles of atomic absorption spectrometry.
1. Emission spectra from a hollow cathode lamp.
2. Sample absorption of resonance line.
3. Spectra after absorption.
4. Monochromator isolation of resonance line.
5. Detector recording the resonance line diminished by sample absorption.

of the resonance line reduced in intensity by absorption in the sample. The natural width of the resonance line is about 10^{-4} Å and in spite of line broadening by the Doppler effect, and others related to field and pressure effects which vary with element and temperature, absorption of radiation still takes place over a very narrow spectral region, typically about 10^{-2} Å.

Numerous publications are available describing atomic absorption spectroscopy for example, Ramirez-Munoz (5), Slavin (6), L'vov (7), Angino and Billings (8), Willis (9), Smith and Winefordner (10), Price (11), Welz (12), while recently Kirkbright and Sargent (13) have provided an excellent account of atomic absorption and fluorescence spectroscopy; the Society of Analytical Chemistry (14) publishes annual reports describing all aspects of methods together with an up-to-date bibliography. The main components of an atomic absorption spectrometer follows.

Light Source

The most commonly used light source is the hollow cathode lamp as illustrated in Figure 62. The lamp consists of two electrodes (anode and cathode), in which the cathode consists of the element of interest sealed in a glass tube, with a facing window which is transparent to the resonance line of the element of interest. The cathode of most lamps consists of a single element, although by using mixtures of elements multielement lamps can be constructed but often suffer from problems of preferential sputtering or selective emission of one element; hence, the most useful mixture will consist of elements with similar emission characteristics. The lamp is filled with an inert gas, such as argon, at 1 to 5 torr; lamp sensitivity can be improved by using different filling gases; for some elements neon can result in an increase in sensitivity of about 3.

A discharge is established at the cathode by application of 300-400 v across the electrodes and the inert filling gas atoms are ionized, producing ions which bombard the cathode causing sputtering and displacement of atoms in an excited state and as a result the emission of line spectra, of which resonance lines are prominent, for the element of which the cathode is made.

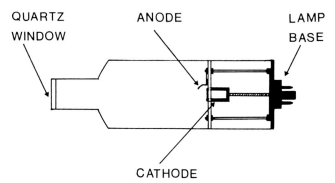

Figure 62: Typical hollow cathode lamp in which the element is contained within the hollow cathode.

Modern lamps provide a steady source of light, low noise levels, low drift, clean characteristic spectra of high intensity, and no spectral interference.

When using lamps it is essential to subject them to a period of warm-up, which varies depending upon the element from a few minutes to several hours, in order to establish equilibrium between the cathodic temperature and population of analyte atoms. Under normal conditions of operation, lamps should be run at about 70 percent of the recommended maximum current and usually have a useful life of about five hundred hours. In order to reduce the loss of time when changing lamps, multi-turret lamp holders are available in which several lamps may be held in the conditioned state. Demountable lamps are also available in which the material of the cathode may be changed. Such systems require high vacuum techniques but an advantage lies in an ability to continuously recirculate the filler gas through a cleaning system thus removing contaminants in the atomic gas cloud and preventing the build-up of sputtered metal on the window of the lamp.

Very high intensity lamps are also available in which the cloud of atoms in the vicinity of the cathode are excited by a secondary discharge and contribute to the resonance radiation. Quite clearly problems will arise if the material of the cathode is easily vapourised and is deposited on the window. This problem can be overcome by the use of electrodeless discharge lamps, which consist of a sealed tube containing a few milligrans of a metal or volatile salt in an inert atmosphere of filler gas such as argon or neon; excitation is provided by radio or microwave frequencies, hence the need to purchase both a lamp and a power supply. The outputs of these lamps are more intense and stable than most hollow cathode lamps, and with lower noise levels for many elements; such lamps have proven to be useful for analysis of volatile elements such as As, Sb, Pb, Se, and Te.

Another type of source is the laser, consisting of a coherent unidirectional beam of radiation with a large amount of energy contained in a small spectral region; of the many types of laser available, the tuneable laser is of interest as it allows the selection of laser output in the form of a "line" whose half width may be selected through a given wavelength range thus providing sharp lines at a metal absorption line.

Sample Atomisation

Flames

One of the most critical components in liquid atomic absorption spectrometry is the atom cell and associated parts in which a liquid sample is atomised into the flame. Standard burner heads, using premixed gases, are of the slot type, as illustrated in Figure 63, provide long narrow flames, and are made of corrosion resistant materials such as titanium; single slots are common but multiple slots, such as the three-slot air-acetylene head, are useful for samples containing high levels of dissolved solids. Atomisation of the sample does not take place until all the solvent phase has been removed, hence a stage of desolvation is required in which the droplets are dried. Very careful adjustment of the composition of the oxidising gases and control of flow rate and height of flame are essential; the spectral light source is fixed to provide the maximum of radiant power reaching the detector but the distribution of the analyte in the flame is not homogeneous, in fact much remains to be learned of the chemistry of flames; the successful use of flames is largely empirical.

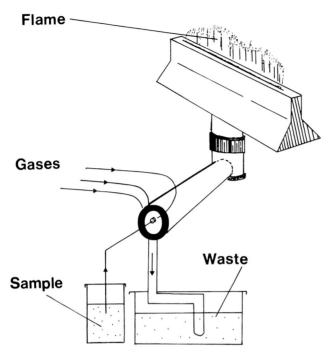

Figure 63: Typical laminar type burner used in atomic absorption spectroscopy.

Commonly used gases are various mixtures of air, acetylene, oxygen, nitrous oxide, nitric oxide, and propane, and modern instruments use automated gas mixing and ignition systems. Only a small amount of the sample drawn up into a capillary tube is aspirated, the rest is removed through a drain into a suitable waste receptacle. The mixture of analyte solution and gases are premixed in a chamber coated with an inert plastic coating. Although pneumatic nebulisation is commonly used, ultrasonic techniques can provide a very fine spray with particles less than 1 μ in diameter, thus reducing the energy required to convert the aerosol into an atomic vapour.

Apart from the use of samples dissolved in simple acids, various types of organic solvents can be satisfactorily nebulised; hence it is practical to carry out preconcentration chemistry and extract single or particular groups of elements into small volumes of organic solvent thus overcoming some of the matrix problems. For biological materials a commonly used extractant is ammonium pyrrolidine dithiocarbamate (APDC) in methyl isobutyl ketone (MIBK). Although direct nebulisation of samples into flames may be used for many applications, it is preferable to use some form of preconcentration technique. Wendt and Fassel (15) report the use of an induced coupled plasma (see page 258) as an atom reservoir in atomic absorption spectrometry although Greenfield, McGeachin, and Smith (16) see little advantage, if any, in its use. Nevertheless, plasma sources offer the possibility of extended linear ranges of calibration curves and longer residence times in the flame compared to conventional gas flames.

In biological work it is often only possible to obtain very small amounts of sample but these can be analysed by use of the boat-in-flame or the Delves cup (17) technique. A few microlitres of sample are placed in a nickle, molybdenum, or stainless steel cup and then dried at about 140°C on the edge of a premixed gas flame and then inserted into the flame; vapour arising from the sample then passes into a cylindrical tube made of quartz positioned at a higher level in the flame. The axis of the tube is aligned to the radiation emitted by the hollow cathode lamp and illuminates the sample vapour in the tube for longer times than those experienced with open flames, thus enhancing sensitivity. Clearly the technique is most suitable for volatile species such as Pb, Bi, Se, and Ga; a critical review of the technique has been published by Clark, Dagnall and West (18), while Cernik and Sayers (19) have used the technique for the direct assay of lead in blood by pricking a finger and absorbing the blood on filter paper from which a disc is punched for direct insertion into the cup.

Electrothermal Atomisation

Electrothermal or flameless atomisation techniques have become very popular because they can be used for very small quantities of sample and in many instances detection limits are some fifty to a thousand times better than those obtained with conventional flames for a variety of matrixes. Flameless techniques should be considered as complimentary to, rather than competitive with, flames. The basic technique consists of placing an aliquot of a sample on a surface, such as graphite, or a high melting point metal ribbon, such as tantalum, and rapidly heating in an enclosed space such that, in the ideal case, the whole sample is volatilised and atomised, hence providing an environment in which the residence times for the atoms is long compared to flames. Light from a source, such as a hollow cathode lamp, is passed through the atomised sample; light transfer into and out of the chamber is accomplished through optically clear windows. Early systems consisted of furnaces such as the L'vov oven (7), but modern systems using graphite tubes are based upon the Massmann (20) heated graphite analyser (HGA), which is now commercially available in several designs. Most systems consist of a pyrolitic graphite tube, for example about 55 mm long with an internal diameter of 8 mm, which is water cooled and can be used up to a temperature of 3500°C. An aliquot of a sample, typically 10-50 μl, is injected through an entry port with a micropipette into the tube. The sample is then subjected to a sequence of heating stages; first the sample is dried and solvent removed at temperatures of between 100 and 150°C for a period of about twenty-five seconds; the residue is then charred and ashed over a temperature range of 150-300°C, typical times for this stage are about fifty to sixty seconds; then follows a period of rapid heating between 1900 and 3000°C over a period of less than ten seconds when the residue is atomised into the space of the tube. As the tube is water cooled new samples may be rapidly injected; a typical ramp heating profile is illustrated in Figure 64. An example of some typical flameless devices and others are illustrated in Figure 65. Because of the speed of atomisation, data are usually calculated from measurements of peak height rather than peak area but there is a need to develop methods for the integration of peak area.

The literature abounds with measurements of elements present in biological materials determined by this technique; for example Perry, Koirtyohann, and Perry (21) described assay for Cd in blood and urine; lead in whole blood has been described

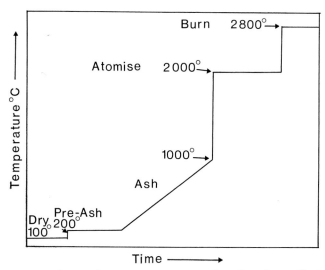

Figure 64: Typical ramp heating sequence used in flameless cells. Note that the ashing process takes place under reducing conditions because of the use of a nonreactive carrier gas.

by Fernandez (22) and is illustrated in Figure 66, comparing data obtained by a micro-HGA technique and a macro-extraction method; also Cu, Cd, and Zn in human liver has been described by Evensen and Anderson (23). In spite of the apparent simplicity of assay by this technique, and the possibility of determining very low levels of trace elements in a variety of biological materials, it is associated with some serious problems which are often ignored in the quest for data. In spite of careful manufacturing processes the performance of each individual graphite tube has to be determined; when about fifty injections have been made the tube will usually start to show a decrease in response, and often by the time two hundred injections have been made the tube may have to be replaced. The rate and nature of tube degradation depends upon several factors but is largely influenced by the nature of the samples analysed. The rate of gas flow through the tube must be controlled in order to optimise residence times for atoms and remove the sample cloud before chemical processes take place impairing the element signal recorded at the detector. One commercial instrument, the

Perkin-Elmer HGA 21000, utilises improved gas flow within the tube and an auxiliary inert flow outside the tube to prevent oxidation of the atom cloud as it is being removed from the system.

Electrothermal atomisation is associated with various types of matrix effect; often the matrix effects are more pronounced than when encountered in flames. Of particular importance is the formation of salts (for example NaCl), broad band absorption, and light scattering effects—which can be difficult to correct. Ramp-heating can result in the loss, or partial loss, of an element or compound during the drying or ashing stage, and constitutes a source of error. The so-called ashing stage takes place under reducing conditions, hence the efficiency and type of ashing will differ from that of conventional oxidising systems. During the process of drying samples, chemical fractionation of the sample can occur which may form a fractionated substrate from which rates of atomisation of elements may vary. The response curves for electrothermal atomisation are usually restricted to a narrow range of concentrations over which a linear response is obtained and become a very important consideration when using the standard addition technique of analysis. Limited linearity of response is directly related to the high density of particles in the atom cloud, which may be compared with a wide linearity of response, over very large concentrations, associated with low density flames such as those obtained with plasmas. Finally, because the sample is atomised over a short period of time, it is essential to use optical background correction systems in order to provide some degree of correction related to short term instabilities.

When analysing for very volatile elements, such as Hg, Bi, Sb, Se, and As, the elements can usually be efficiently removed from the sample matrix by simple heating in the case of Hg or through generation of volatile hydrides using pellets or a solution of sodium tetrahydroborate. Rooney (24) has described the determination of bismuth in blood by this technique and obtained a value of 0.0245 μg Bi/ml $\pm$ 0.0288 μg/ml^{-1} with a range of 0.007 to 0.18 μg Bi/ml^{-1}.

The next stage in atomic absorption analysis concerns processing of the light emerging from the source region by isolation of spectral lines of interest by a monochromator (which serves to isolate a resonance line of interest from other lines originating

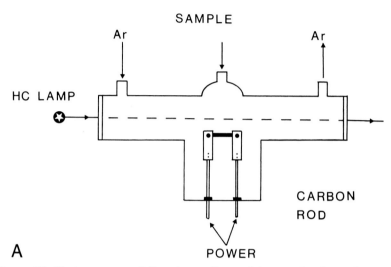

Figure 65: Various types of flameless cells used in atomic absorption spectroscopy. A: Carbon rod system in which about 2 μl of sample containing perhaps 10^{-12} g of an element is pipetted onto the carbon rod and rapidly heated by passage of an electric current to 2500°C.

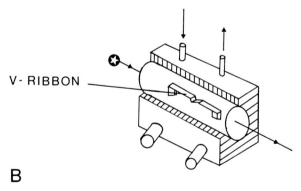

Figure 65B: A similar system but using a tantalum ribbon, which can receive between 1 and 25 μl of solution and which is heated to about 3000°C. Useful element detection range 10^{-10} to 10^{-13} g.

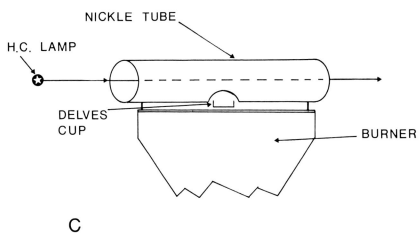

C

Figure 65C: Delves Cup method in which a solution is placed in a cup, evaporated to dryness slowly, and then passed into a laminar flame for atomisation.

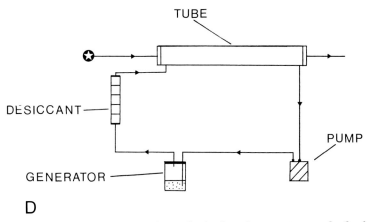

D

Figure 65D: A nonflame method in which the element in a volatile form, produced by chemical reaction in a generator, is passed into a tube along the axis of which it is irradiated by the emission from a hollow cathode lamp. By using a closed recirculating system, the volatilised element can be retained in the light path for a long time compared to those methods using flash atomisation. The method is particularly useful for mercury, arsenic, and selenium.

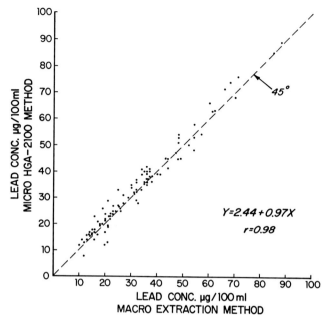

Figure 66: Comparison between the determination of lead by the micro heated graphite analyser (HGA) technique and conventional atomic absorption spectroscopy following extraction of lead from a sample. From Fernandez, F. J.: Micromethod for lead determination in whole blood by atomic absorption with use of the graphite furnace. *Clin Chem*, 21(4):558, 1975.

from the source spectrum, molecular emission, and other background continua originating from the atom cell). Most modern instruments use a grating instead of the conventional prism as the dispersing element and with improvements in holographic recording of diffraction gratings very considerable improvements in resolution are possible. Cresser, Keliher, and Wohlers (25) have described the use of echelle gratings in atomic absorption analysis for multielement analysis by using a continuum source as a single illumination source and achieving adequate spectral resolution by use of the grating. If gratings having high resolution are used, it is possible to achieve separation of the spectral lines associated with individual isotopes of the elements.

Final detection of the spectra of interest is usually carried out by use of photomultiplier detectors, although considerable advances are taking place in the use of silicon photodiode and photon counting techniques and the use of vidicon tubes as multi-element detectors (Ganjei et al., 4). For many purposes, particularly when using flames, single beam spectrometers are acceptable but double beam systems are essential for many applications. In a double beam spectrometer, as illustrated in Figure 67, the light emitted from the source lamp is alternately passed through and past the flame and the ratio of the two intensities determined, thus allowing for compensation for variables such as drift in the intensity of the lamp emission with time and various forms of electronic fluctuations. A further improvement in overcoming problems related to light scattering, such as the presence of a flux of particles in the flame or in the sampling beam, is provided by the use of a deuterium lamp for background correction. The magnitude of radiation scatter effects varies with wavelength; radiation from a hollow cathode lamp source and a deuterium UV lamp source is directed alternatively through the atom cloud by means of a rotating beam switch. By focussing the radiation from the deuterium lamp onto the same focal spot in the sampling area as the hollow cathode lamp radiation, a sig-

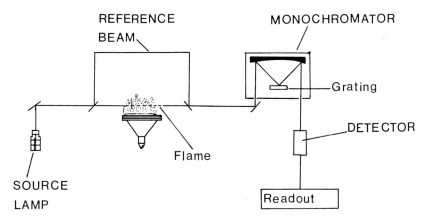

Figure 67: Double beam atomic absorption spectrophotometer. In this system the sample beam passes through the flame where absorption of the element of interest takes place, then the sample beam, together with the reference beam, are focussed on the grating monochromator.

nal is obtained consisting of the true atomic absorption and background scatter, but the broad continuous radiation from the deuterium lamp will not be significantly absorbed and will provide a measure of background scatter. By using appropriate instrument settings, the background scatter can therefore be automatically eliminated.

Some atomic absorption spectrometers, provided they have a good monochromator, can be used in the emission mode by direct aspiration of a liquid into a flame. In emission measurements the burner and head are the same as used in atomic absorption measurements; nitrous oxide-acetylene flames are preferred rather than air-acetylene except for the alkali metals. A wavelength scanner is required if changes in baseline occur from sample to sample, but a fixed wavelength is appropriate when the analytical line can be isolated and when the baseline is constant for different samples.

ATOMIC FLUORESCENCE SPECTROSCOPY

When a species becomes excited by the absorption of radiation and is elevated from the ground state energy to some higher energy level, it then tends to return rapidly to the original ground state; in this process the energy of the absorbed photon is dissipated, usually in the form of heat. Some molecular species return to the ground state but are preceded by degradation to lower energy excited states with the emission of heat; photons produced as a result of these transition processes from the second excited state then pass to the ground state. Photons produced by the second excited state have lower energies and hence longer wavelengths than the excited photons, and the overall process is called fluorescence emission. The analytical use of fluorescence is described as follows (West, 26).

$$F = Q \, I_o \, \epsilon_A \, l \, c \, a \, p$$

where F = the analyte fluorescence signal
Q = the quantum efficiency of the fluorescence mechanism
I_o = the intensity of the selected resonance line

ϵ_A = the atomic absorption coefficient corresponding to the probability that absorption will take place for those particular lines and species

l = the absorption path length through a population of atoms

c = the concentration of element atoms

a = a conversion factor including an efficiency of nebulisation and atomisation in a flame

p = a geometric factor relating the fraction of emitted radiation which can be accepted by the detector

The main components of instrumentation in fluorescence analysis are illustrated in Figure 68. From the above equation it is seen that the strength of the analyte signal is proportional to the in-

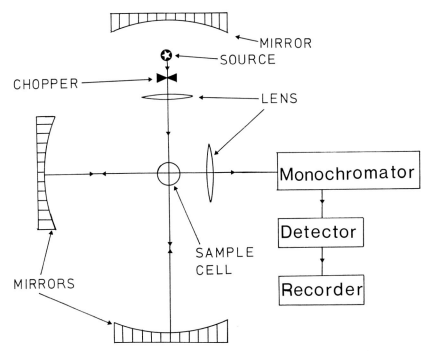

Figure 68: Schematic diagram illustrating the main components of an atomic fluorescence system. The source can consist of an electrodeless discharge lamp, xenon arc lamp, or a laser; as the emission spectra consists of a few lines, only a simple monochromator is required.

tensity of the source, hence the greater the intensity of the excitation source the better the sensitivity of detection. Requirements of a good fluorescence source are the presence of very intense lines, electrodeless lamps are preferred and multielement analysis can be achieved.

Recently attention has been directed towards the use of continuum sources for multielement analyses. Because the light source is not in direct line with the detector, it is possible to use very strong sources, such as the xenon arc. Johnson, Plankey, and Winefordner (27) have described the use of a 500 W xenon arc lamp chopped at 50 Hz as an excitation source for simultaneous assay of eighteen elements. Atomic fluorescence spectrometry has been developed by Winefordner and his co-workers, for example see Winefordner and Vickers (28), as a very versatile and sophisticated method of analysis, particularly suitable for volatile elements, but the method has tended to be associated with laboratory rather than commercial development. Montaser and Fassel (29) have described the use of inductively coupled plasmas in atomisation cells for atomic fluorescence spectrometry; detection for Cd, Zn, and Hg was found to be superior to atomic emission. A comparison of advantages and disadvantages of atomic fluorescence spectroscopy has been given by Kirkbright and Sargent (13); some advantages are that sensitivity is related to light intensity and not temperature, instruments are not usually complex, and there is high sensitivity in the far ultraviolet range; some disadvantages are that quenching is affected by gas species in the cell, light is scattered, and there is limited linearity of response.

Atomic absorption spectroscopy has become an established tool in medical studies, discussed by Christian and Feldman (30), while Van Ormer (31) has described its application in toxicology. Data illustrating the sensitivity of detection for a number of elements by various methods are given in Table LIV. The best data tend to be obtained for between 15 and 65 percent absorption, and detection limits are considered to be those signals which represent values of twice background levels. It is important to note that detection levels based upon simple solutions

TABLE LIV

SOME DETECTION LIMITS FOR ATOMIC ABSORPTION AND
ATOMIC FLUORESCENCE SPECTROMETRY*

Element	Atomic Absorption Flame Methods $\mu g\ ml^{-1}$	Atomic Absorption Nonflame Methods picogram	Atomic Fluorescence $\mu g\ ml^{-1}$
As	0.2	1	0.1
Ca	0.002	0.4	0.02
Cd	0.005	0.08	0.000001
Co	0.005	2	0.005
Cr	0.005	2	0.05
Cu	0.005	0.5	0.001
Fe	0.005	3	0.008
K	0.005	2	—
Mg	0.0005	0.04	0.001
Na	0.005	0.2	—
Pb	0.01	7	0.01
Se	0.5	10	0.04
Sn	0.06	2	0.1
Zn	0.002	0.03	0.00002

* Detection limits will depend in practice upon type of instrument, operating characteristics and overall chemistry of a sample.
[1 $\mu g = 10^{-6} g$; 1 picogram = $10^{-12} g$]

of single elements may have little bearing on real samples because of interferences. Today the tendency is often to attempt analyses on solutions of samples without any chemical processing by direct aspiration or atomisation of aliquots in thermal devices. It is for the analyst to provide sufficient evidence to confirm the quality of such data. Often, and in spite of the need for more effort and skill, it is far better to carry out some type of element preconcentration, remove major sources of interference, and obtain signals from elements free from element interferences. In atomic absorption spectroscopy, it is often better to use flame sources rather than electrothermal atomisation; the latter technique is very attractive and can be very advantageous but is subject to a variety of sources of error, and the overall analytical time involved to obtain acceptable data can be considerably longer than when using flame atomisation.

In the thermal excitation methods described so far, it is clear that for most purposes sensitivity is not a problem; in the passage of a few years we have seen considerable improvements in commercially available instruments, detection limits have gone from microgram per millilitre to nano- or picogram levels per millilitre of sample solution. With advances in electrothermal techniques, we have passed from the need to use 50-100 ml sample solutions to the direct insertion of 1-50 microlitres of solution. All these advances are very exciting, but there has not always been a similar improvement in providing exceptionally clean working conditions, which are essential for trace element analysis. It is worth noting that in atomic absorption spectroscopy good ventilation is essential as flame sources can emit some 3 kw of heat; in order to adequately remove the volatilised species from the environment of a laboratory, air flow rates of some 100 ft^3/min are required. Quite clearly, the air should be free of particulate matter or alternatively the source flame, or heating area, should be supplied with a protective zone of a laminar-introduced, purified, inert gas supply in order to isolate the laboratory air from the area in which atomisation processes are taking place.

Until considerable improvements are made in the total philosophy of trace element analysis, from sampling through sample preparation and analysis, with particular reference to the exclusion of contaminants and the preparation of adequately clean reagents, the full advantage of these very sensitive methods of analysis will not be realised. Unfortunately, commercially available systems rarely consider the total problem of analysis and tend, at least initially, to illustrate the advantages of their products with reference to a few simple matrixes. In a similar manner the vast number of publications which appear in the analytical literature are often concerned with a particular matrix or the special development of an instrumental procedure, often requiring sophisticated equipment or very high technical skill.

REFERENCES

1. Woodson, T. T.: A new mercury vapour detector. *Rev Sci Inst*, 10:308, 1935.
2. Walsh, A.: The application of atomic absorption to chemical analysis. *Spectrochim Acta*, 7:108, 1955.

3. Meret, S. and Henkin, R. J.: Simultaneous direct estimation by atomic absorption spectrophotometry of copper and zinc in urine and cerebrospinal fluid. *Clin Chem, 17*:369, 1971.
4. Ganjei, J. D., Howell, N. G., Roth, J. R., and Morrison, G. H.: Multielement atomic spectrometry with a computerised vidicon detector. *Anal Chem, 48*:505, 1976.
5. Ramirez-Munoz, J.: *Atomic Absorption Spectroscopy.* Amsterdam, Elsevier, 1968.
6. Slavin, W.: *Atomic Absorption Spectroscopy.* New York, Wiley, 1968.
7. L'vov, B. V.: *Atomic Absorption Spectroscopy (1966).* Israel, Prog Sci Trans, 1969.
8. Angino, E. E. and Billings, G. K.: *Atomic Absorption Spectrophotometry.* New York, Elsevier, 1968.
9. Willis, J. B.: Atomic absorption, atomic fluorescence and flame emission spectroscopy. In Robinson, J. W. (Ed.): *Handbook of Spectroscopy.* Cleveland, Ohio, CRC Press, 1974.
10. Smith, R. and Winefordner, J. D. (Eds.): *Spectrochemical Methods of Analysis.* New York, Wiley, 1971.
11. Price, W. J.: *Analytical Atomic Absorption Spectrometry.* London, Heyden and Sons Ltd, 1972.
12. Welz, B.: *Atomic Absorption Spectroscopy.* (1st English Ed.): Verlag Chemie Int Inc, New York, 1976.
13. Kirkbright, G. F. and Sargent, M.: *Atomic Absorption and Fluorescent Spectroscopy.* New York, Academic Press, 1974.
14. *Atomic Absorption Abstracts.* Society of Analytical Chemistry, London. (Annual Publication)
15. Wendt, R. H. and Fassel, V. A.: Atomic absorption spectroscopy with induction-coupled plasmas. *Anal Chem, 38*:337, 1966.
16. Greenfield, S., McGeachin, H. McD., and Smith, P. B.: Plasma emission sources in analytical chemistry, III. *Talanta, 23*:1, 1976.
17. Delves, H. T.: A micro-sampling method for the rapid determination of lead in blood by A.A.S. *Analyst, 95*:431, 1970.
18. Clark, D., Dagnall, R. M., and West, T. S.: The occurrence of multiple peaks in the determination of various elements by the "Delves sampling cup" method. *Anal Chem Acta, 60*:219, 1972.
19. Cernik, A. A. and Sayers, M. H. P.: Determination of lead in capillary blood using a paper punched disc by atomic absorption technique. *Br I Industry Med, 28*:392, 1971.
20. Massmann, H.: Vergleisch von atomabsorption und atomfluorezenz in der graphit kuvette. *Spectrochim Acta, 23B*:215, 1968.
21. Perry, E. F., Koirtyohann, S. R., and Perry, H. M.: Determination of cadmium and urine by graphite furnace atomic absorption spectrophotometry. *Clin Chem, 2114*:626, 1975.
22. Fernandez, F. J.: Micromethod for lead determination in whole blood

by atomic absorption with use of the graphite furnace. *Clin Chem* 21(4):558, 1975.
23. Evenson, M. A. and Anderson, C. T.: Ultra micro analysis for copper and zinc in human liver tissue by atomic absorption spectrometry and the heated graphite tube furnace. *Clin Chem,* 21(4):537, 1975.
24. Rooney, R. C.: Determination of bismuth in blood and urine. *Analyst,* 101:749, 1976.
25. Cresser, M. S., Keliher, P. N. and Wohlers, C. C.: Echelle Grating Spectrometers in Analytical Chemistry. *Anal Chem,* 48:333A, 1976.
26. West, T. S.: Atomic fluorescence and atomic absorption spectrometry for chemical analysis. *Analyst,* 99:886, 1974.
27. Johnson, D. S., Plankey, F. W., and Winefordner, J. D.: Multi-element analysis via computer-controlled rapid-scan atomic fluorescence spectrometer with a continuum source. *Anal Chem,* 47:1739, 1975.
28. Winefordner, J. D. and Vickers, T. J.: Flame spectrometry. *Anal Chem,* 46:192R, 174.
29. Montaser, A. and Fassel, V. A.: Inductively coupled plasmas as atomisation cells for fluorescence spectrometry. *Anal Chem,* 48:1490, 1976.
30. Christian, G. D. and Feldman, F. J.: *Atomic Absorption Spectroscopy Applications in Agriculture, Biology and Medicine.* New York, Wiley, 1970.
31. Van Ormer, D. G.: Atomic absorption analysis of some trace metals of toxicological interest. *Journal of Forensic Sciences,* 595:20, 1975.

High Temperature Atomic Emission Sources

High temperature atomic emission sources may collectively be called plasma emission sources, first described by Gerdien and Lotz (1); some of the first analytical applications were by Margoshes and Scribner (2) and Korolev and Vainshtein (3). In multielement analysis, by simultaneous or sequential detection, it is necessary to couple the fast speeds with which light signals can be processed with an ideal stable emission thermal source. The multielement capability of emission spectroscopy requires systems capable of exciting as many lines as possible characteristic of the chemical elements, with the minimum of spectral overlap and interference. Conventional premixed gas flames provide analytically useful temperatures of between 1850°K (air-coal-gas) and about 4600°K (oxygen-cyanogen) into which aspirated liquids, and some very fine solids, may be injected and atomised.

Flame spectra usually consist of a few simple lines because of low excitation energies, and it is not possible to excite all the elements at once, particularly those of a refractory nature. Further, flames are complex systems and require well-defined conditions for operation; many types of chemical reactions take place in flames, the rate and type depending upon the overall chemical matrix of the sample. In emission spectroscopy, flames are not hot enough to excite all elements simultaneously and other methods of excitation, such as the arc or spark, are often used; hence a considerable amount of the energy is lost in heating the electrodes. Arc and spark sources require careful control and in the case of the arc are subject to considerable variability in volatilisation of samples because of wandering, or selective, arcing across some parts of the electrode. A plasma source is a gas in which greater than 1 percent of the atoms and molecules are ionized and into which a very large quantity of energy can be transferred providing very high temperatures of between 4700°K to 16000°K depending upon the type of plasma. It is important to note that plasma flames represent a volume of high temperature regimes which are not subject to changes in temperature gradients with time, hence it is possible to inject a sample, usually a liquid, into a fairly even high temperature field for volatilisation.

An optimum source should be versatile, providing high sensitivity of detection in order that trace and minor elements may be determined together with major elements; the system should also be free from changes in the rate of spectral emission with time and the system should be extremely stable. Other desirable requirements are that injection of samples should be easy and instrumentation should be as simple as possible in order to be commercially viable and should not have to rely upon an individual's technical expertise. In recent years plasma instruments have become commercially available consisting of a source system coupled to an optical emission spectrometer, with fast computer systems for data retrieval and evaluation. A schematic dia-

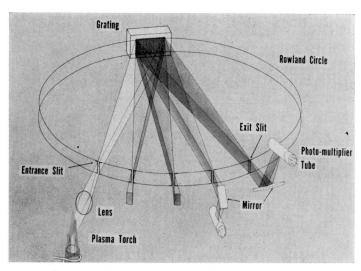

Figure 69: Schematic diagram of a typical atomic emission system using multichannel polychromators. Photograph provided by Professor V. A. Fassel.

gram of a typical atomic emission multichannel polychromator system is illustrated in Figure 69.

A plasma is formed when the current density to be dissipated on the surface of an electrode increases to a level at which it cannot be carried by radiation and conduction alone. By restricting the plasma to a narrow tube a jet is formed; because of the confinement, current density increases to give rise to a so-called thermal-pinch effect followed by a magnetic-pinch, which together result in the development of very high temperatures. The net result of these advances has been a major breakthrough in element analysis when it is further realised that this type of system provides one constant thermal source which is very stable and simple to operate; true simultaneous multielement analyses can be carried out under one set of operating conditions; detection of elements is usually equal or superior to conventional atomic absorption techniques and the dynamic range is vastly superior. With a single calibration line it is possible to obtain a linear response from nanogram to gram levels of the elements present in solutions; analytical curves for the determination of Cr and Ni in steel over the range of 0.001 to 50 percent are illustrated in Figure 70. Because of the high temperatures, inter-element effects are minimal, and while the technique is mainly restricted to liquid samples, finely divided solids can also be analysed. The development of this field of analysis is fascinating and the reader is referred to the papers by Reed (4), Fassel and Kniseley

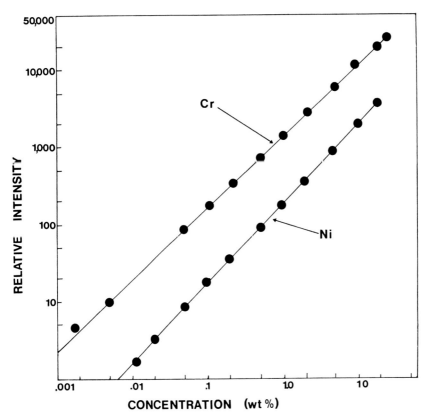

Figure 70: An example of the linear response obtained by inductively coupled plasma-atomic emission spectroscopy for Cr and Ni in a sample of steel. Reprinted with permission from Butler, C. C., Kniseley, R. N., Fassel, V. A. Inductively coupled plasma-optical emission spectrometry: Application to the determination of alloying and impurity elements in low and high alloy steels. *Anal Chem*, 47:827, 1975. Copyright by the American Chemical Society.

(5), Greenfield, McGeachin, and Smith (6), Butler (7), Boumans and de Boer (8), Fassel (9), and Kirkbright, Ward, and West (10) for details.

A plasma may be formed in the normal manner between two conducting electrodes; however, improving the sensitivity for many elements is subject to some matrix effects, chemical interferences, and the possibility of contamination from the electrodes. Many of the problems can be overcome by using very dilute solutions, thus reducing matrix effects but with the loss of some sensitivity. Liquids are injected into the arc by nebulisation through a capillary tube; solids require special techniques such as impregnating one electrode with the sample, but such techniques are not very convenient. The temperature of the arc is about 5000°C, which is sufficient to vapourise and dissociate most materials. Typical detection limits for many elements are 0.1-0.5 μg element/ml; for Be, Ca, Cr, Fe, Li, Ni, and Y, 0.001-

Figure 71: Spectraspan III high dispersion, echelle grating plasma instrument (manufactured by Spectrametrics Inc., Andover, Massachusetts, United States; photograph supplied by Techmation Ltd, United Kingdom). Assay by photoelectric or photographic mode for simultaneous determination of up to twenty different elements.

0.008 μg/ml; and for Bi, C, Ir, Nd, Pr, Se, Ta, Th, and W, between 1-15 μg/ml.

One commercially available instrument, the Techmation-Spectrametrics Spectraspan IV, is illustrated in Figure 71; samples are atomised in a d.c. argon plasma, as shown in Figure 72, which operates at low power (40 v, 75 amps) but produces temperatures as high as 10000°K; wavelength dispersion is achieved by use of a high dispersion echelle grating monochromator with a focal length of only 0.75 m, which provides a two-dimensional wavelength control and high dispersion of all wavelengths from 1900 to 8000 Å in a 4 × 5 inch area. Detection can be by photographic plate, photomultiplier tubes, or vidicons for simultaneous detection of all elements or as a direct reader with twenty end-on photomultiplier tubes.

In radio frequency plasmas, it is possible to dispense with the conventional electrodes by direct coupling to a gas; there are

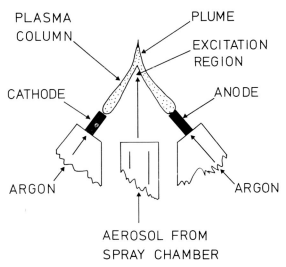

Figure 72: An illustration of the Spectrametrics, Spectraspan III d.c. argon plasma which operates at a low power (40 v, 7.5 amps) and produces temperatures as high as 10,000°K. The geometry of the plasma, formed by two intersecting streams of argon, traps the sample aerosol momentarily at the apex of the inverted V. Courtesy of Techmation Limited, United Kingdom and Spectrometrics Inc, Andover, Massachusetts, United States.

three main types of electrodeless plasmas: (a) inductive coupling, (b) capacitance coupling, and (c) microwave.

Inductive Coupling

In an inductively coupled plasma source, energy is transferred from a coil carrying an oscillating current to the electrodeless plasma contained in a quartz tube. As the plasma is an ionized gas it is electrically conducting, hence oscillatory electric currents are induced into it by the electromagnetic field created by the coil. In order to protect the quartz tube from melting, a flow of coolant gas is introduced tangentially at high velocity and the sample is introduced axially (for high frequency plasmas), as illustrated in Figure 73, together with an Applied Research Laboratories Ltd. ICP instrument. Although there are many types of plasma available, those described by Fassel (9), Fassel and Kniseley (5), and Scott et al. (11) are mainly used for element assay.

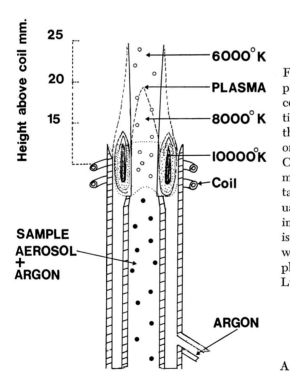

Figure 73: Diagram depicts typical inductively coupled plasma configuration. Photograph shows the Applied Research Laboratories, Ltd. Inductively Coupled Plasma Instrument; the plasma is contained within the box situated to the right of the instrument and the plasma is viewed through a glass window (photograph supplied by A. R. L., Ltd., Luton, England).

A

Detection limits for such systems are excellent for almost all the elements; typical detection limits for many elements are about 1-10 nanograms ml^{-1}, for example Be 0.04, Ca 0.002, Mg 0.01, Sr 0.002, Sb 200, As 90, Au 40, Ta 70 and U 30; comparative data illustrating improvements of detection with time are given in Table LV. Typical temperatures in such plasmas are 9000-10000°K using argon cooling gas at a flow rate of 10-15 litres/min and sample introduction in a flow of argon gas at about 1.5 litres/min.

As aspirated samples pass up into the central eddy current "tunnel," they are subjected to heating through convection, conduction, and radiation; even the presence of low ionization materials, a major problem for flames, provides little inter-element effect. In the core of the plasma, vapourisation and atomisation occurs, but because of the intense continuum emission in this re-

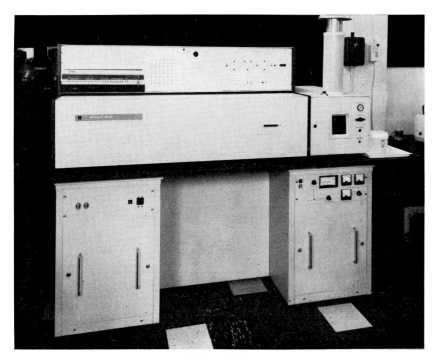

Figure 73B

TABLE LV*

IMPROVEMENTS IN DETECTION LIMITS FOR ASSAY BY
INDUCTIVELY COUPLED PLASMA SPECTROMETRY
(μg/ml)

	1964 Greenfield et al.[†]	1965 Wendt Fassel[‡]	1965 Greenfield et al.[§]	1969 Dickinson Fassel[‖]	1971-72 Fassel[¶]	1975-76* Boumans-de Boer** Olson-Haas-Fassel[††]
Al	50	3	0.5	0.002	0.002	0.0004[††]
As	—	25	—	0.1	0.04	0.002[††]
Ca	1	0.2	0.005	—	0.00007	0.0000001**
Cd	—	20	—	0.03	0.002	0.00007[††]
Co	—	—	0.2	0.003	0.003	0.0001[††]
Cr	20	0.3	—	0.001	0.001	0.00008[††]
Cu	10	0.2	0.01	—	0.001	0.00004[††]
Fe	—	3	0.05	0.005	0.005	0.00009**
La	—	50	—	0.003	0.003	0.0001**
Mg	5	2	0.03	—	0.0007	0.000003**
Mn	10	1	0.05	—	0.0007	0.00001[††]
Ni	5	1	—	0.006	0.006	0.0002**
P	—	10	0.8	0.1	0.04	0.015**
Pb	—	—	—	0.008	0.008	0.001[††]
Si	—	3	—	—	0.01	—
Sn	—	50	4	—	0.3	0.003**
Sr	—	.09	—	0.00002	0.00002	0.000003**
Ta	—	16	—	0.07	—	—
Th	—	40	—	0.003	—	—
V	—	—	0.1	0.006	—	0.00006**
W	—	3	—	0.002	—	0.0008**
Zn	—	30	—	0.009	0.002	0.0001[††]
Zr	—	15	—	0.005	—	0.06**

* From Fassel, V. A.: *Current and Potential Applications of Inductively Coupled Plasma-Atomic Emission Spectroscopy in the Exploration, Mining and Processing of Materials.* Plenary lecture at International Symposium on Analytical Chemistry in the Exploration, Mining and Processing of Materials, Johannesburg, Republic of South Africa, August 24, 1976.

† Data from Greenfield, S., Jones, I. L., and Berry, C. T. *Analyst, 89:*713, 1964.

‡ Data from Wendt, R. H. and Fassel, V. A. *Anal Chem, 37:*920, 1965.

§ Data from Greenfeld, S., McGeachin, McD., and Smith, P. B.: Plasma emission. Sources in analytical spectroscopy III. *Talanta, 23:*1, 1976.

‖ Data from Dickinson, G. W. and Fassel, V. A. *Anal Chem, 41:*1021, 1969.

¶ Data from Fassel, V. A.: Hasler Award Address, Tenth National Meeting, Society for Applied Spectroscopy, St. Louis, Missouri, United States, Oct. 20, 1971; Eleventh National Meeting, St. Louis Society for Applied Spectroscopy, Dallas, Texas, United States, Sept. 14, 1972.

** Data from Boumans, P. W. J. M. and De Boer, F. J. *Spectrochim Acta, 30B:*309, 1975.

†† Data from Olson, K. W., Haas, W. J., and Fassel, V. A. *Anal Chem,* in press, 1978.

gion, it cannot be used for analysis; instead use is made of higher zones in the thermal plume where core temperatures are about 6500°K and are about twice as hot as those obtained with the hottest gas flame namely $N_2O-C_2H_2$; unlike gas flames the free ions are formed in a chemically inert environment. Although the plasma instrument is still being improved, it has reached a state in the analysis of liquids which seems to be distinctly superior to any other method for multielement analysis. Improvements are continually taking place; for example, by employing ultrasonic nebulisation techniques, detection limits for some elements can be improved by a factor of ten (Fassel, private communication).

Although an ultimate requirement in element analysis is for an instrument that is free from any form of inter-element interference, plasma instruments go a long way in attempting to reach this goal; because of high sensitivity very high dilutions of samples can be made thus reducing common effects associated with major ions. One type of interference concerns the problem of background or stray light, which has been discussed by Larson et al. (12); in simple terms the spectrum of a major element may be so intense that it "spreads" across a wide band and contributes continuum at least up to 10 nm from the element line, such that when multiphotodetectors are used there is spectral overlap. Not all of this is due to line broadening; some arise from imperfections in the optical system giving rise to stray light, which can be reduced by using recently developed gratings in conjunction with very narrow bandpass filters positioned in front of polychromator detector or at the exit slit of a monochromator.

Capacitance-Coupled Plasmas

This type of plasma is produced between a pair of electrodes which act as a capacitance in a tuned circuit; an aerosol is introduced into an air gap between the two electrodes; and breakdown occurs as normal air is at a low enough potential. Aspirated aerosols of samples are introduced by several techniques, such as around the electrode and into a brush discharge or more efficiently by axial injection through the material of the discharge tip. The flames produced in this plasma are nonequilibrium types, and difficulties can be experienced in the volatilisation of refractory materials unless several kilowatts are dissipated in the source. The emission intensity of an element will depend upon the concentration of other elements in the matrix with which the excited sample element species may combine; like flames, band spectra may constitute a serious source of interference. Detection limits for many elements are better than 0.02 μg ml^{-1} and some show improvements on those obtained with an air-acetylene flame.

Microwave Flames

In this electrodeless system, the microwave energy is coupled to a stream of gas contained within a nonconductive tube at atmospheric pressures, and containing gases such as argon, helium, or nitrogen into which aqueous aerosols are injected from flame nebulisation devices; Skogerboe and Copeland (13) have recently reviewed microwave plasma emission sources in spectrometry. The system has been widely used for microsamples: detection limits of 10^1-10^{-3} ng/ml can be obtained with 10 μl solutions. Easily ionized elements give rise to matrix effects; this problem arises because of the present need to inject the sample solution into the peripheral zone of the plasma, while the light emitted from the central portion of the plasma is used for analysis; as a result of lateral diffusion through the plasma, changes in the composition of the sample occur. This type of problem can be partly overcome by improvements in nebulisation and techniques for the generation of aerosols. However, the resultant spectra tend to be simple and arise from ground state transitions; a comparison of microwave single electrode plasma (SEP) and radiofrequency inductively coupled plasma (ICP) by Larson and Fassel (14) confirm that easily ionizable elements produce severe changes on the line emission of analyte elements in SEP compared with negligible or small effects in the ICP.

Microwave discharges have been used for the analysis of gases and even isotope assay, which is based upon the shifting of atomic lines by as much as 1 Å of molecular bands caused by the different masses of the isotopes; this technique has been useful in nitrogen studies using ^{15}N labels. Using this method, Nemets and Petrov (15) describe the simultaneous determination of the isotopes of nitrogen, hydrogen, carbon, and oxygen.

In recent years, microwave plasmas have been used as detectors for gas chromatography; organic effluents from a gas chromatograph are passed into the plasma, and elements present in the various fractions eluted from a column can be determined by emission spectrometry. Provided that the spectrometer is equipped with a direct reading system, multielement measurements can be made; so far attention has been directed mainly to H, D, F, Cl, Br, I, S, and P, with detection limits of less than 1 ng/sec of sample flow and for O and N about 5 ng/sec; Dagnall, West, and Whitehead (16) have determined volatile metal chelates; Houpt and Compaan (17) levels of organic mercury in foods; Talmi and Andren (18) selenium in environmental materials; Talmi and Bostick (19) alkylarsenic acids in pesticides. A commercially available instrument, the Applied Chromatography Systems Ltd MPD 850, is schematically illustrated in Figure 74 and uses a helium microwave plasma source with quantitative response to organic and inorganic constituents eluted from gas chromatographic columns. The output from a gas chromatograph is divided into two equal streams, one of which passes through the chromatograph detection system, the other through the microwave-sustained helium plasma. As the microwave generator is not capable of initiating the plasma, the helium is excited by a high frequency tester incorporating a Tesla coil; when microwave power is applied, the plasma is self-sustaining. In the plasma, organic matter is totally fragmented into atomic species which are excited to produce atomic emission spectrum lines.

Up to fourteen photomultiplier tubes and twelve amplifiers (or integrators) can be conveniently located in the instrument, enabling twelve elements to be determined simultaneously from a programme of fourteen. The output of this instrument enables the normal organic trace from the gas chromatograph to be recorded together with the inorganic trace from the plasma source, as illustrated in Figure 74; from atomic ratios, empirical formulae of compounds can be calculated provided that some correction is applied, when required, to allow for differences in sensitivity. Because of the versatility and high resolving power of modern gas chromatographs and the specific nature for element detection, this instrument provides a simple means for identifying and quantitative assay for inorganic-organic compounds. Using a ratio frequency electrodeless discharge cell, Runser and Frank (20) have determined an absolute detection limit of 8.2 μg for DDT in soils even in the presence of 10000 ppm of sodium chloride.

Having outlined, in brief terms, the advantages of using plasma sources, it is worth noting that apart from the simplicity of analysis it is carried out very quickly. If samples are brought into solution by using batch techniques then, as each analysis takes about one minute and if about forty elements are determined simultaneously, we are talking about a throughput of

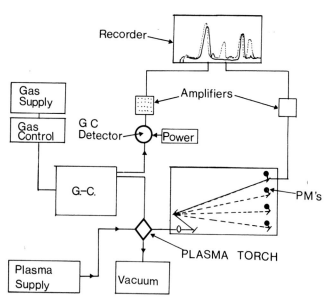

Figure 74A: A diagram illustrating the main components of the Applied Chromatography Systems Ltd, United Kingdom. Organic Analyser MPD 850. The final recorder trace contains an output from both the gas chromatograph and the plasma detectors.

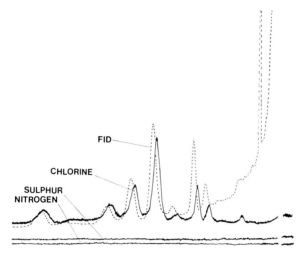

Figure 74B: G-C Flame Ionization Detector (FID) and element traces for an organochlorine compound. G-C column 3% SE 30%, temperature 200°C, and total chlorine content 107 ng/μl.

about 480 samples per day and obtaining data for some 20,000 individual element assays. Such instruments therefore have a role to play in central laboratories and especially those concerned with monitoring. Clearly the total number of sample measurements made each day will usually be less than I have indicated, as adequate numbers of blanks and standards will also have to be run. If confronted with a problem of possible sample matrix effects, the speed at which analyses can be made under identical running conditions enable quite detailed studies of matrix to be made, which are not normally attempted with other instruments because of time and labour involved. With these instruments, standardisation plays an important role, and exceptionally clean working conditions are essential for preparative steps; in order to carry out analyses it is advisable to carefully control the temperature and humidity of rooms housing such instruments. Fully equipped induction coupled plasma or d.c. arc plasma instruments are between 30 and 50 percent more expensive than fully comprehensive atomic absorption spectrometer, but they are far easier to operate, relatively free from any major matrix effect, and can be used for simultaneous multielement analysis. For many problems it is becoming more important to consider groups of elements rather than single elements and for this reason such instruments are likely to play a very important role in biological research, particularly as there is scope for combining them with a variety of detectors for simultaneous organic analysis.

REFERENCES

1. Gerdien, H. and Lotz, A.: Uber eine Lichtquelle von sehr hoher Flächenhelligkeit. *Wiss Veröffentl Siemens-Konzern*, 2:489, 1922.
2. Margoshes, M. and Scribner, B. F.: The plasma jet as a spectroscopic source. *Spectrochim Acta, 15*:138, 1959.
3. Korolev, V. V. and Vainshtein, E. E.: The use of a plasma generator as an excitation source in spectrographic analysis. *J Anal Chem USSR, 14*:731, 1959.
4. Reed, T. B.: Plasma torches. *Inter Sci and Technol, 42*, June 1962.
5. Fassel, V. A. and Kniseley, R. N.: Inductively coupled plasma-optical emission spectroscopy. *Anal Chem, 46*:1110A, 1155A, 1974.
6. Greenfield, S., McGeachin, McD. H., and Smith, P. B.: Plasma emission sources in analytical chemistry, I, II, III, *Talanta, 22*:1, 1975; *Talanta, 22*:553, 1975; *Talanta, 23*:1, 1976.
7. Butler, L. R. P., Human, H. G. C., and Scott, R. H.: Electrical flames. In Robinson, J. W.: *Handbook of Spectroscopy*, vol I. Cleveland, Ohio, CRC Press, 1975.
8. Boumans, P. W. J. M. and de Boer, F. J.: Studies of flame and plasma torch emission for simultaneous multielement analysis. Preliminary Investigations. *Spectrochim Acta, 27B*:391, 1972.
9. Fassel, V. A.: *Current and potential applications of inductively coupled plasma (ICP)—atomic emission spectroscopy (AES) in the exploration, mining and processing of materials.* Plenary Lecture, Int Symp on Anal Chem in the Exploration, Mining and Processing of Materials, Johannesburg, R.S.A., August 24, 1976.
10. Kirkbright, G. F., Ward, A. F., and West, T. S.: The determination of sulphur and phosphorous by atomic emission spectrometry with an induction coupled high frequency plasma source. *Anal Chem Acta, 62*:241, 1972.
11. Scott, R. H., Fassel, V. A., Kniseley, R. N., and Nixon, D. E.: Inductively-coupled plasma optical emission analytical spectrometry. A compact facility for trace analysis of solutions. *Anal Chem 46*:1, 1974.
12. Larson, G. F., Fassel, V. A., Winge, R. K., and Kniseley, R. N.: Ultra analyses by optical spectroscopy: the stray light problem. *Appl Spectroscopy, 30*:384, 1976.

13. Skogerboe, R. K. and Copeland, T. R.: Microwave plasma emission spectrometry. *Anal Chem, 48*:6111A, 1976.
14. Larson, G. F. and Fassel, V. A.: Comparison of interrelement effects in a microwave single electrode plasma and a radiofrequency inductively coupled plasma. *Anal Chem, 48*:1161, 1976.
15. Nemets, V. M. and Petrov, A. A.: *Spektrosk Tr Sib Soveshch,* 4th ed. N. A. Prilezhaeva, Moscow, Izd. "Nauka," 1969.
16. Dagnall, R. M., West, T. S., and Whitehead, P. W.: Use of the microwave-excited emissive detector for gas chromatography for quantitative measurement of inter-element ratios. *Anal Chem, 44*:2074, 1972.
17. Houpt, P. M. and Compaan, H.: Une nouvelle méthode spectrographique (émission) pour l'identification de traces de matières organiques contenant des halogènes et du mercure dans les fractions obtenues par chromatographique en phase gazeuse. *Analusis, 98*:647, 1973.
18. Talmi, Y. and Andren, A. W.: Determination of selenium in environmental samples using gas chromatography with a microwave emission spectrometric detection system. *Anal Chem, 46*:2122, 1974.
19. Talmi, Y. and Bostick, D. T.: Determination of alkylarsenic acids in pesticide and environmental samples by gas chromatography with a microwave emission spectrometric detection system. *Anal Chem, 47*:2145, 1975.
20. Runser, D. J. and Frank, C. W.: Plasma emission detection of chlorinated pesticides in inert matrixes. *Anal Chem, 48*:514, 1976.

Radioactivation Analysis

Since the discovery of radioactivity by Becquerel in 1896, a symbiotic relationship has existed between analytical chemistry and the progress of nuclear science. Today various types of radioactivation analyses have become established as capable of providing very reliable data but, because of the need to have access to a nuclear reactor or other sources for inducing radioactivity in materials, the methods are not available for general use.

In general terms, radioactivation analysis concerns the bombardment of a sample with nuclear particles, such as neutrons, gamma photons, or alpha particles; as a result, stable isotopes of many elements become radioactive, emit characteristic signals in the form of radioactivity, and (through measurements of the intensity of the emissions) the concentration of an element in a sample may be determined. The method can be very sensitive and is applicable to both single and multielement analyses. For details of the many methods and techniques the reader is referred to the many excellent publications on the subject, for example Smales (1), Bowen and Gibbons (2), Adams and Hoste (3), McKay (4), and Crouthamel, Adams, and Dams (5).

One of the most common sources of nuclear irradiation is provided by nuclear reactors, as illustrated in Figure 75, which provides a source of thermal and fast neutrons. Briefly the core region of a nuclear reactor (or pile) contains a super-critical mass of ^{235}U or ^{239}Pu in which controlled fission takes place and neutrons are emitted. Neutrons emitted from the core region then pass through a moderator, formed of a low-atomic-weight material such as graphite or heavy water, where they become thermalised and have energies of $\approx$ 0.4 ev which are most suitable for use in radioactivation analysis. The nuclear processes in the core region are controlled by use of a series of removable rods composed of a material which strongly absorbs neutrons, such as cadmium. The graphite-moderated pile contains a series

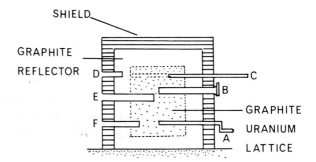

Figure 75: Sketch of a nuclear reactor (pile) illustrating the main features used in neutron activation analysis.
A. Pneumatic tube through which samples can be fired into and then returned to a distant laboratory on site.
B. Sample irradiation hole.
C. Control rod.
D. Thermal neutron beam.
E. Pile neutron beam.
F. Recessed experimental hole.

of holes into which samples, contained in cans, can be placed and subjected to a flux of neutrons. General requirements for irradiations follow:

1. Samples are sealed in capsules which can withstand the neutron flux without deterioration; for fluxes of $\approx 10^8\text{-}10^{12}$ neutrons cm²/s rigid polyethylene is suitable, while for higher fluxes quartz is used. The encapsulating materials should be free of elements which can be activated, or the half-life (see later) should be short so that after removal most of the induced activity will have decayed. The same requirement applies to the can into which the sealed samples are placed; they are normally made of pure aluminium which has a short half-life.
2. For biological samples it is preferable to irradiate in a thermally cool part of the reactor to prevent decomposition and the production of gaseous compounds, which would result in the buildup of an internal pressure and constitute a risk of an explosion when the capsules are opened.
3. Activation analysis is a comparative method of analysis and involves comparing radioactivity induced in a standard with that from a sample. Therefore an irradiation can will contain—
  a. a sample, in a homogenised form such that elements are randomly distributed in relation to the neutron flux;
  b. a standard, which can consist of a material of the same composition as the sample and in which the concentration of elements of interest are known; alternatively an aliquot of an element in a pure form or as a simple compound can also be used;
  c. a neutron flux monitor, in order to determine any variation in neutron flux through the container. This becomes important if a sample contains an element with a high neutron absorption cross section and can result in the shielding of adjacent samples;
  d. if the sample, prior to irradiation, has been chemically processed, an adequate number of "blanks" are included to determine levels of contaminants.

Today neutron fluxes from 10^8 to 10^{14} neutrons/cm²s are readily available and irradiation times range from about one minute to several days, depending upon the type of activation reaction required. Apart from thermal neutron reactors, other sources of neutrons include the following:

1. Neutrons produced from mixtures of light elements bombarded by alpha particles, for example radium-beryllium sources, can provide fluxes of $\approx 1 \times 10^7$ neutron sec^{-1} Curie^{-1}. Radionuclides which undergo spontaneous fission, for example californium 252, can provide fluxes of $\approx 3 \times 10^9$ neutrons sec^{-1} mg^{-1} element. Both types of sources are portable and can be contained within shielded containers to reduce radiation exposure; the former has limited use in activation analysis of biological material because of low output of neutrons but the latter is of interest particularly when more material becomes available; neutron moderation is usually performed by means of a paraffin or water shield.
2. High energetic neutrons can be produced by various nuclear reactions which have analytical applications; for example 14 Mev neutron generators are commercially available and provide an output of $\approx 2 \times 10^{11}$ neutrons sec^{-1} and is particularly suitable for short-lived radionuclides.
3. Some nuclear reactors are particularly suited to nonneutron activation; electron accelerators or betatrons, often available in hospitals, can provide high fluxes of photons capable of much higher penetration of matter and are particularly suited for the analysis of light and some heavier elements, such as lead, which cannot be easily determined in biological materials by neutron actuation.
4. In the analysis of surfaces, or thin samples, charged particles produced in cyclotrons have several analytical applications but difficulties arise because of inhomogeneities in samples, surface contamination, and loss of incident energy with depth of penetration into a sample.

When a nucleus, in the ground state, acquires energy in the

form of excitation energy, any excess is emitted in the form of electromagnetic radiations of high energy such as gamma radiation (γ), electrons (β^-), positrons (β^+), alpha particles (α), or X-rays; energy is emitted in the form of spectrum bands rather than sharp lines, hence the need to use techniques capable of high resolution of similar energies.

Using thermal neutron irradiation as an example of activation analysis, a sample containing the stable elements is irradiated in a flux of neutrons, and selected interactions take place between the bombarding particle and the nuclei of atoms in the sample; for example, in the n, γ reaction the target nuclei absorb a neutron and give off a gamma ray. As a result of the activation of the stable nuclide by neutron bombardment, the amount of activated product formed during the irradiation is directly proportional to the amount of the parent isotope; hence measurement of the induced activity of the product provides a measure of the total concentration of the parent element; the method does not discriminate between elements present in different chemical forms.

The basic equation for neutron activation analysis is as follows:

$$W = \frac{AM}{\sigma f \phi (1 - e^{-\lambda t}) \times 6.02 \times 10^{23}}$$

where W = the weight of the element irradiated in grams
A = the induced activity present in the sample at the end of the irradiation expressed in disintergrations/sec
ϕ = the flux of neutrons which have passed through the sample, expressed as neutrons/cm^2/sec, and incident onto the target
σ = the activation cross section for a specific nuclear reaction (cm^2)
f = the fractional abundance of the isotope of the element to be determined
M = the atomic weight of that element
λ = the decay constant of the induced radionuclide (sec^{-1})
t = the total time taken for the irradiation (sec)

The most convenient technique of neutron activation analysis involves simultaneous irradiation of sample and standard, and element concentrations are determined according to the following relationship:

$$\frac{\text{weight of element in unknown}}{\text{weight of element in standard}} = \frac{\text{activity of element in unknown}}{\text{activity of element in standard}}$$

Therefore:

$$\text{weight of element in sample} = \frac{(\text{activity in sample}) \times (\text{weight of element in standard})}{(\text{activity in standard})}$$

Sensitivity of detection can be improved by increasing the neutron flux or the length of irradiation, provided that a saturation level of induced activity is not reached when the rate of production of activity is balanced by the same rate of loss by radio-decay processes. In relation to the rate of production of radionuclides by thermal neutron irradiation, the following features are of importance: scattering, reflecting the extent to which the incident neutron beam is scattered when impinging upon or passing through a target; absorption, a measure of the extent to which particles are lost when passing through a target; activation cross section, responsible for the yield of a product in a target; and the total cross section of a reaction, which is the sum of effects for all processes occurring as an incident beam passes through a target. Following irradiation, the quantity of interest to be measured is the activity induced in the target which is characteristic of the element of interest; examples of some typical sensitivities are given in Table LVI.

An attribute of radioactivity is that of radioactive decay, which is characteristic for a particular radionuclide. The rate of radioactive decay is determined in terms of the concept of the half-life for decay, which is the time required for the activity of a radionuclide to decrease by a factor of 0.5; radioactive decay may be expressed as follows:

$$A = A_0 \, e^{-\lambda t} \tag{1}$$

where A = activity at time t
A_0 = activity at time t = 0

t = time expressed in seconds
λ = decay constant for a radioisotope

Therefore the time (T) required for the activity to decrease by 0.5 of its initial value is given by equation (2):

$$T = \frac{\log_e 0.5}{\lambda} = \frac{0.693}{\lambda} \tag{2}$$

Equation (1) may then be rewritten as follows:

$$A = A_0 \, e^{-0.693 \, t/T} \tag{3}$$

Therefore,

$$\log_e \frac{A}{A_0} = -0.693 \, \frac{t}{T}$$

A graphical plot of $\log_e \frac{A}{A_0}$ against time will, for a radionuclide associated with a single decay product, be an expression of the half-life for radioactive decay, and the graph will have the form of a straight line as illustrated in Figure 76. The rate of radioactive decay is used to identify particular radionuclides

TABLE LVI
SENSITIVITY FOR ELEMENTS BY NEUTRON ACTIVATION ANALYSIS*

Sensitivity g	Elements
$10^{-13} - 10^{-12}$	Dy, Eu
$10^{-12} - 10^{-11}$	Au, Mn
$10^{-11} - 10^{-10}$	Hf, La, Sm, V
$10^{-10} - 10^{-9}$	Ag, Al, As, Ba, Co, Cu, Er, Hg, K, Na, Pr, Sb, Sc, U, W, Yb
$10^{-9} - 10^{-8}$	Cd, Ce, Cs, Ge, Mo, Nd, Pt, Sr, Ta, Th
$10^{-8} - 10^{-7}$	Bi, Ca, Cr, Mg, Ni, Rb, Se, Te, Ti, Tl, Zn, Zr
$10^{-7} - 10^{-6}$	Pb
Less than 10^{-6}	Cl, Fe, Nb, S, Si, Sn

* This data can only be considered as a general guide to sensitivities which depend upon irradiation time, neutron flux, and type of detector system used. Because biological materials consist predominantly of the light elements, problems of self-shielding and gamma-ray attenuation do not arise. However, because of the high activity induced in major constituents such as Na, Mg, Al, Cl, K, and Br, it is usually essential to allow short-lived products (Mg, Al) to decay and to increasingly rely upon instrumental methods with high energy resolution in order to resolve adjacent energies. In spite of these problems, it is usually practical to determine about twenty-five elements by this method (including As, Ca, Cr, Cd, Co, Cu, Fe, Hg, K, Mg, Mn, Na, Se, Sb, Zn, and V) while other important biological elements such as P and F require special techniques.

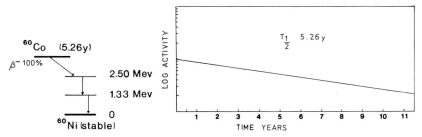

Figure 76: Radioactive decay plot and decay scheme for the radionuclide ^{60}Co.

which can also be confirmed by measurement of the energy of the emission.

A single element consisting of several isotopes will give rise to a complex decay system as illustrated in the case of barium in Figure 77; using graphical techniques individual components of

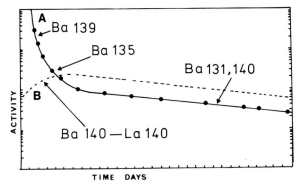

Figure 77A: The composite radioactive decay plot for natural barium radionuclides and uranium fission product ^{140}Ba. In the determination of the uranium content of bone, it is possible to select one of several fission products as a measure of uranium levels. However, when selecting barium, natural bone also contains stable isotopes of barium, which also gives rise to a number of barium radionuclides; hence, in the separation of radiochemically pure barium, the isotopes originating from the fission of uranium are contaminated by other barium radionuclides. The initial part of the curve arises because of the presence of several barium nuclides with different half-lives. When only two decay components are present they can be separated graphically, as illustrated in Figure 77B.

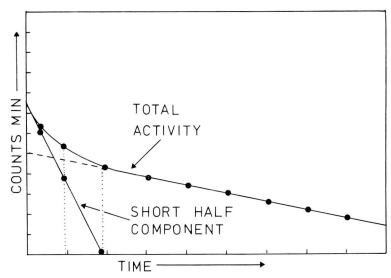

Figure 77B: The decay scheme for barium separated from pure uranium; the initial part of the curve illustrates a growth feature due to the production of the daughter product ^{140}La from the parent ^{140}Ba, followed by a straight line decay as a result of a state of equilibrium between daughter and parent, i.e. the short-lived daughter product (half-life 40h) is being produced at the same rate that it decays; hence, the slope of the decay line is determined by that of the longer lived parent which decays with a half-life of 12.8 days. In order to determine the uranium content of a sample containing natural barium, the separated barium is left for about five days and then the ^{140}La which has reached maximum activity is separated as a measure of uranium content.

the decay curve can be identified particularly when formed of a short- and long-lived component.

Radiation is detected through its ability to ionize matter. Detectors produce an electrical signal which may originate from the production of a voltage pulse (counters), a light pulse (scintillation detectors), or as a result of electron transfer processes across junctions (semiconductor detectors); some commonly used detectors are illustrated in Figure 78. Most detectors have a very high efficiency of response arising from the passage of ionizing radiation but not all the radiation will be detected; for example radiation is emitted from a radionuclide in all directions, hence

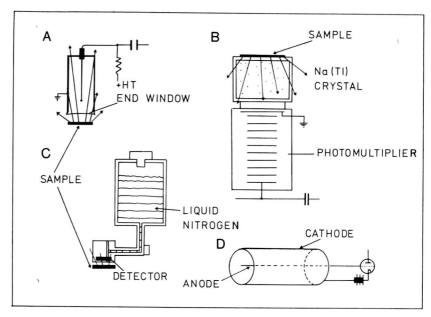

Figure 78: Some typical detectors used for radioassay.

A. Geiger-Müller counter. The detector is filled with argon at reduced pressure and is operated at high voltages producing constant and large pulses irrespective of the extent of the primary ionization.

B. Thallium activated sodium-iodide detector crystal. The absorbed incident radiation is converted to light pulses, which are detected by a photomultiplier tube. The crystal is hydroscopic and is therefore sealed in an aluminium can. Efficiency for energy conversion is about 12 percent, and the crystal has a resolution of about 5-10 percent. In order to improve the detection of highly penetrating gamma activities, large crystals are required. For most analytical purposes a 3 × 3 inch crystal is sufficient.

C. Semiconductor solid state detector. Charged particles lose energy by interactions with the detector through processes involving lifting electrons from valence bands. In order to enhance the absorption of photoelectrons of gamma radiation, the detectors are made of high Z materials, such as germanium, into which lithium is diffused. The lithium occupies interstitial positions in a solid solution of germanium and acts as donors. The detector is kept at liquid nitrogen temperatures in order to reduce diffusion of the lithium.

D. A proportional counter which can either be filled with a radioactive gas or by use of a window of low Z material in order to allow the entry of radiation with the minimum of absorption; a typical gas used in these counters is 90 percent argon and 10 percent methane. The cathode is usually made of carbon; in the case of very weak energies, materials can be coated onto the inside wall of the cathode thus overcoming problems of absorption.

a detector consisting of a flat surface can only receive, in the ideal case, half of the radiation, i.e. 2π. The efficiency of counting is therefore a question of geometry, absorbing properties of detector materials, and the rate of loss of ionizing radiation in passage through the detector. All detectors will respond to natural background radioactivity but this can be reduced by using anticoincident shields which basically consist of an additional detector surrounding the sample detector; external signals passing first through the shield and then registered by the sample detector are rejected by electronic circuits. Some nuclear decay processes are associated with several particles, for example the simultaneous emission of a β particle and a γ ray; coincidence counters can be used which will only respond to the coincident registering of a β and γ signal, hence particular nuclides can be identified in a mixed source. Today most measurements are made using scintillation (sodium iodide crystals activated with thallium) crystals or lithium drifted semi-conductor detectors.

METHODS OF ANALYSIS

Radiochemical Separation

The simplest technique involves the separation of a radionuclide free from any interference from other nuclides followed by counting the radioactivity and confirming radiopurity by determining the rate of radioactive decay. For such measurements, all that is required is a simple Geiger counter which responds to ionization but without energy discrimination, although absorbers can be placed between the sample and detector to remove some types of radiation. With the rapid development of instrumental methods of analysis, this simple technique has tended to be ignored, yet is capable of providing very accurate data. If the radionuclide of interest has a short half-life, then chemical processing has to be carried out quickly thus restricting the technique to those who have access to an irradiation facility on site; this is further required, since soon after irradiation the total radioactivity of the sample will be high and adequate facilities are essential to reduce ionizing radiation exposure to which the worker is subjected. While some separation schemes may take several hours to perform, others can be carried out in minutes; for details of selected radiochemical procedures the reader is referred to the National Academy of Sciences, National Research Council, Nuclear Science Series, United States, which has prepared a series of short monographs of selected procedures, for example *The Radiochemistry of Copper* (Dyer and Leddicotte, 6). Following an irradiation the element to be determined will be mainly nonradioactive and only a small number of atoms will be radioactive; the stable element now has an "inbuilt" radiotracer, and hence, in the instance of a trace element, if a known mass (for example a few milligrams) of the element of interest is added to solutions of samples and standards, they may be processed by conventional chemical techniques to isolate the desired element in a defined chemical form which is then weighed to determine the overall chemical yield of processing. The only requirement is that both carrier and element are present in the same chemical form so that isotopic exchange is assured. The presence of a large mass of stable element, called a carrier, is usually of no concern to the final radioassay which will only detect radioactivity species; quantitative separations are not required as any loss can be corrected for from the measurement of recovery; further, additions of stable elements do not result in contamination as they will not be active and therefore will not be detected. In radiochemical processing two other types of processes are commonly used:

1. Holdback carriers—large amounts (mgms.) of an element added to the active solutions in order to reduce, by dilution, the separation of the element to be determined during some types of chemical processing.
2. Scavengers—addition of elements by which (through processes of precipitation) unwanted elements and associated radionuclides may be removed, for example the addition of iron followed by hydroxide precipitation to remove elements such as Fe, Al, Sc, Mn, Cu, Zn, Pb, Ti, and leave the alkaline metals in solution.

Providing that the radioactive half-life of the element of interest is sufficiently long, radiochemical separations can be carried out in laboratories distant from a source of irradiation, and for many purposes quite simple and inexpensive counting techniques, such as Geiger-Muller counters, are quite satisfactory for quantitative analyses.

Instrumental Methods of Analysis

In these techniques emphasis is placed upon an ability to identify radionuclides on the basis of distinctive energies by use of high resolution detectors coupled to multichannel energy discriminating spectrometers. The two detectors in common use are NaI (Tl) scintillation crystals and Ge (Li) semiconductor detectors; the latter is capable of very high resolution as illustrated in Figure 79, and multielement analysis can often be performed after irradiation without any separation chemistry.

Adequate energy resolution is only part of the problem. For most biological samples following irradiation the intense activity arising from major elements, such as Cl, Na, Mg, K, Br, and I, is often so high that detectors become saturated and emissions from other elements cannot be identified. One solution is to leave the sample until most of the short-lived intense activity has decayed; this is acceptable provided that the nuclides of interest do not themselves have short half-lives, in which case they will have decayed before a measurement can be made. The energies normally determined in activation analysis are gamma rays which range from about 50 to 3,500 kev. The resolution of a germanium detector is between 1 and 4 kev; hence if a multichannel gamma spectrometer is used with a pulse height spectrum, covering about 0 to 4,000 kev, some four hundred photopeaks could, in theory, be resolved from one another but, because of the complexity of the energy spectrum, overlap of energies occurs. In practice the total number of energy peaks for which measurements are possible is between fifty and about one hundred; energy overlap is reduced for energies above about 500 kev. Often there is a choice of several gamma rays associated with an element and one free from overlap can be selected while final verification of purity can be determined by measure-

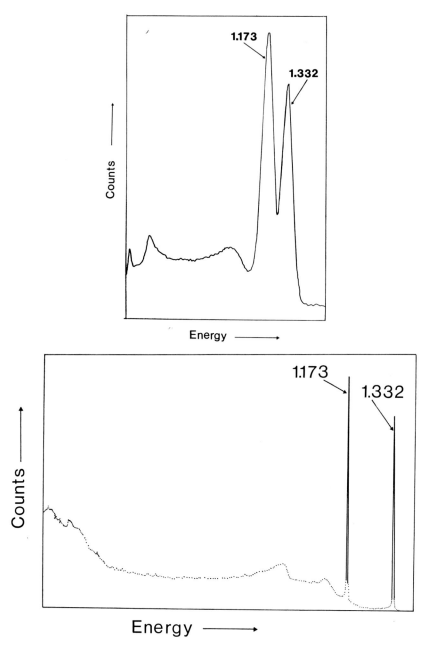

Figure 79: A comparison of the resolving power of a sodium iodide crystal (top) and a germanium detector (lower) using ^{60}Co. The Ge(Li) detector has a very high resolving power, and therefore, in instrumental neutron activation a very large number of energies arising from different radionuclides can be resolved; hence multielement analysis is possible. In the case of the Na(Tl) crystal, inferior resolution results in severe overlap of adjacent energies.

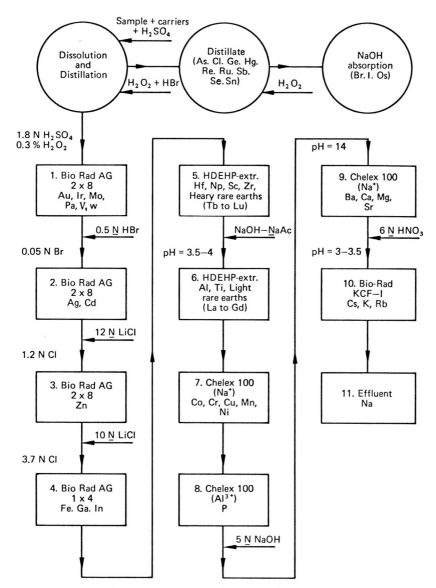

Figure 80: An illustration of a complex radiochemical separation scheme used in neutron activation analysis for multielement assay, which can be used in fully automated separation systems. See Figure 82. Figure reproduced by permission of the AB-Atomenergi, Sweden (based upon the method of Samsahl, K.: High speed, automatic radiochemical separations for activation analysis in the biological and medical research laboratory, *Sci Total Environ*, 1:65, 1972).

ment of the radioactive half-life. In the analysis of biological materials, each type of tissue will present different problems and in practice a combination of techniques is advocated, selection being determined by the particular demands of the sample. Many elegant techniques exist to overcome particular problems characteristic of biological materials; for example in order to overcome the problem of the intense activity arising from sodium after activation, it is possible to remove most of the radioactive sodium by precipitation with butanol and hydrochloric acid or to pass the sample solution through a hydrated antimony pentoxide column, described by Girardi and Sabbioni (7), which selectively retains sodium ions together with tantalum. Chromatographic methods have a wide application and are widely used in various chemical separation techniques; of particular interest are the many ion-exchange techniques, many of which can be automated for the separation of individual elements and groups

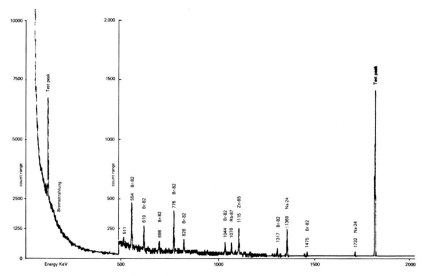

Figure 81: Spectrum of irradiated NBS-Liver sample before chemical separation. Sample size: 25 mg. Irradiation: Neutron flux 7×10^{-13} n cm^{-2} sec^{-1}. Cooling time: 10D. Counting time: 200 sec. Detector: 90 cm^3 Ge(Li) well-type. (A well detector has the form of a cylinder along the vertical axis of which a hole is present into which samples are placed thus approaching 4π geometry.) Reproduced by permission of G. V. Iyengar.

of elements. A general scheme for radiochemical processing described by Iyengar and Samsahl (8) is given in Figure 80 and an illustration of the separation of Cd, Au, Mo, Ag, and W in N B S Bovine Liver is illustrated in Figure 81. For the optimum analysis of biological materials it is essential to remove, as quickly as possible, those unwanted elements which are easily activated; if they remain within the sample the gamma ray emission interacts with the detector to give rise to scattering phenomena which produces a whole sequence of energies less than the maximum energy, hence obscuring the presence of other elements; sodium will also give rise to a characteristic signal from the ^{24}Na 1368 kev and on the lower energy side of the spectrum, a continuum which usually obscures most other energies.

Samsahl, Wester, and Landström (9) have developed sophisticated automated methods of analysis; biological samples are digested by reaction with nitric acid vapour followed by complete

Figure 82A: General view of apparatus.

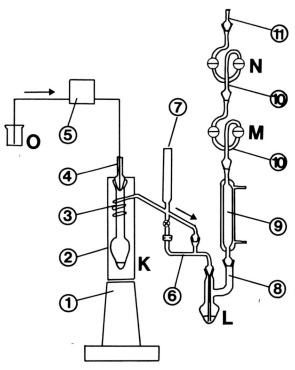

Figure 82B: The distillation unit.

Item No.	Description
1	Electroburner, Horo, 500 W
2	Chimney
3	Sample flask
4	Reagent inlet tube
5	Pump, Ismatec Mini-Micro-2
6	Connecting tube
7	Air inlet tube
8	Distillation flask
9	Condenser
10	Absorption, tube
11	Suction connection

mineralisation in Teflon bombs at 150-160°C under a pressure of 3-12 kg/cm². Ion exchange separations are performed with an automatic ion-exchange separator, now commercially available from AB Atomenergi, Studsvik, Sweden; a scheme for assay of biological materials is illustrated in Figure 82; values for the recovery of many elements with this system are given in Table LVII. An excellent review of selective ion media is obtainable from Carlo-Erba (10) and includes both organic and inorganic materials. Advantages of an automated system are that (a) it overcomes problems of radiation dose to workers, (b) very complex chemical separations can be carried out, (c) manpower requirements are small, and (d) separations can be carried out over long periods of time in a reproducible manner.

On occasions, particular nuclear processes can be used for the

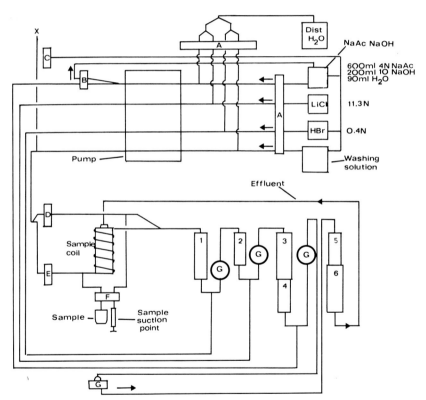

Figure 82C: Flow diagram for SA-7100
A,B,G = Double valves
C,D,E,F = Single valves
1-6 = Ion-exchange columns

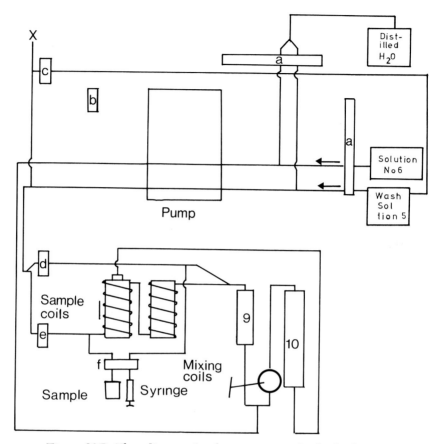

Figure 82D: Flow diagram for the separation of volatile elements.
  A,B,C Double valves
  C,D,E,F = Single valves
  G = Mixing coils
  9, 10 = Ion-exchange columns

Figure 82: The Automatic Ion Exchange Apparatus SA-7100 designed by Dr. K. Samsahl (see Figure 80) and commercially available from Studsvik AB Atomenergi, Nyköping, Sweden.

The Automatic Ion Exchange Apparatus SA-7100 developed by Dr. K. Samsahl gives a separation of thirty elements in approximately ten minutes on eight ion exchange columns. The reproducibility is very good, and the apparatus may also be used for elements with comparatively low yields with good accuracy due to the automatic nature of the process.

Instrumental Methods

Three simultaneous runs are possible with a 30-element set-up of the apparatus. To analyse only a few elements the number of ion exchange columns in the set-up may easily be reduced.

The apparatus may be used for many different types of samples provided that it is possible to dissolve them in a suitable way. The dissolution procedure is mainly intended for organic samples and is performed in five to ten minute depending on the sample composition. For inorganic samples, particular dissolution procedures are necessary.

APPARATUS

A general view of the SA-7100 is shown in the photograph. The lower cabinet contains the ion exchange system and the top cabinet contains a peristaltic pump and the solutions necessary to perform the separation. The construction materials are mainly PVC and other plastics due to the corrosive effects of the solutions.

The lower cabinet has 15 grooves for ion exchange columns. The heights of these have been chosen to take ion exchangers of different lengths. The ion exchangers consist of polythene tubes with the resins between two plastic filters. Tightness at both ends is ensured by stopper with O-rings. According to the flow diagram, the unit has sample coils for collecting the sample before separation and mixing coils to ensure efficient mixing of sample and added reagent solutions. The unit is also equipped with four valves, D, E, F, and G, and a sample inlet. Sample coils, valves, and sample inlet are designed to take six samples at a time.

Space is reserved inside the cabinet for lead bricks to shield the operator from radiation from the final eluate, which contains rather large amounts of ^{24}Na and ^{32}P. To achieve both convenient access and sufficient protection, in case of leakage, or accidental breakage of tubing or joints due to excess pressure, the apparatus is surrounded by PVC walls and sliding removable Plexiglas doors.

The top cabinet has an Ismatec® peristaltic pump to pass reagent solutions and samples through the ion exchange columns in a predetermined ratio. It also contains storage flasks for reagent solutions and the regulating valves A, B, and C according to the flow diagram. An electric clock for shutting off the pump automatically and a panel for pressure control are located on the front.

An Ismatic® thirteen-channel pump with fixed speed is the standard version of pump. The dimensions of tygon tubings have been selected to suit the time schedule for the separation scheme. For special purposes, e.g. when extremely short-lived nuclides have to be determined, a faster separation may be important. With a thirteen-channel pump having a variable speed it is possible to reduce the separation time to five minutes which is the minimum time prescribed by the capability of the ion exchange resins used. It is, however, not recommended to increase the speed of the solution flow if a complete system of 8 ion exchangers are to be used due to the pressure build-up in the system.

If required, a twenty-five channel pump with fixed or variable speed may be installed to serve two separation cabinets.

The distillation set included in the SA-7100 is used to perform wet incineration of the sample and distillation of volatile elements (As, Sb, Sn, Se, and Hg). The set-up of the distillation unit is according to the sketch. A 500 W electroburner and a small peristaltic pump are necessary for the distillation.

MEASUREMENT

After the separation the columns are removed from the ion exchange system. A gamma spectrum of each element may be measured directly on the column if a 20 mm well-type Na(Tl)I crystal is available.

assay of specific elements. For example, as a result of nuclear fission, following neutron irradiation, a series of delayed neutrons are emitted which can be used for assay of fissionable nuclides; in the case of natural materials analysis can be made for uranium and provides a very simple method which approaches "black box technology." A sample is irradiated for one minute in a fast return tube (rabbit) of a reactor and then pneumatically ejected into a neutron detector. After a lapse of one minute all emitted neutrons are detected (efficiency of detection is about 10%) and then compared with standards processed in an identical manner; the concentration of uranium can be calculated by simply comparing the total neutron counts of standard and sample. For normal biological materials, the technique does not suffer from any problems of interference and provided a suitable sample transfer facility is available at a reactor, the total analysis time for each sample is about four minutes; with a thermal neutron flux of about 5×10^{12} n cm^{-2} sec^{-1}, 1 µg of uranium will give about four hundred counts and a typical background arising from the container in which the sample is irradiated is only about four counts. The technique has been described by Amiel (11) and Hamilton (12), while Hamilton (13) has described the technique for the analysis of uranium in normal human tissues and various types of diet.

Uranium provides a further example of the utilisation of particular properties of radioactivity by measurement of alpha particle emissions. One method, namely alpha-particle autoradiography, concerns the detection of alpha particles emitted from a surface which can be recorded by placing the surface of a sample in contact with a photographic emulsion; the passage of the alpha particle through the emulsion leaves a trail of radiation damage which can be developed, through photographic techniques, to form a track composed of silver grains and can be observed with a conventional light microscope. The technique is restricted to alpha emitters such as U, Th (together with daughters of the natural radioactive series), and the transuranium elements, for example, Pu; apart from assay for total levels of an alpha-emitting nuclide, the technique provides valuable infor-

A. EXAMPLES OF RECOVERY VALUES FOR ION-EXCHANGE PRECONCENTRATION FOR TRACERS ADDED TO BIOLOGICAL SAMPLES

Element	Nuclide Used	Energy Measured MeV	Mean Value of Yield %	Standard Deviation of a Single Value	Standard Error of Mean Value	Number of Determinations
Ag	Ag^{110}	0.66	90	7	2	8
As	As^{76}	0.55	90	6	2	6
Ba	Ba^{131}	0.22	90	10	4	7
Br	Br^{82}	0.77	92	7	2	10
Ca	Ca^{47}	1.31	98	5	2	8
Cd	Cd^{115}	0.52	93	5	2	10
Ce	Ce^{141}	0.14	97	7	2	7
Co	Co^{60}	1.33	99	9	3	11
Cr	Cr^{51}	0.32	97	2	1	7
Cs	Cs^{134}	0.60	96	2	1	6
Cu	Cu^{64}	0.51	98	7	2	10
Fe	Fe^{59}	1.29	91	6	2	7
Ga	Ga^{72}	0.83	90	3	1	7
Hf	Hf^{181}	0.48	92	4	2	6
Hg	Hg^{203}	0.28	96	7	3	8
In	In^{114m}	0.19	70	6	2	9
K	K^{42}	1.53	93	6	2	10
La	La^{140}	1.60	91	7	3	6
Mn	Mn^{56}	0.84	92	2	1	6
Mo	Tc^{99m}	0.14	97	5	2	6
Na	Na^{24}	1.38, 2.76	96	5	2	7
P	P^{32}	Bremsstrahlung	92	5	2	7
Rb	Rb^{86}	1.08	97	4	1	6
Sb	Sb^{124}	0.60	97	2	1	7
Sc	Sc^{46}	1.12	95	5	2	5
Se	Se^{75}	0.26	85	5	2	9
Sm	Sm^{153}	0.10	92	5	2	5
Sr	Sr^{85}	0.51	95	4	2	6
Th	Pa^{233}	0.31	79	14	5	7
U	Np^{239}	0.106	84	5	2	10
W	W^{187}	0.68	90	5	2	6
Zn	Zn^{65}	1.11	100	3	1	11

B. PRECISION OF ASSAY USING AUTOMATED GROUP SEPARATIONS ON BOWENS KALE

Values expressed in counts per 100 mg standard.

As	5.21 · 10³ 5.75 5.35	5.44 · 10³ ± 0.28 · 10¹	Mn	6.30 · 10³ 5.73 5.33	5.79 · 10³ ± 0.48 · 10³
Br	2.73 · 10⁵ 2.82 2.91	2.82 · 10⁵ ± 0.07 · 10⁵	Mo	3.42 · 10³ 3.68 3.64	3.58 · 10³ ± 0.14 · 10³
Ca	3.15 · 10⁴ 2.99 2.65	2.93 · 10⁴ ± 0.26 · 10⁴	Na	7.32 · 10⁴ 7.45 7.50	7.42 · 10⁴ ± 0.09 · 10⁴
Cd	2.36 · 10³ 2.16 2.10	2.21 · 10³ ± 0.14 · 10³	P	1.42 · 10⁵ 1.32 1.35	1.36 · 10⁵ ± 0.05 · 10⁵
Ce	4.40 · 10³ 4.53 4.07	4.33 · 10³ ± 0.24 · 10³	Rb	7.55 · 10³ 7.50 7.69	7.58 · 10³ ± 0.10 · 10³
Co	1.00 · 10⁴ 1.02 1.09	1.04 · 10⁴ ± 0.05 · 10⁴	Sb	5.58 · 10³ 4.36 6.18	5.37 · 10³ ± 0.93 · 10³
Cr	3.48 · 10³ 3.38 3.59	3.48 · 10³ ± 0.10 · 10³	Sc	2.82 · 10³ 3.04 2.92	2.93 · 10³ ± 0.11 · 10³

Cs	3.59 · 10³ 4.08 3.76	3.81 · 10³ ± 0.25 · 10³	Se	7.90 · 10² 9.34 8.96	8.73 · 10² ± 0.75 · 10²
Cu	3.23 · 10³ 3.34 3.34	3.30³ · 10 ± 0.06 · 10³	Sm	3.36 · 10³ 3.04 3.27	3.19 · 10³ ± 0.13 · 10³
Fe	2.32 · 10³ 2.22 2.57	2.37 · 10³ ± 0.18 · 10³	Sr	6.00 · 10⁴ 5.90 5.25	5.72 · 10⁴ ± 0.41 · 10⁴
Ga	5.88 · 10³ 5.38 5.40	5.55 · 10³ ± 0.22 · 10³	Th	1.62 · 10³ 1.61 1.10	1.44 · 10³ ± 0.30 · 10³
Hg	4.31 · 10³ 4.82 4.46	4.53 · 10³ ± 0.26 · 10³	W	2.73 · 10⁴ 2.73 2.75	2.74 · 10⁴ ± 0.01 · 10⁴
K	8.60 · 10³ 9.21 9.54	9.12 · 10³ ± 0.48 · 10³	Zn	1.41 · 10⁴ 1.42 1.41	1.41 · 10⁴ ± 0.01 · 10⁴
La	7.30 · 10³ 7.86 7.28	7.48 · 10³ ± 0.33 · 10³			

* Reprinted with permission from Samsahl, K., Wester, P. O., and Landström, O.: An automated group system for the simultaneous determination of a great number of elements in biological material. *Anal Chem*, 40:184-185, 1968. Copyright by the American Chemical Society.

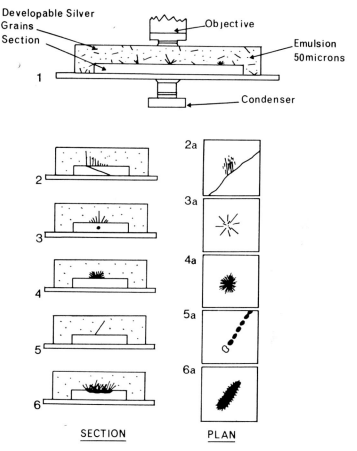

Figure 83: Autoradiography using alpha particle sensitive nuclear emulsions.
1. Diagram illustrating the system of observation using a conventional light microscope and a dry × 40 fluorite objective.
2. Track pictures observed for an active alpha particle deposit along a fracture plane in a solid. As the distance between the source of the particle and the detector increases, because of the passage of the alpha particle through the sample, so the recorded length of the track decreases because of absorption of energy in the sample.
3. Track distributions for a point source buried below the surface of a sample illustrating effects of sample absorption. Measurement of the angle of emergence of the tracks provides information of the depth below the surface of the point source.
4. Track features formed by a point source on the surface of a sample.

mation to describe the distribution of such emitters from surfaces, as illustrated in Figure 83. Alpha-particle autoradiography is most commonly used for studying natural alpha particle emitting materials, and sensitivity is limited by background tracks and a fading of the latent image produced in the emulsion before it is developed. For most practical purposes, exposures of about three months are possible and it is possible to determine levels of uranium at a concentration of about 0.1 ppm.

The use of track phenomena was greatly enhanced by the analytical use of induced tracks from fissionable materials by Fleischer, Price, and Walker (14, 15) and Fleischer et al. (16) for assay of uranium and the transuranium group of elements. Briefly the detector can consist of a simple plastic, such as polycarbonate, cellulose nitrate, glass, or micas. A sample, in the form of a thin section of biological tissue or a powder, is placed in contact with the detector and both are irradiated (together with standards) in a flux of thermal neutrons. As a result of the irradiation, fission is induced in fissionable nuclides (for example uranium and plutonium) and the fission fragment penetrates the detector resulting in a very narrow zone of radiation damage. Following an irradiation, the detector is removed from the sample and placed in a suitable solvent which will dissolve away the radiation damaged zones at a rate faster than solution of the bulk material; for most plastics and cellulose nitrate, sodium hydroxide is used, and hydrofluoric acid is used for glass and micas. As a result of etching, the narrow tracks of radiation damage become enlarged and can be readily viewed using a con-

---

5. Illustration of a single track. 5a illustrates micro-features of a track which consist of a series of separated silver grains.
6. Track features observed for alpha particles emerging from the surface of an opaque body.

In these illustrations the sample section has been covered with a layer of liquid emulsion, thus preserving relationships between origin of tracks and features in the sample. Alternatively, a glass plate coated with an emulsion can be used, but it is then essential to provide some form of position orientation in order to match up the detector with features in the sample; this can be accomplished by marking the surface of the sample with an ink containing an alpha emitter.

ventional optical microscope at low magnification; the enlarged tracks appear black because of total internal reflection of the light scattered along their length. By comparing the total number of tracks produced by a sample with those from a standard,

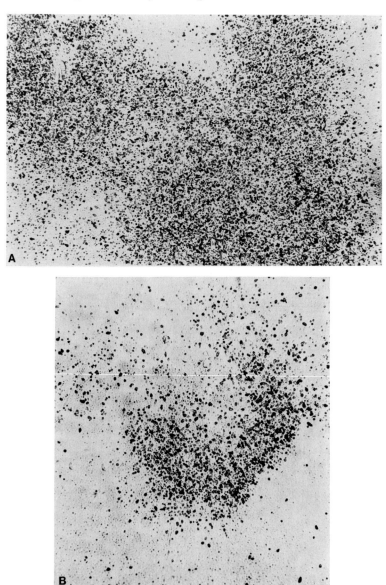

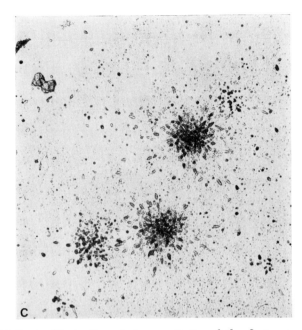

Figure 84A-C: An illustration of the sensitivity of the fission track method for uranium in normal human bone which has been irradiated with about 10^{18} n cm^{-2}. From Hamilton, E. I.: The concentration and distribution of uranium in human skeletal tissue. *Calc Tiss Res*, 7:150, 1971. Courtesy of Springer-Verlag, Heidelberg.

A: High density of tracks associated with cortical bone and low density with trabecular bone (reduced one-half from ×150).

B: Tracks from an embayment of trabecular bone in cortical bone (reduced one-half from ×150).

C: Tracks originating from point sources in bone (reduced one-half from ×150).

quantitative assay for fissile nuclides can be made. Hamilton (17) has described the use of this technique for the assay of uranium in normal human blood while Hamilton (18) describes a further application for determining the concentration and distribution of uranium in human skeletal tissues. Some features observed for the natural distribution of uranium in human bone are illustrated in Figure 84A which clearly shows deposition on endosteal surfaces; in Figure 84D illustrations are presented for fission tracks produced from samples of urine for

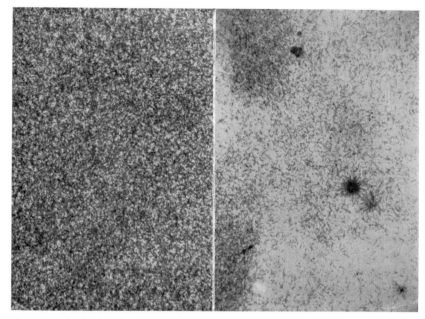

Figure 84D: An application of the fission track method for the determination of fissionable nuclides in human urine. *Left:* Uranium fission tracks from a twenty-four hour sample of urine after chemical separation of uranium. Note the high density of tracks resulting in an almost total "burn-out" of the detector. Length of tracks 20 μm. *Right:* ^{239}Pu fission tracks from a twenty-four hour sample of urine after chemical separation of plutonium. Note the presence of a "hot-spot" approximately 60 μm in diameter caused by the aggregation of plutonium on a particle.

workers exposed to uranium and plutonium. Following the collection of a twenty-four hour urine sample, water was removed by freeze drying and uranium and plutonium separated by chemical techniques followed by evaporating aliquots of the solutions onto polycarbonate detectors which were then irradiated together with standards. Figure 84D illustrates a very dense population of tracks from uranium and plutonium; in both cases the workers were exposed to a variety of different salts of both elements, although oxide and nitrate forms were mainly used. This technique has many applications in the nuclear industry: preliminary studies using room air filters, for determining the mass of fissile material suspended in working environments, have been compared with filters placed near the nose, with nasal wipes and samples of sputum to investigate the extent to which such nuclides enter man; when solid particles of fissile material are present they are easily recognised by the appearance of "hotspots," the diameter of which is related to particle size.

Further examples of the fission track technique are illustrated in Figure 85 for the distribution of ^{239}Pu in the femur of a six-week-old rabbit which had been injected with 1.29 μCi of ^{239}Pu. Because of its very low content of uranium, polycarbonate is the detector of choice for fissile nuclides and the fission tracks are easily recognized. However, this detector will also record damage caused by alpha particles together with a variety of etch features related to nuclear processes, often caused by the small, fast component of the largely thermal neutron flux used for irradiations. By superimposing the etched detector onto the original sample, it is possible to identify the origin of the tracks in relation to biological features. This type of study can also make use of etchable features directly related to the biological matrix and which are recorded in the detector; thus, in some circumstances the record preserved in the detector is sufficient to recognise biological features such as cell walls. Hamilton (18) has discussed some advantages in tissue recognition using etch-pit features while Green, Howell, and Thorne (19) have used this technique for the accurate localisation of ^{239}Pu in bone.

A further interesting application of the track technique is the production of tracks from alpha particles in cellulose nitrate without using neutron irradiation. In this technique the sample is placed in contact with a sheet, or film, of cellulose nitrate and then exposed for a suitable period of time in a box filled with aged dry nitrogen and surrounded by about 4 cm of lead to reduce contributions from the cosmic flux. Figure 86A illustrates tracks produced in sheet cellulose nitrate from ^{210}Po, while in Figure 86B the tracks are produced by irradiation of the detector with ^{241}Am, but in this instance thin films of detector were prepared on glass slides by evaporation of the cellulose nitrate dissolved in suitable solvents containing plasticisers. An advantage of this technique is that the thickness of the detector can

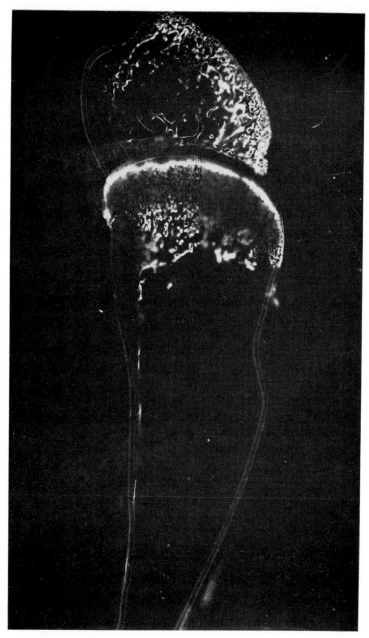

Figure 85: Figures illustrating the high resolution obtained by the fission track method for the femur of a rabbit; animal injected with ^{239}Pu. A: Contact print of femur. The bone was set in Araldite®, sectioned, and then polished. Detector, sheet polycarbonate 0.5 mm thick, Mag ×7. Using very low magnification the distribution of tracks in the trabecular bone can be easily seen, in fact the general distribution is visible with the naked eye. The highest density of tracks is associated with the surfaces of trabecular bone.

Instrumental Methods 299

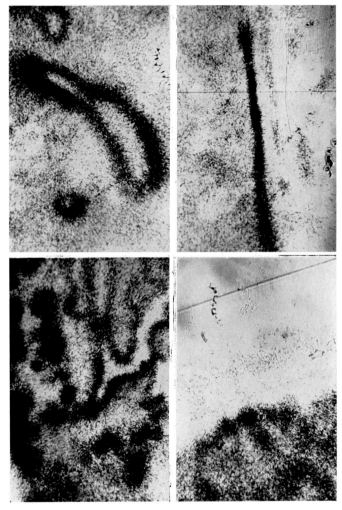

Figure 85B: Top left: Labelling of Haversian systems. Top right: Tracks associated with inner linings of trabecular bone. Bottom left: Heavy label associated with the inner wall of shaft cortical bone and a lower density of tracks associated with general cortical bone. Bottom right: Articular surface of femur head illustrating trabecular label and very low track density associated with the covering layer of cartilage. Track length about 10 microns.

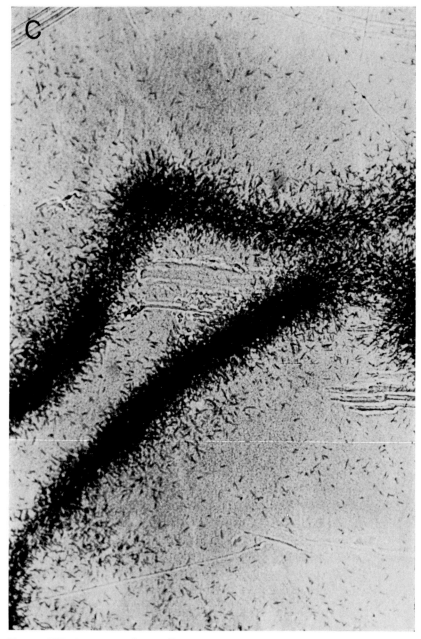

Figure 85C: Dense track population associated with the trabecular surfaces. Track length about 10 microns.

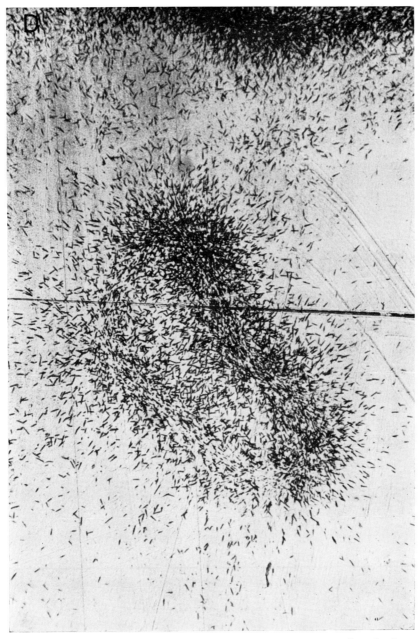

Figure 85D: Tracks associated with a Haversian system. Track length about 10 microns.

be controlled and any track features caused by background activity, or those caused by ageing, are absent at the start of the exposure. As the maximum energy loss of an alpha particle occurs toward the end of the track length, the etchable tracks are produced at various depths in the detector depending upon their energy. In the case of ^{241}Am, tracks are produced just below the surface of the detector and fully developed tracks have dimensions of 4×1.5 μm. The dual use of cellulose nitrate for detecting both fission events and alpha particles is illustrated in Figure 86C using a mixed ^{250}Cf–^{252}Cf source. The fission tracks are present from the surface of the detector and are approximately 20×2 μm in size; Figure 86C illustrates the fission tracks,

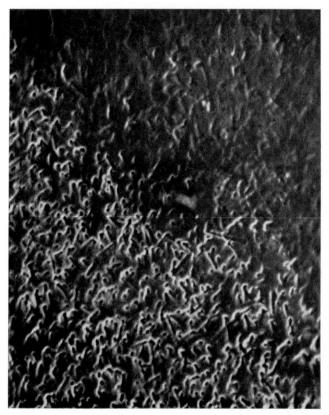

Figure 86A: Alpha tracks (4×1.5 μm) from ^{210}Po.

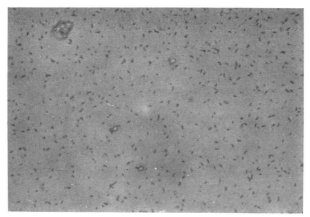

Figure 86B: Alpha tracks from a thin electrodeposited sample of ^{241}Am.

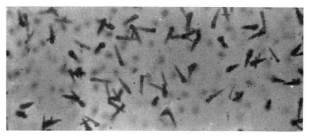

Figure 86C: Fission tracks produced from a californium source (20 × 2 μm) produced on the surface of the detector. The larger diameter of the track occurs at the surface as it has been subjected to maximum etching.

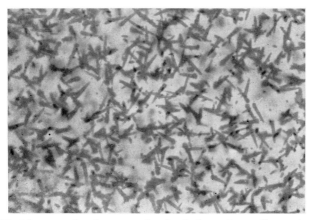

Figure 86D: Photograph taken at a depth of about 2 μm illustrating the first production of alpha particle tracks as small dark "spots."

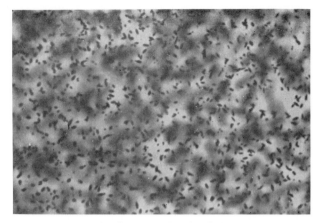

Figure 86E: Same as D but taken at a depth of about 10 μm illustrating the detection of well-formed alpha particles (5 × 3 μm) with ghost pictures of the surface fission tracks still visible.

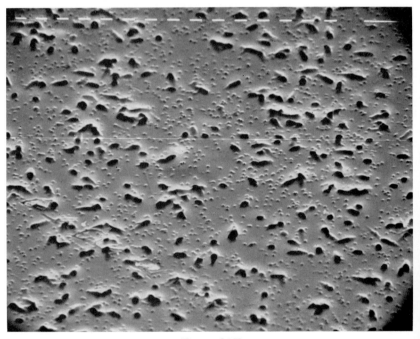

Figure 86F

Instrumental Methods

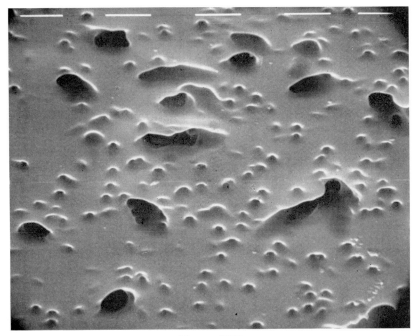

Figure 86G

Figure 86F, G: The detection of fission tracks and alpha particles in cellulose nitrate.

which often have side protuberances presumably arising from various nuclear processes associated with fission. Figure 86D is of track features produced about 2 μm below the surface of the detector, although the faint outline of the fission tracks can still be clearly seen, together with a new etch feature represented by small dark spots which have been formed by some alpha particles; by focussing deeper into the detector at a depth of about 10 μm, a new population of tracks becomes evident (see Fig. 86E) formed by alpha particles associated with alpha decay of the californium source and have dimensions of about 5 × 3 μm. The possibility is currently being investigated of using the depth at which alpha tracks are produced, together with the diameter or volume of the track, as a measure of alpha particle energy, hence, an application of such features to identify the energy of alpha particles and the nuclide from which they have originated, thus providing a potentially simple, and cheap, technique for determining the abundance of various alpha-emitting nuclides in biological and environmental materials. Apart from simplicity, and the lack of any sophisticated electronic equipment associated with conventional alpha particle spectrometers, track techniques have one particular advantage, namely that long exposures, for example six months or one year, may be used which are not possible with conventional electronic instrumentation because of instability of electronics with time and problems associated with background activity. The levels of alpha particle activity in many biological and environmental samples are very low and with conventional techniques it becomes essential to pool large amounts of material in order to obtain sufficient activity for measurement. Ignoring the possibility of spectroscopy by this technique, sensitivity of detection for the assay of alpha particle activity can be enhanced by dispersing alpha emitters, or chemically separated species, in liquid cellulose nitrate and then casting thin films which can be readily etched; a possible advance in technique is the use of pressure etching through thick films, although at present information is lacking concerning the nature of etch processes in solid systems.

Apart from naturally occurring alpha emitters of the uranium and thorium natural decay series, alpha activity induced by neutron irradiation of biological material can also be used for the assay of boron and lithium; in a further application Hamilton (20) bombarded samples of animal bone with 30 Mev alpha particles produced in a cyclotron and through the $(\alpha, 2n)$ reaction on isotopes of lead to form ^{210}Po, the alpha particles are observed in cellulose nitrate after etching in 6.2 N NaOH for two minutes at 55°C. The threshold of specific ionization for the creation of etchable tracks in cellulose nitrate is about 3 Mev; hence in order to register the 5.3 Mev alpha particle from ^{210}Po, an aluminium absorber or an air gap between the sample and detector is essential to record etchable tracks at the surface of the detector. The validity of this technique was confirmed by irradiating similar samples in an electron linear accelerator providing 30 Mev electrons and through the reaction ^{204}Pb $(\gamma, n) \rightarrow$ ^{203}Pb the 0.28 Mev ^{203}Pb activity, with a half-life of fifty-two hours, was detected using a 3 by 3 inch NaI (Tl) crystal; an activity of 85 cpm μg^{-1}Pb was obtained.

Autoradiographic techniques can also be used for γ and β emitters using appropriate fine grain photographic emulsions such as those used in X-ray radiography (Kodak No Screen X-ray Film). In studying the distribution of elements, considerable information can be obtained from animals injected with radionuclides in different chemical form. An example is given in Figure 87D for determining the distribution of cesium in the rat following I.V. injection while in Figure 87E an illustration is provided for ^{147}Pm. In this technique, a whole rat is frozen and total body thin sections are taken with a sledge microtome and then exposed to X-ray photographic paper; after development the precise distribution of the radionuclide can be examined and the relative concentrations in different parts of the body can be determined by microdensitometry or by removing parts of the tissue and counting the radioactivity. In furthering our knowledge of the metabolic pathways of the chemical elements, such techniques compliment stable element investigations.

Activation analysis can also be applied to in vivo studies of man (see Anderson et al., 21; Boddy et al., 22) by activation of the whole body with 14 Mev neutrons, followed by whole body counting of induced activities. Using such techniques, it is possible to determine total and exchangeable sodium in the body and to investigate total body calcium and features associated with changes in body calcium metabolism, for example osteoporosis, hormone calciton, bone wasting diseases, and calcium balance during long term hemeodialysis; other elements which may be studied by whole body activation are chlorine, aluminium (from activation of phosphorous), and nitrogen.

It is evident that activation analysis and related techniques offer tremendous advantages over many other methods of analysis for the assay of the total mass of an element present in a sample. In instrumental neutron activation analysis of biological samples, possibly one of the most important features is the reduction of extraneous contamination; for example, in the simplest case all that is required is the removal of a wet sample of tissue, encapsulation in an irradiation vial, followed by irradiation, after which contamination by stable elements, even those to be determined in the samples, is of no consequence. Tech-

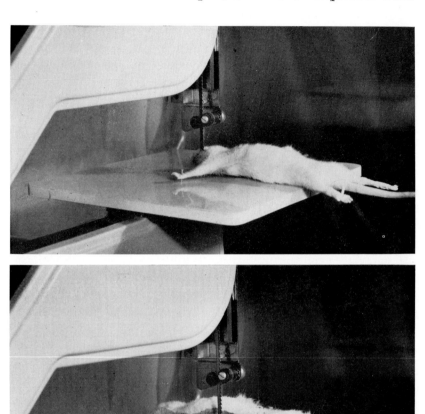

Figure 87: An illustration of techniques for the preparation of whole body sections of the rat for autoradiography. The rat is (A.) gradually frozen in liquid nitrogen and (B.) cut along the central line using a band-saw.

Instrumental Methods

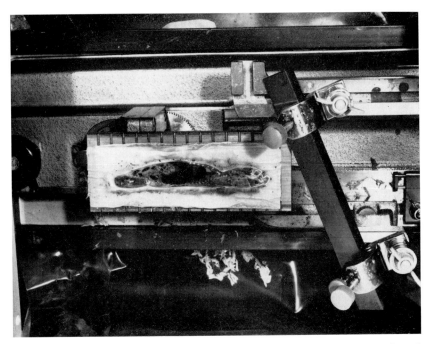

Figure 87C: A whole body section about 0.5 cm thick is removed and frozen onto a metal slab in a sledge microtome with tissue paper soaked in water; the microtome is kept in a standard chest type deep freeze. The surface layers are skimmed off to remove saw cuts and at this stage detailed tissue structures can be easily seen in almost natural colours.

Figure 87D: Approximately fifteen micron sections are cut and removed on Scotch® tape, freeze dried, and then exposed to an appropriate photographic paper detector. In this illustration the animal had been injected with ^{137}Cs and exposure made with Kodirex No Screen X-ray Paper. The white areas represent sites of labeling. (See Ullberg (23) for development of technique.)

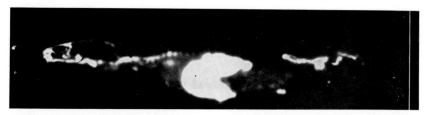

Figure 87E: The distribution of ^{147}Pm in sagittal sections of the rat after I.V. injection. ^{147}Pm localized in skull, vertebrae, caecum and large intestine.

niques available today are very sensitive and multielement analysis is possible for about thirty to forty elements, although chemical processing to some degree is usually required. There is now available a wide range of instruments and associated detectors, which are complimentary to one another; most of the published work on biological materials is concerned with thermal neutron activation which is not suitable for assay of many important light elements, such as calcium, sulphur, magnesium, silicon, and phosphorous, and for some heavier elements such as lead. Increasingly, other techniques, such as energy dispersive X-ray spectrometry and atomic absorption spectrometry, are being used for these elements but are often standardised by measurements obtained by particular activation techniques.

Neutron activation has provided very large numbers of analyses for very rare elements, in both the natural environment and man. Some, such as scandium and indium, have been analysed simply because the nuclear properties of these elements are very amenable for activation and very small masses can be determined accurately. In the case of scandium, there is cause to look at the published data very carefully, as it is rare for this element to become chemically separated from the more abundant element aluminium; hence if scandium levels in biological tissues maintain the characteristic Al/Sc ratio (see p. 56) common to environmental materials, such as soil and dust, scandium may be of value in tissues, such as the lung, for identifying particulate matter resulting from inhalation of air-borne mineral debris.

Non-Activation Radiocounting Techniques

Apart from the heavy, naturally occurring radionuclides, the human body contains several other radioactive isotopes; of particular interest to clinical studies is ^{40}K which can be measured directly using standard detectors located over particular parts of the body or by whole body counting. As a result of nuclear tests and the release of effluent from nuclear reactors now distributed throughout the natural environment, the human body and items of diet contain a variety of radionuclides, some of which can be measured in the body, for example ^{131}I in the thyroid and ^{137}Cs distributed through the body. With the availability of high-resolution anticoincidence gamma ray spectrometers, multiradionuclide analyses are possible on a variety of materials, and extremely low levels of many radionuclides may be determined, although levels found in normal populations are far below those of concern to health. Although whole body studies require special conditions, radio-bioassay of human tissues, particularly the liver, bone, and lung, can provide very useful information for the metabolic pathways of many elements as traced by various radionuclides in nuclear fallout; for example, ^{58}Co, ^{60}Co, ^{54}Mn, ^{59}Fe, ^{65}Zn, ^{106}Ru, ^{110}Ag, ^{95}Zr, ^{95}Nb, ^{131}I, ^{134}Cs, and ^{137}Cs. Other commonly present radionuclides, such as ^{3}H, ^{14}C, and ^{85}Kr, together with ^{238}Pu and ^{239}Pu, which require special radiocounting techniques.

Gamma ray spectrometers and other basic counting equipment are therefore essential items in modern experimental biological research as, apart from activation analysis, such equipment may be used for single or multi-tracer experiments, in order to study metabolic behaviour of the elements. However, one word of caution is required in relation to the performance of stable elements, namely the need to determine that the radiotracer is in the same chemical form as the stable element for which it is used as a tracer.

STANDARDISATION AND HARMONISATION PROCEDURES

At the onset of the nuclear period, it soon became apparent that nuclear debris had entered the natural environment and was available directly or through the food chain to man; hence it became essential to develop accurate techniques of analysis in order to determine hazards. Although the very early data was often crude, this situation did not last long and soon, through programmes of international interlaboratory standardisation, suitable materials were made available, notably through the auspices of the International Atomic Energy Agency, Vienna, Austria; more recently this agency has entered the field of stable element standards. The success of this programme can be traced through the various reports of the agency and associated laboratories. Radionuclides present a potential hazard to man and it soon became essential to obtain reliable data from various laboratories scattered throughout the world. Some of the analyses required chemical separations procedures, while others could be carried out by direct counting.

Although the quality of data has improved over the years, examination of the record and the current state of the art provides clear evidence of the difficulties of environmental analysis. Consider the relatively simple problems of determining the unique character of radioactivity present in a variety of materials, and then visualise the far greater problems presented by trace, minor, and even major element assay for the stable elements which have not really been subjected to such detailed analytical examination as applied to the radionuclides. For the stable elements we are only now entering an era in which the many problems are being examined in sufficient detail, and it is only through programmes of international quality harmonisation that any acceptable progress is likely to be made. At the same time we must be careful not to become enraptured with the selectivity of assay provided by many methods, neither must we assume that an element is hazardous to man unless acceptable biological evidence of detrimental change is available. The scandium and indium story associated with activation analysis has its counterpart in stable element analysis; the analysis of Cu, Zn, and perhaps Pb in vast numbers of samples at tremendous cost and expenditure of manpower has possibly provided little return in terms of improving the health of man. Certainly in the case of Zn, hazards to man do not really exist, and even for lead, clinical studies by very sensitive indicators such as neurological response to elevated levels are inconclusive. Until far better links are forged between epidemiology and clinical practice, and the complex problem of total analysis for levels is improved, potential hazards to man, indeed if they even exist at all, are not likely to be identified, but some elements may become prominent simply because of ease of assay by a particular method or technique. The availability of advanced instruments for radioassay and the existence of irradiation sources in hospitals provides, in conjunction with medical expertise, the required structure for advancing our knowledge concerning the metabolism of all the chemical elements found in man.

REFERENCES

1. Smales, A. A.: Radioactivation analysis. *Annual Rept Chem Soc London, 46*:285, 1949.
2. Bowen, H. J. M. and Gibbons, D.: *Radioactivation Analysis.* Oxford, Clarendon Press, 1963.
3. Adams, F. and Hoste, J.: Non-destructive activation analysis. *GEC Atomic Energy Review, 4*:113, 1966.
4. McKay, H. A. C.: *Principles of Radiochemistry.* London, Butterworths, 1971.
5. Crouthamel, C. E., Adams, F., and Dams, R.: *Applied Gamma-Ray Spectrometry,* 2nd ed. Int Ser Monographs on Anal Chem, 41, Oxford, Pergamon, 1970.
6. Dyer, F. F. and Leddicotte, G. W.: *The Radiochemistry of Copper.* Washington, D.C., National Academy of Sciences, National Research Council, Nuclear Science Series, 1961.
7. Girardi, F. and Sabbioni, E.: Selective removal of radiosodium from neutron activated material by retention on hydrated antimony pentoxide. *J Radioanalyt Chem, 1*:169, 1968.
8. Iyengar, G. V. and Samasahl, K.: Recovery of ion-exchange resens and partition chromatographic supports from large scale radiochemical separations. *J Radioanalytical Chem, 25*:47, 1975.
9. Samsahl, K., Wester, P. O., and Landström, O.: An automated group separation system for the simultaneous determination of a great number of elements in biological materials. *Anal Chem, 40*:181, 1968.
10. *Selective Ion Retention Media.* Carlo Erba, Divisione Chimica Industriale Pub, Sem No. 620, Milano, Italy, 1971.
11. Amiel, S.: Analytical application of delayed neutron emission in fissionable elements. *Trans Israel AEC,* Rep 1A-621, 1961. See also *Anal Chem, 32*:1683, 1962.
12. Hamilton, E. I.: The determination of uranium in rocks and minerals by the delayed neutron method. *Earth Planetary Sci Lettr, 1*:77, 1966.

13. Hamilton, E. I.: The concentration of uranium in man and his diet. *Health Physics, 22*:149, 1972.
14. Fleischer, R. L., Price, P. B., and Walker, R. M.: Nuclear tracks in solids. *Sci Am, 220*:30, 1969.
15. Fleischer, R. L., Price, P. B., and Walker, R. M.: Tracks of charged particles in solids. *Science, 149*:383, 1965.
16. Fleischer, R. K., Alter, H. W., Furman, H. W., Price, P. B., and Walker, R. M.: Particle track etching. *Science, 178*:255, 1972.
17. Hamilton, E. I.: Uranium content of normal blood. *Nature, 227*:501, 1970.
18. Hamilton, E. I.; The concentration and distribution of uranium in human skeletal tissues. *Calcified Tiss Res, 7*:150, 1971.
19. Green, D., Howell, G., and Thorne, M. C.: A new method for the accurate localization of ^{239}Pu in bone. *Phys Med Biol, 22*:733, 1977.
20. Hamilton, E. I.: New technique for determining the concentration and distribution of lead in materials. *Nature, 231*:524, 1971.
21. Anderson, J., Battye, C. K., Osborn, S. B., Tomlinson, R. W. S., Fry, F. A., and Newton, D.: Production of a uniform neutron fluence in a man-like phantom and determination of total-body sodium in humans by neutron activation analysis. In *Nuclear Activation in the Life Sciences, 1972*. Vienna, IAEA, 1972.
22. Boddy, K., Holloway, I., Elliott, A., Glaros, D., Robertson, I., and East, B. W.: Low-cost facility for partial-body and total-body in vivo activation analyses in the clinical environment. In *Nuclear Activation in the Life Sciences, 1972*. Vienna, IAEA, 1972.
23. Ullberg, S.: The technique of whole body autoradiography. Cryosectioning of large specimens: Science tools. *LKB Instrument J*, Special Issue, 1977.

X-Ray Fluorescence Spectrometry

X-ray fluorescence spectrometry is a method suitable for rapid qualitative and quantitative analysis of the chemical elements present in a wide variety of matrixes at major, minor, or trace levels of abundance; the physics underlying measurements are well known and with modern instrumentation measurements are easily made. Depending upon the type of instrument, single, sequential, or simultaneous analyses can be undertaken. Excellent accounts of the practical use of X-ray spectrometry have been described by Jenkins and de Vries (1) and Jenkins (2).

The X-ray region of the electromagnetic spectrum lies between 0.1 and 100 Å. When an element is bombarded with electrons in the X-ray region, a spectrum is obtained, as illustrated in Figure 88 for the bombardment of a gold target by electrons produced from an incandescent tungsten filament. The spectrum consists of the continuous spectrum with a well-defined minimum wavelength (maximum frequency) corresponding to an electron

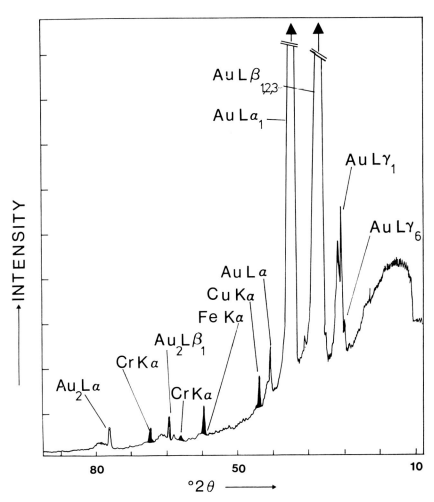

Figure 88: Characteristic X-ray spectra obtained with a gold target. Tube conditions: 80 kv, 124 mA. Detector crystal LiF 200 cut. Lines from Cu, Fe, and Cr arise from impurities in the gold target and from internal parts of the instrument.

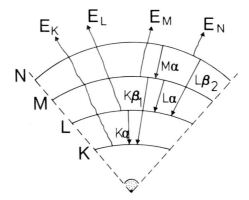

Figure 89: Transitions involving orbital rearrangement following ejection of one or more electrons as a result of excitation; for example

$$[E]_{K_a} = E_K - E_L.$$

losing all its energy in a single collison with a target atom. The longer wavelengths correspond to a more gradual loss of energy as the electrons are gradually decelerated and constitute the bremsstrahlung. Superimposed upon the continuous spectrum are a series of emission lines characteristic of the target element, together with any element impurities present in the target, which correspond to the quantum of radiation emitted when an electron changes energy levels. The radiation results from the rearrangement of orbital electrons of the target element following ejection of one or more electrons in the excitation processes as illustrated in Figure 89. For example if an electron from the K shell is ejected, the atom becomes unstable, but stability is regained by single or multiple transitions from outer orbitals. As a result of the transfer of an electron, the atom moves to a less energetic state and radiation is emitted at a wavelength corresponding to the difference in the energy between the initial and final states of the transferred electron.

Moseley (3) measured the frequency of corresponding characteristic line spectra for a number of elements and noted a linear relationship between the square root of frequency and atomic number; a few gaps existed in the original graph due to the fact that some elements had not been discovered. The actual process of electron changes are complex but may be likened to cascade phenomena which continue until the energy of the atom is lowered to a value corresponding to that found for normal electron vibrations in outer orbitals. Tables are available for the principal lines of X-ray emission together with values for relative intensities and identification of lines which are interfered with by the presence of other elements. Apart from the additional energy an atom contains after an electron jump from one energy shell to another, energy may also be used in processes of reorganising electron distributions within the atom resulting in the ejection of one or more electrons from the outer shell; this is called the Auger effect (see page 356). Restricting comments mainly to commercially available systems, the following are available:

The Sequential Wavelength Dispersion Spectrometer

This instrument consists of a source of X-rays produced in a sealed tube which contains a heated tungsten filament as a source of electrons. The electrons are accelerated through a large potential drop and strike a metal anode of high purity metal, such as tungsten, molybdenum, or gold, cooled by recirculating water. The anode provides the site of energy conversion processes, and although a considerable amount of the energy is in the form of heat, about 1 percent of the total applied power emerges through a thin end window of the tube as useable X-rays. In some instruments the end window is dispensed with and continuous pumping systems are used in order to provide a high spectral output over a large range of wavelengths which is required to excite the light elements.

The polychromatic radiation from the X-ray tube then impinges upon the surface of the sample, giving rise to characteristic fluorescent radiation from the various elements present in the sample which then passes through a collimator in order to provide acceptable angular resolution. Dispersion of the polychromatic radiation arising from the sample is achieved by use of the specific diffraction properties of crystals or diffraction gratings. Most spectrometers use flat crystals upon which the collimated beam from the sample impinges, such that diffraction

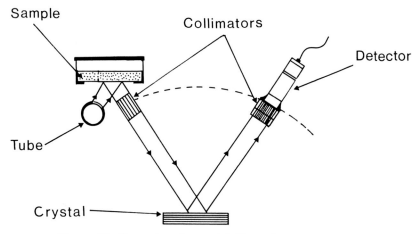

Figure 90: Geometry of a crystal dispersive spectrometer.

is achieved according to the Bragg relationship, namely $n\lambda = 2d \sin \theta$, where λ is the wavelength of the radiation diffracted through an angle θ by planes in the crystal of spacing, d, and n is an integer. A variety of single inorganic crystals are available, also various types of synthetic organic and organic-inorganic types providing detection for elements from boron to uranium; in most instruments, several crystals are housed in single instruments in aligned sledges and can be selected according to requirements. A second collimator is placed at an angle 2θ to the primary collimator after which the X-rays are detected through their ability to ionize matter. Commonly used detectors are Geiger counters, scintillation crystals, or proportional counters in conjunction with pulse height analysis in which a narrow range of wavelengths is isolated on the basis of energy. A schematic illustration of a dispersion spectrometer is given in Figure 90.

Multiple Crystal Simultaneous Wavelength Dispersion Instruments

These instruments are designed for rapid analysis for a specific number of elements usually in defined matrixes; each crystal and detector is optimised for a specific element in order to obtain the best accuracy of assay in the minimum of time. The basic components are the same as used in sequential spectrometers with the exception perhaps of curved rather than flat crystals for diffraction. This type of instrument has been widely used in quality control for on-line analysis in the metallurgical, ceramic, and cement industries where repetitive analyses of similar materials are required.

Energy Dispersive Spectrometers

In the conventional X-ray spectrometer, discrimination between the different energies of X-rays is obtained by using a crystal. In nondispersive spectrometers, the crystal is replaced by a single solid state detector which receives all the radiation, and provides a means of energy discrimination. Energy dispersive spectrometers have been described by Birks (4), Kneip and Laurer (5), Statham (6), Servant, Meny, and Champigny (7), and Frankel and Aitken (8). The energy dispersive spectrometer provides a gain in efficiency of ×100 or more, by eliminating the diffraction crystal and collimators, thus bringing the detector close to the source. Standard X-ray tubes can be used as a source of X-rays; alternatively, radioisotope sources have become very popular particularly in portable instruments. Maximum fluorescence excitation is obtained by using low energy gamma rays or X-ray emitters with energies suitable for exciting elements of interest, hence for the analysis of a wide range of elements several sources are required as illustrated in Table LVIII. Problems of safety limit the size of a radioactive source to a few tens of millicuries with total photon outputs in the range of 10^7-10^8 photons per sec, which may be compared with 10^{10}-10^{12} photons per sec from an X-ray tube. Because of the characteristics of radioactive decay, the intensity of the source will decrease with time according to the radioactive half-life of the radionuclide; for example a ^{55}Fe source with a half life of 2.7 years will show a loss in intensity at a rate of 0.35 percent per week.

Following excitation of the sample, the secondary radiation is then collimated and impinges directly onto the surface of the detector; the most commonly used detector is the lithium-drifted silicon, Si (Li), type which is kept at the temperature of liquid

TABLE LVIII
SOME COMMERCIALLY AVAILABLE RADIOISOTOPE SOURCES FOR NONDISPERSIVE X-RAY FLUORESCENCE SPECTROMETRY

Source	Half-Life	Useful Radiation	Highest Atomic Number Usefully Excited (K X-rays)
^{55}Fe	2.7y	Mn, K X-rays, 5.6 kev	24 (Cr)
^{3}H/Zr	12.3y	Bremsstrahlung 2-12 kev Zr, L X-rays, 2 kev	30 (Zn)
^{109}Cd	1.3y	Ag, K X-rays, 22 kev	43 (Tc)
^{147}Pm	2.6y	Bremsstrahlung, 10-100 kev	60 (Nd)
^{241}Pm	470y	Gamma-rays, 59.6 kev Np, L X-rays, 11-22 kev	69 (Tm)
^{153}Gd	236d	Gamma-rays, 100 kev Eu, K X-rays, 42 kev	88 (Ra)
^{57}Co	270d	Gamma-rays, 122 kev Fe, K X-rays, 6.4 kev	98 (Cf)

nitrogen in order to reduce the mobility of the lithium diode and reduce electronic noise. The signal from the detector is then passed to a preamplifier and finally into some form of pulse height analyser where energies are recorded simultaneously. The main components of an energy dispersive spectrometer are illustrated in Figure 91. Energy dispersive systems are compact and lack the delicate and complex mechanical parts and crystal assemblies associated with conventional dispersive instruments but there are some notable disadvantages:

1. Resolution is some fifty times poorer than with a crystal spectrometer, which means that additional techniques are required in order to resolve adjacent energies which becomes an important factor in multielement work.
2. Si (Li) detectors can only accept a limited activity which is usually less than a count rate of $\approx$ 20,000 counts per second for the whole of the spectrum, including scattered primary radiation and continuum, which in some instances may contribute to almost all of the observed count rate.
3. Much of the commercial literature associated with energy dispersive systems emphasises their use in multielement analyses, but they have limited application for natural sam-

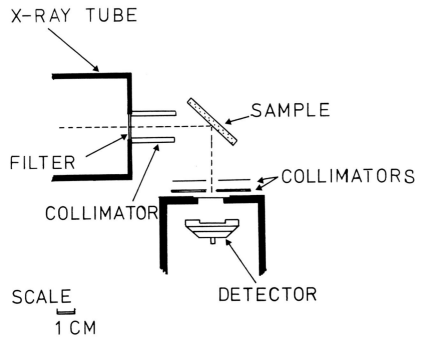

Figure 91: Geometry of an energy dispersive spectrometer.

ples of biological materials usually because of poor sensitivity for many trace elements; in the analysis of bulk samples a realistic detection limit is between 1 and 100 ppm but is very dependent upon the bulk composition of the sample. However, in many analyses energy dispersive spectrometers are competitive with crystal instruments and have been shown to be very satisfactory for the analysis of air particulates collected on filter papers as described by Camp et al. (9).

In recent years a new technique, based upon particle-induced X-ray emission (PIXE), offers promise of high sensitivity for simultaneous multielement analysis and has been described by Deconnick, Demortier, and Bodart (10), Johansson et al. (11) and Katsanos et al. (12). Briefly, the source of excitation is provided by a proton beam produced by an accelerator, such as the van de Graff, and X-ray detection is achieved with an energy dispersive Si (Li) detector; typical proton beams of between 1 and 5 Mev are used although, in order to reduce effects of bremsstrahlung, radiation energies of about 2 Mev are most suitable. Modern hospital complexes are often equipped with accelerators and there are usually no serious difficulties in providing space at the beam exit to irradiate materials, hence this type of technique is of particular interest to clinical studies.

Developments in the field using various types of irradiation sources are rapid; powerful monochromatic X-ray sources are required in order to reduce the background under the fluorescent energies and hence improve detection for elements present at levels of a few ppm or less; strong diffracting graphite crystals have been developed as a means of selecting a single energy from a metal anode X-ray tube, while the use of electrons circulating in storage rings and synchrotrons have achieved a considerable reduction in background count rates.

Methods of Analysis

X-ray fluorescence analysis is based upon an identification of the characteristic lines of an element present in a mixture of the elements, all of which to various degrees produce their own characteristic lines. Once a characteristic line has been identified and isolated, quantitative analysis can be carried out by external or internal standardisation techniques (Jenkins and de Vries, 1; Stern, 13).

As X-rays pass through matter they are attenuated depending upon thickness and density of the material, in relation to X-ray wavelength. Therefore absorption as a matrix effect is concerned with depth of penetration of the exciting radiation and the ability of excited species produced below the surface of the sample to emerge and be detected. In a biological matrix some of the elements present will preferentially absorb radiation leading to enhancement and quenching effects. As X-ray spectrometry is concerned with the penetration of matter by X-rays, then clearly grain size is important. The ideal case is a sample which has been fused to form a homogeneous media with reference to the distribution of the elements fixed in a homogeneous matrix; the worst case would be a loose, heterogeneous, sized powder for which both penetration and emergence of radiation will be dependent upon grain size. For most purposes, the chemical form of the elements in a sample is of no importance but in special cases the method can be used to differentiate between some elements present in different valency states.

The first requirement for X-ray fluorescence analysis is that the sample be ground to a fine powder and that the exciting beam irradiate a representative area; this can most satisfactorily be achieved by grinding the sample so that it will at least pass a four hundred mesh sieve. The only satisfactory way of achieving a homogeneous mix is to grind the sample and then pass the total sample through a series of graded sieves until the total mass is recovered through the finest sieve. This process does present problems of contamination from grinding equipment and loss of sample in the form of fine dust. By experiment, on defined matrixes, the total time taken for homogenisation can often be reduced. Careful attention should be paid to the manufacturer's instructions concerning maximum grinding load; if vials are overloaded, efficiency of grinding is greatly reduced because of compaction. It is also important that the sample be dry, otherwise it becomes smeared over the internal surfaces. In biological materials the main problems concern an adequate destruction of cell constituents before they become compacted; efficiency of grinding a single material varies considerably, for example cartilage or connective tissue can be difficult to grind in agate mortars. Following conventional drying techniques, for example in an oven or freeze dryer, many biological materials readily take up moisture while lipids may become encrusted with a hardened surface enclosing "wet" material. One standard technique for sample preparation is the compaction of the material into the form of a tablet, as illustrated in Figure 92, which fits into the source of the instrument. When the tablet technique is used, inspection of the surface will clearly indicate whether or not homogeniety has been achieved; the surface should have an even colour and be free from any structural features, such as slivers of compressed connective tissue. This simple technique is very

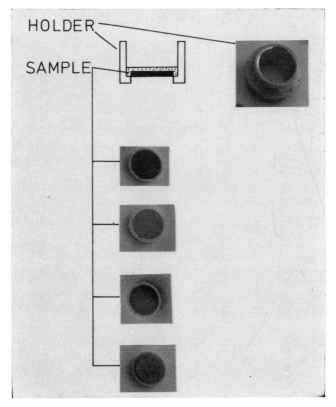

Figure 92: Sample holder used in XRF and examples of some types of sample tablet, including examples of sample deterioration.

attractive and although the exciting beam does heat the sample slightly and causes surface deterioration of the tablets, with care they can be considered as permanent items and may be stored for further use.

Following storage, most biological materials undergo some type of deterioration as a result of expansion due to absorption of water; however, such samples can be dried and pressed again before use. It is possible to prepare a tablet in a hydraulic press without any other treatment, in which case both surfaces can be analysed in order to determine homogeneity and provide some indication of surface contamination. Although stainless steel dies are used for tablet making, they rarely result in contamination if they are kept clean and highly polished; if contamination does occur then the surfaces of the die can be protected with an inert ultrapure material such as tantalum or platinum foil. If a limited mass of sample is available the tablet can be backed with an inert matrix such as cellulose or borax. Leoni and Saitha (14) have described techniques for microsample analysis requiring $\approx$ 40 mg of material, compared with conventional amounts of 2 to 3 grams. When analysing ashed or organic-free material, the sample can be fused in various low temperature fluxes which are then cast in the form of tablets; this technique improves homogeniety and also allows the addition of standards or buffers to modify the matrix and can overcome some types of interelement effect. From my experience in the analysis of biological materials, I prefer to homogenise the untreated sample in a liquid nitrogen mill followed by freeze drying and the preparation of unbacked tablets.

Although loss in sensitivity for many elements occurs, another popular technique is the preparation of samples as thin films, thus reducing or eliminating matrix effects. When attempting to make thin films of particulate material, physical fractionation of constituents can occur but is reduced if the samples are finely ground. The thin film technique is useful when analysing liquids which can be evaporated onto suitable element-free substrates, although difficulties can arise because of selective migration of a constituent usually to the edge of the substrate. Liquids can also be absorbed onto materials such as cellulose nitrate, dried, and then homogenised, followed by preparation of a tablet. Using commercially available holders, liquids can also be analysed directly, but because they are retained in a capsule with a mylar end-window, sensitivity is lost; when analysing liquids it is common practice, because of evaporation, to use an air path, which results in a further loss of sensitivity, or, alternately, a helium path can be used for which there is little absorbance of radiation.

In attempts to employ instrumental methods of analysis, most of the major problems concern sensitivity and complex matrix effects, which limit the number of elements which can be analysed in a single matrix. Improvements in instrument design and sophisticated electronic systems usually yield only marginal rewards. Using wavelength dispersive X-ray spectrometers as an example, the range of elements which can be analysed can be dramatically increased, together with sensitivity, if preliminary chemical separations are performed. It is beyond the scope of this monograph to consider this matter in detail but the following serve as illustrations:

1. The selective extraction of single or groups of elements on ion exchange resins which are then made into tablets; alternatively various types of ion exchange filter paper are commercially available thus providing a thin source. Apart from the usual precautions concerning purity of reagents, the ion exchange material has to be free of the elements of interest. This can usually be accomplished for exchangeable sites by elution with suitable reagents but elements present in the structural body of the resin may be difficult to remove.
2. The use of liquid extraction techniques to separate and concentrate elements into small volumes which can then be absorbed onto suitable substrates, for example borax or powdered cellulose nitrate.
3. Selective precipitation of elements, such as with sodium diethyldithiocarbamate or chelation by 8-hydroxyquinoline, followed by preparation of tablets or the preparation of thin films on filter paper or carbon substrates.

With current advances in analytical chemistry a great variety of selective extraction techniques are available and through their use interfering elements can be removed and trace elements concentrated in small volumes when sensitivity of detection is usually no longer a problem. If such techniques are not used, then conventional X-ray spectrometry analysis is restricted to major and minor elements of biological matrixes together with some trace elements such as Rb, Br, Zn, Cu, Mn, and Ti. Although clinical interest appears to centre upon trace elements, because of the interrelations which exist between the chemical elements, quantitative assay for the major elements may be rewarding in studies of disease and health; for example a change in the level of a major element by a few percent may be far more significant than a change of several hundred percent for a trace element.

Various types of X-ray fluorescence analysis play an important role in determining levels of elements in a variety of materials. For example, levels of lead in printing ink of children's comics, described by Hamilton (17), are illustrated in Figure 93 and can cause concern for children exhibiting pica. Some X-ray spectra are illustrated in Figure 94 for a low abundance element (Cu) and in Figure 95 for a minor trace element (Zn) in human liver and blood, determined by conventional crystal dispersive spectrometry on freeze dried tablets of samples. Evans has described techniques for the analysis of light elements in plants and faecal materials; Flint, Lawson, and Standil (15) determined levels of Na, K, Ca, V, Mn, Fe, Ni, Cu, Zn, and Br in human serum; a mean value of 1.07 ppm Fe was obtained for healthy

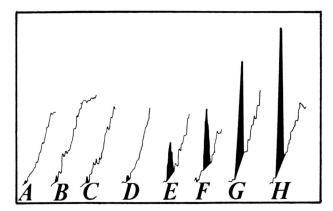

Figure 93: X-ray fluorescence spectra for trace levels of lead in various coloured inks used in printing children's comics. Key for colour of ink: A = black; B=navy-blue; C=light blue; D=white; E=scarlet; F=red; G=green; H = yellow. In H the intensity of the lead peak corresponds to about 2,000 ppm Pb. Apart from children who exhibit pica most small children suck various objects, including paper, and hence for some types of print quite considerable quantities of lead can be ingested, but the biological availability of such lead is not known.

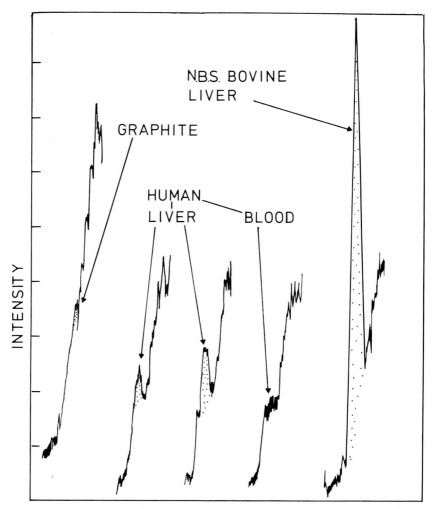

Figure 94: XRF (crystal) scans for levels of copper in a graphite blank, in two samples of human liver, in human blood, and for NBS Bovine liver.

blood serum from 1,180 individuals for which individual values ranged from 0.52 to 2.01 ppm Fe. High levels of Br and Cu were observed for reticulum cell sarcoma which has also been reported by Cowgill (16) for a tumor in a nocturnal prosimian. In clinical studies, XRF analysis can be used to determine iron levels in blood in relation to hemoglobin and requires only about

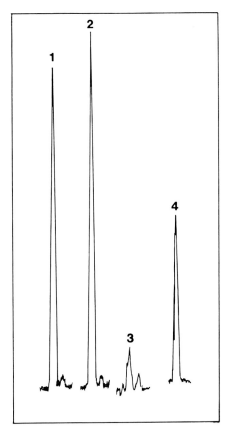

Figure 95: Illustrations of XRF (crystal) zinc spectra for various freeze-dried tissues. 1 and 2=human liver; 3=human blood; 4=NBS Bovine liver.

25 µl of sample absorbed onto a filter paper. Indirect methods of analysis can also be useful. For example, the determination of ammonia-nitrogen can be made by using Nesslers reagent and absorbing the product on filter paper: the ratio of Hg, I, and ammonia are present in fixed proportions, detection limits are excellent as assay is made using Hg or I X-ray lines; 14 µg of ammonium-nitrogen corresponds to ≈ 400 µg of mercury.

In recent years, very significant improvements in XRF methods have taken place. Giaugue, Garrett, and Goda (17), using energy dispersive X-ray fluorescence, utilise relationships of X-ray energy between adjacent major element absorption edges; both the reciprocal of the specimen mass absorption coefficient and the spectral background intensity vary linearly with the intensity of the Compton scattered excitation radiation, providing an internal standard and a measure of spectral background intensity; assays are reported for twenty-six trace and two major elements in geochemical materials with an accuracy of ± 5 percent or better.

Dyson and Simpson (18) have used 2.5 Mev protons from an accelerator and have determined the following levels in normal livers (μg/g wet weight): Fe, 288; Cu, 8.1; Zn, 54.5; Br, 0.8, and Mn, 0.8.

Wobrauschek and Aiginger (19), using an energy dispersive spectrometer system in a total reflection mode of operation, determined manganese at a concentration of 0.2 ppm in a 5 μl volume. Aiginger, Wobrauschek, and Brauner (20) describe further techniques and methods for assay, at the nanogram level, using Bragg reflected polarised X-rays. Crystal and energy dispersive analyses can be attached to scanning and transmission electron microscopes, thus providing assay for the micro distribution of elements in their sections; Geiss and Huang (21) describe the analysis for elements present in areas between 250 and 2,000 Å in diameter.

Using modern instruments in combination with ultraclean laboratory facilities for preconcentration of elements, X-ray fluorescence is a very simple, sensitive, and versatile method for element analysis. When dealing with thin film techniques, most problems of absorbance and matrix effects are eliminated and linear relationships between intensity of signal and concentration can be obtained. However, the limitation of such techniques rests with the purity of substrates.

Within the general array of commercially available instruments for element analysis, X-ray fluorescence spectrometry is a reliable workhorse. For dried biological samples it is suitable for the analyses of major and some minor elements; for particular elements, analysis after preparation of tablets can be rapid, for example strontium in bone can be determined at a rate of about 200 measurements per day. The scope of the method is dramatically increased by preconcentrating elements but requires ultra clean working conditions. A major problem concerns matrix effects even for samples having similar compositions; modern instruments are available which incorporate small computors for matrix correction for defined matrixes, while individual laboratories can develop special items of software for particular materials. Instruments are now available for automated analysis, allowing selection of appropriate crystals, and are equipped with automatic sample changers which can contain a large number of samples; the only item which is not usually automated is the selection of X-ray tubes with different target elements.

REFERENCES

1. Jenkins, R. and de Vries, J. L.: *Practical X-Ray Spectrometry.* London, Macmillan, 1967.
2. Jenkins, R.: *An Introduction to X-Ray Spectrometry.* New York, Heyden and Son, Ltd., 1976.
3. Moseley, H. G. J.: The high frequency spectra of the elements. *London, Edinburgh and Dublin Philosophical Magazine,* 26:1024, 1913.
4. Birks, S.: *X-Ray Spectrochemical Analysis,* 2nd ed. New York, Wiley, 1969.
5. Kneip, T. J. and Laurer, G. R.: Isotope excited x-ray fluorescence. *Anal Chem,* 44:57A, 1972.
6. Statham, P. J.: A comparative study of techniques for quantitative analysis of the x-ray spectra obtained with a Si (Li) detector. *X-ray Spectrometry,* 5:16, 1976.
7. Servant, J. M., Meny, L., and Champigny, M.: Energy dispersive quantitative x-ray microanalysis by scanning microscopy. *X-ray Spectrometry,* 4:99, 1975.
8. Frankel, R. S. and Aitken, D. W.: Energy x-ray dispersive emission spectroscopy. *Appl Spectroscopy,* 24:557, 1970.
9. Camp, D. C., Van Lehn, A. L., Rhodes, J. R., and Pradzynski, A. H.: Intercomparison of trace element determinations in simulated and real air particulate samples. *X-ray Spectrometry,* 4:99, 1975.
10. Deconnick, G., Demortier, G., and Bodart, F.: Application of x-ray production by charged particles to elemental analysis. *Atomic Energy Review,* 13:367, 1975.
11. Johansson, T. B., Van Grieken, R. E., Nelson, J. W., and Winchester, J. W.: Elemental trace analysis of small samples by proton induced x-ray emission. *Anal Chem,* 47:855, 1975.
12. Katsanos, A., Xenoulis, A., Hadjiantoniou, A., and Fink, R.: An external

beam technique for proton induced x-ray emission analysis. *Nucl Instr and Met, 137*:119, 1976.
13. Stern, W. B.: On trace element analysis of geological samples by x-ray flourescence. *X-ray Spectrometry, 5*:56, 1976.
14. Leoni, L. and Saitha, M.: X-ray fluorescence analysis of powder pellets utilising a small quantity of material. *X-ray Spectrometry, 3*:74, 1974.
15. Flint, R. W., Lawson, C. D., and Standil, S.: The application of trace element analysis by X-ray fluorescence to human blood serum. *J Lab Clin Med, 85*:155, 1975.
16. Cowgill, U. M.: Elemental analysis of a tumor from a nocturnal prosimian with special emphasis on bromine. *Sci Total Environ, 7*:63, 1977.
17. Giauque, R. D., Garrett, R. B., and Goda, L. Y.: Energy dispersive X-ray fluorescence spectrometry for determination of twenty-six trace and two major elements in geochemical specimens. *Anal Chem, 49*:1012, 1977.
18. Dyson, N. A. and Simpson, A. E.: Studies of normal and diseased human liver tissue by proton-induced X-ray emission spectroscopy. *Phys Med Biol, 21*:853, 1976.
19. Wobrauschek, P. and Aiginger, H.: X-ray total reflection fluorescence analysis. In *Measurement, Detection and Control of Environmental Pollutants.* Vienna, IAEA, 187, 1976.
20. Aiginger, H., Wobrauschek, P., and Brauner, C.; Energy-dispersive fluorescence analysis using Bragg-reflected polarized X-rays. In *Measurement, Detection and Control of Environmental Pollutants.* Vienna, IAEA, 187, 1976.
21. Geiss, R. H. and Haung, T. C.: Quantitative X-ray energy dispersive analysis with the transmission electron microscope. *X-ray Spectrometry, 4*:196, 1975.

Mass Spectrometry

Over the past forty years the mass spectrometer has become established as an important tool for the analysis of inorganic, inorganic-organic, and organic substances at major, minor, and trace levels. Mass spectrometers usually involve high capital costs, the need for computor data retrieval systems, a very high degree of technical skill, and some degree of art for optimum performance. Principles and applications of mass spectrometry have been described by Beynon (1), McDowell (2), Reed, (3), White (4), Ahreans (5), and Watson (6).

A mass spectrometer is an instrument in which a beam of ions, generated from the substance to be analysed by electrothermal processes, is sorted out to produce an ion spectrum in which ions are displayed according to their mass to charge ratios and which provides a record of the relative abundance of each ion species present. Most measurements are concerned with the measurement of positive ions, although for special purposes negative ions can also be determined. Through mass spectrometry, the existence of stable isotopes of the chemical elements was discovered and also the realisation that the chemical properties of an element are determined by its atomic number and not its atomic weight.

Restricting remarks to the analysis of elements, there are two types of instrument; the mass spectrometer, in which the ion signals are detected electrically and then amplified by electronic systems before being recorded, and the mass spectrograph, in which the ion beam is detected and recorded on photographic plates.

There are three ways in which an ion beam can be focussed, and basic principles used in ion optics are analagous to those used in light optics:

1. Direct focussing—the concentration of a beam of ions all of the same mass in which the ion beam is focussed for a range of different initial directions when all ions in the beam are moving at the same speed.
2. Velocity focussing—the ion beam is focussed when it contains ions travelling with a range of different speeds but which are all moving in the same initial direction.
3. Double focussing—an ion beam of varying initial speed and direction is brought into focus.

For the analysis of the chemical elements, two types of instrument are in general use; the direction-focussing thermal ionization spectrometer, which is usually concerned with the assay of individual elements and their isotopes, and the spark source mass spectrograph, a double focussing machine for the simultaneous detection of all elements by photographic detection or more recently the sequential detection of the elements by scanning through the mass range and using electrical detectors to record ion intensities. The former is an absolute method of analysis while the latter usually involves comparing ion intensities from samples with internal or external standards.

THERMAL IONIZATION MASS SPECTROMETRY (SURFACE IONIZATION)

The basic instrument, illustrated in Figure 96, consists of a source region in which ions of the elements are produced by evaporating an element in a suitable form onto a metal filament, for example tantalum, tungsten, or rhenium, as illustrated in Figure 97, which is then heated by passage of an electric current; when a substance is evaporated from a heated surface, there is a probability that it will evaporate to produce positive ions. The use of triple filaments provides more careful control of the evaporation process and improves the ionization and production of positive ions. In filament volatilisation processes, the chemical form of the sample is critical; with very few exceptions it is essential to separate the element of interest into a pure form of known chemical composition.

An objective in thermal ionization is the production of a steady stream of positive ions. Some compounds have a low ionization potential; for example, cesium chloride produces few ions compared with cesium sulphate, because of the higher volatility of the chloride which evaporates from the filament before atomic decomposition occurs. The efficiency of ion production, and stability of ion beams, is dependent upon intimate relationships between the thermochemistry of the surface of a filament and form of the deposited compound. Highly volatile compounds and elements such as mercury and some organic forms of elements are best analysed by passing the vapour into an electron bombardment source of a gas analysis mass spectrometer, where the source of ionization is provided by incandescent filaments. Because thermal ionization mass spectrometry involves very exacting chemical separations, ultra-clean laboratory facilities are essential and all operations must be carried out without

Instrumental Methods

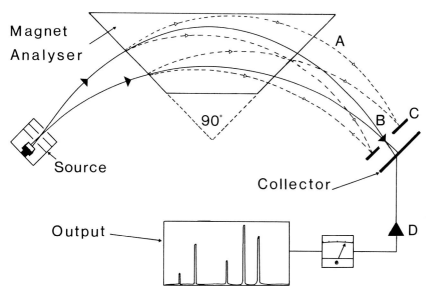

Figure 96: Basic ion-optics of a thermal source mass spectrometer. Through magnetic scanning, the required ion mass can be focussed at point B and adjacent, but unwanted ions are grounded on the slit at C; the focussed ion then impinges upon a collector and then the signal is amplified (D) before passing to some form of recorder.

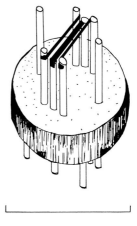

Figure 97: An illustration of a triple filament assembly; the sample can either be placed on the central filament and heated by passage of an electric current directly, or alternatively can be placed on the central filament and ionization achieved by heating the side filaments.

1 CM.

the introduction of contaminants. Initially the problem seems very daunting; for example, the separation of a few micrograms of an element from one gram of a sample in a pure form and without any contamination. Thermal ionization mass spectrometry provides invaluable experience in the field of trace element analysis. There is a satisfaction in taking a sample of bone, dissolving it, separating out the lead, precipitating or electroplating the micro amounts of separated lead so that it can be seen, and then observing the emission of the isotopes through the detection system. Although most elements of the Periodic Table which have more than one isotope can be determined by this method, the tendency has been for work to be limited to those elements for which very accurate data are required. Of all the analytical methods this is possibly one that can be described as an absolute method and may be compared with gravimetric analysis.

In order to reduce loss of ions by collision processes, the source and ion flight path is maintained under high vacuum, typically 10^{-8} to 10^{-10} torr. Ions produced in the source are passed through a series of slits into the tube region which is placed in a high magnetic field where the isotopes of the elements are separated. The simplest detector allows the ion beam to impinge upon a metal collecting electrode which is connected to earth through a high ohmic resistor; the potential drop across the resistor is then a measure of the ion current. A typical collector is the Faraday cage which can consist of a single or double collector in which two ion beams are recorded simultaneously; comparison of the difference between both signals by null balance techniques permits exceptionally accurate measurements to be made.

Alternatively detectors, such as the electron multiplier, can be used which can have high sensitivity permitting, in some instances, the detection of single ions. The signal from the detector is then amplified and recorded on chart paper or by using digital systems interfaced to computors. The basic techniques of surface ionization spectrometry are largely based on very precise forms of art in the separation of the element and controlling its emission from a filament. Once the required techniques have been evaluated, instruments can be automated and human errors removed. The total time taken for manual analysis can be very long compared with other methods; excluding the chemical processing, between eight and fifteen isotope assays can usually be made each day. A considerable impetus to surface ionization mass spectrometry was provided by the demands of geochronologists who required very accurate data for the isotopic composition of Rb, Sr, Pb, and U; a general description of the relevant techniques has been presented by Hamilton (7). Tera and Wasserburg (8) have described the precise isotopic analysis of lead in picomole and subpicomole quantities; as little as 8×10^{11} Pb atoms can be measured with a level of Pb contamination reduced to 4×10^{10} atoms; lead ions were emitted from a silica-zirconia gel deposited on a filament, which marks a considerable improvement over the use of PbS which requires about 10^{-6} gram of Pb.

Biologists and clinicians have been mainly concerned with organic mass spectrometry, which possibly covers more than 90 percent of all literature on mass spectrometry, but more recently the value of element assay by mass analysis has been shown in various environmental studies.

Spark Source Mass Spectrometry

Spark source mass spectrographs are mainly used for the analysis of solids, although liquids if frozen can also be analysed; the design of a basic spectrometer is illustrated in Figure 98. Samples of homogeneous powders, mixed with a conducting matrix such as graphite (10%), are formed into two electrodes between which an electric spark is passed. Samples can also be prepared in a single electrode and sparked against a pure graphite electrode; for the analysis of surfaces, providing that they are electrically conducting, a pure graphite electrode can be used as a counterprobe to provide data for the distribution of the elements on the surface. Spark sources have an energy spread of about 1 kv, hence double focussing instruments are required in order to provide sufficient resolution of the isotopes.

Further, by nature the spark is intermittent, hence it is essential to use a system which will integrate ion beams with time. A commonly used spark is the pulsed radio-frequency type with a voltage frequency amplitude of up to 100 kv. As a result of sparking, the sample is volatilised at the site of spark impact and elements brought into the spark plasma are ionized as single, together with some multiple, ions. The ratio of singly to doubly charged ions has been shown to be constant for all elements. Within a factor of about three, the efficiency of excitation for most elements is the same; hence with the exception of very vola-

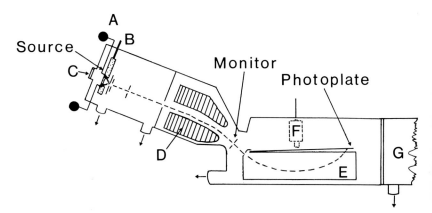

Figure 98: Main features of a spark source mass spectrograph. The sample is contained within an electrically conducting electrode mounted in the source; through a manipulator, A, the electrodes can be positioned to form the appropriate gap for sparking which is then automatically controlled. In order to maintain the best vacuum in the source during sparking, a gold plated cryopump, B, is used and kept cool by a flow of liquid nitrogen supplied in a copper pipe from a dewar. The electrostatic analyser, D, provides a means of energy selection and mass separation is achieved in the magnetic analyser, E. The monitor current provides a means of determining the total ion intensity of the beam prior to entering the magnetic analyser and being resolved and then detected by the photoplate. In the electrical mode an electron multiplier, F, provides direct output of ion intensities by scanning across the mass range using electrostatic scanning. Sequential ion mass scans can also be obtained by magnetic scanning but, in order to overcome hysteresis of the magnet, a Hall probe, positioned between the poles of the electromagnet, is essential. A number of preconditioned photoplates are stored in the magazine, G.

tile materials, the general analytical problem of matrix effect is reduced but nevertheless is still present. During sparking, the temperature of the electrodes will rise and, as the temperature increases with time, selective loss of volatile constituents can occur; further, if the electrode is not homogeneous, selective sparking can occur for some areas which will become depleted in readily volatilised materials.

Following the production of ions in the source, they are then passed through a series of defining slits into an electrostatic analyser. Ions having energies of a selected range then pass into the magnetic analyser sector where mass separation takes place. A monitor collector is positioned between the two analysers, intercepts about 50 percent of the ion beam, and serves to provide a means of adjusting sparking conditions in the source; the integrated output from the monitor also serves as a measure of the exposure laid down on the photographic plate. The ions emerging from the magnetic analyser impinge upon a glass plate coated with a photographic emulsion. Ions covering a mass ratio of 35 : 1 can be focussed simultaneously onto the photographic plate and the magnetic field can be adjusted so that the spectrum is recorded over the range m/e of 7 to 240. For one fixed period of sparking, ions are recorded as a spectrum across the photoplate; the plate can then be racked up in the magazine to accommodate another exposure, and by continuing this process a series of time graded spectra can be recorded as illustrated in Figure 99; it is common practice to increase each exposure period by a factor of three. Although single exposures over a period of at least one hour can be made, most graded exposures cover the range of 1×10^{-3} to 100 coulombs, consisting of about sixteen individual exposures across a photoplate, using a pulse repetition rate of 100 c/s and a pulse length of 100 microsec; during the total time of analysis several milligrams of sample are consumed.

During sparking, the vacuum in the source may rise to about 10^{-6} torr while the rest of the vacuum path is kept at about 5×10^{-9} torr; under such operating conditions limits of detection of 0.001 ppm for most elements can be routinely obtained.

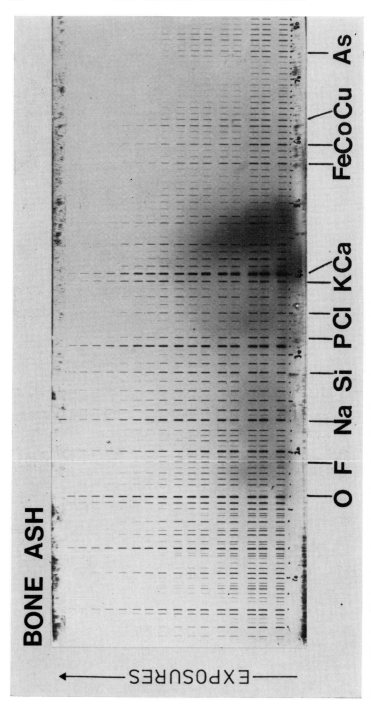

Figure 99: An illustration of a photoplate for ashed bone for 16 graded exposures showing halation effects because of the presence of the major cation—calcium. The electrode consisted of 95 percent bone ash and 5 percent graphite.

Instrumental Methods

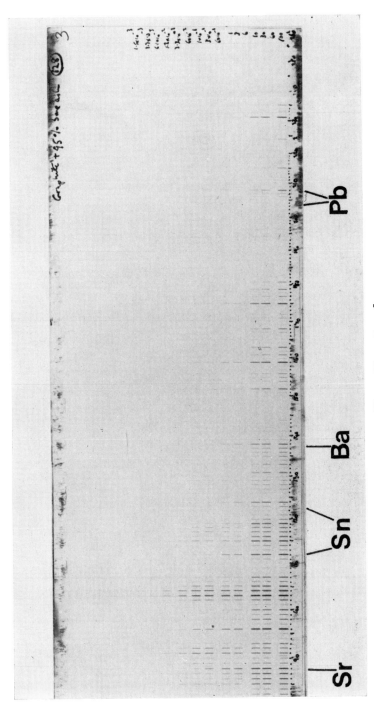

Figure 99 continued.

Following the collection of ions on the photoplate it is then developed and the density of the ion spectra are determined by photodensitometry. Because the difference in relative ionization efficiency of the elements and their compounds is within a factor of three, if the concentration of one element in a sample is known this can, in theory at least, serve as a standard for all the other elements. Matters can be improved considerably if a known mass of an element can be added to a sample as an internal standard; the only basic requirement is that the standard material should exhibit the same degree of ionization and performance as that of the element or elements to be determined. Multiple standard additions for several elements are possible but usually problems arise because of impurities in such mixtures; in practice rare elements are often used as internal standards (such as palladium, lutecium, or silver).

Spark source mass spectrometry provides a very sensitive technique for simultaneous multielement analysis. The sample has to be conducting; nonconductors are mixed with graphite. No chemical processing is required but during process of homogenization of the sample contaminants must not be introduced; the overall chemical matrix of the sample influences the quality of data, and the relative efficiency of ionization can vary with element and chemical form. The method is based upon thermal volatilisation and therefore all the problems of selective volatilisation arise. The problem of relative sensitivity factors for graphite and some biological matrixes has been described by Hamilton and Minski (9); although quite small differences in overall composition can affect relative sensitivity factors from element to element, their significance can be reduced by improving homogenization of the electrodes.

A major problem in biological samples rests with the presence of alkaline metal salts, which give rise to sporadic bursts of volatilised material; hence constant emission from exposure to exposure in not always possible, but which is easily recognised on photoplates. Under such conditions, the monitor current does not reflect a true burn up of the sample and alternative techniques are required; for example, one possibility is the use of an optical spectrometer aligned with the spark and the use of an emission line from the sample spark, e.g. Na, K, Ca, as a measure of sample consumption. The nearest competitor to the spark source mass spectroscopy is optical emission spectroscopy, but spark source spectrometry provides higher sensitivity of detection and is less matrix dependent; by using computor data retrieval systems, a large number of practical correction factors can be made, such as those based upon natural isotopic abundances and relations between single and doubly charged ions.

At present considerable improvements have arisen through the use of multidetector arrays which can now be fitted to most commercial instruments; alternatively a single electron multiplier can be used across the detector surface of which the mass spectrum is scanned and recorded in a sequential mode. In a single scan lasting about eight minutes, using a log-recorder, it is possible to obtain an immediate readout of signals; if processed on line, concentrations can be obtained while the sample is being sparked. Such a technique is an advantage because if an element of interest is detected, very accurate analyses can be made by focussing upon the relevant ion mass and, by repeated scanning, optimum ion counting statistics can be obtained. A disadvantage of photoplate recording is that plates are normally developed at the end of a day and only then can a visual examination be made to determine the levels of elements present.

At the present time electrical detection is less sensitive than the photoplate because of the variable nature of the spark; ion intensities per unit mass of element volatilised with time are not constant, mainly because over short time intervals when an ion mass is being scanned the spark will wander over the electrode and inhomogeneities at the micro level become important. Although the spark source mass spectrometer produces rather simple spectra, this is only possible in the absence of organic materials which must be removed before preparation of the electrodes; in theory it is possible to improve the resolving power of instruments to enhance the separation of the spectral lines from organic contaminants, but this is rarely used. The optimum technique for quantitative data retrieval for both methods of mass spectrometry can be obtained by the technique of isotope dilution analysis, the principles of which are common to several other methods. However, before discussing this technique it is worth noting that excellent data can be obtained from spark instruments by visual comparison of the spectra recorded by photoplates from an unknown sample with a standard, provided that the overall chemical matrix of the two are the same. The visual technique requires considerable skill and is based upon comparing the intensities of lines of the two plates using an illuminated comparator through which the spectral lines are magnified and values of the least detectable exposure observed.

The isotope dilution method can be classed as a tracer technique. The tracer consists of an element whose isotopic composition is different from that found in the natural element; elements enriched in a particular isotope are produced in an electromagnetic separator and are commercially available.

The basic isotope dilution equation developed by Inghram and Hayden (10) may be expressed as follows:

$$R = \frac{C_s S + C_N N}{C'_s S + C'_N N}$$

where R = atom ratio measured mass spectrometrically
C_s, C'_s = atom fractions of the normal and enriched isotope respectively in the tracer (spike)
C_N, C'_N = atom fractions of normal and enriched isotope respectively in the normal element
S = number of moles of enriched tracer
N = number of moles of normal element, that is, the unknown

Since a mole may be defined in terms of the numbers of atoms (one mole contains Avogadro's number of atoms) and since C_s are given as atom fractions and R is measured as an atom ratio, all terms in the above equation are in compatible units. By convention, R is expressed as the ratio of a given isotope with respect to the isotope enriched in the tracer; N can be converted into micrograms by multiplying by the atomic weight, and then to ppm by dividing by the sample weight.

Isotope dilution analysis by mass spectrometry can be applied to any element for which at least two stable isotopes exist. First the enriched tracer is mixed with the sample; in thermal ionization this is accomplished in solutions. In spark source measurements an aliquot of the tracer is added to the sample followed by drying and homogenization by grinding or fusion if the elements to be determined are not volatile; alternatively, a multi tracer can be prepared in a graphite matrix and aliquots homogenized by grinding with the sample. The separation of elements (isotopes) is achieved by the mass spectrometer, hence all that is required is an accurate measurement of the change in isotopic composition of the natural isotopic abundance caused by the addition of the enriched tracer; once complete isotopic exchange between element and tracer has been achieved, chemical recovery of the mixture need not be complete as only ratios between elements are measured. An example of an isotope dilution analysis is given in Figure 100; the method is extremely sensitive for trace element analysis and depends upon knowing the isotopic purity of the added tracer and accuracy of isotope analysis. The chemical purity of the tracer must also be determined and while isotopic purity may be high, other elements, often of abnormal isotopic composition, may be present thus presenting difficulties when used in multiple isotope dilution analysis.

Apart from instruments capable of analysing solids, gas mass spectrometers of simple design are available, and through the use of suitable inlet systems, the composition of volatile species diffusing from body surfaces and the composition of inhaled and exhaled air and stomach gases may be determined. Although seldom used, enriched stable isotopes can be used to investigate the body metabolism for a large number of elements; in this technique the enriched isotope, solid, liquid, or gas, is introduced into the body in a suitable chemical form and following mixing in various body pools aliquots may be sampled and rates at which biological processes take place determined; if total mixing between element and tracer occurs, the method of isotope dilution analysis can be used for mass analysis of elements in various parts of the body. Clearly the mass of administered element

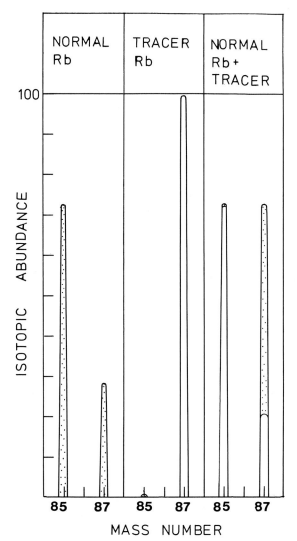

Figure 100: An illustration of the isotope dilution method for the determination of rubidium. The isotopic composition of normal stable rubidium (^{87}Rb, 27.85%; ^{85}Rb, 72.15%) is given first and that for rubidium enriched in the isotope ^{87}Rb second. The tracer rubidium is calibrated against a solution of normal rubidium and a known mass of the tracer is added to a solution of the sample. Rubidium is then chemically separated from the mixture and now consists of the normal rubidium plus the tracer. As the method is only concerned with a measurement of isotopic ratios, quantitative separation of the rubidium is not required. Following measurement of the isotopic composition of the mixture, the enrichment in ^{87}Rb is used to calculate the mass of stable rubidium present in the sample. A requirement for a tracer is that its isotopic composition should be as different as possible from that of the normal composition, hence in this instance ^{87}Rb, which is more sensitive for this type of analysis as it has a lower abundance than the other isotope at mass 85, was selected. In order to optimise the analysis, the final ratio of tracer to stable isotope should be about 1. In this particular analysis, a slight correction is required in order to allow for the small amount of ^{85}Rb present in the tracer.

should not be toxic to an individual, and a sufficient amount must be added in order that an enrichment in isotope ratios can be determined on the sampled aliquots of material. As an example it is feasible, through faeces and urine analysis, to investigate the GI-tract transfer of the rare-earth group of elements, as their abundance in diet and body tissues is low. Analysis of calcium would be more difficult because of the large mass of exchangeable calcium in the body requiring exceptionally large doses to be administered. There are exceptions; for example, when the administered element is in a particular chemical form such that it is only associated with specific biochemical processes which do not involve mixing with the major mass of an element present in the body. When using laboratory animals it is often possible to administer an enriched isotope with a high neutron activation capture cross section for subsequent neutron activation analysis and therefore enhance the detection limits for assay by this technique.

Recently new techniques for mass analysis have been reported by Gray (11), which could overcome the problems of direct analysis of liquids by mass spectrometry. Briefly liquids are nebulised into a d.c. plasma source maintained in an argon carrier gas. In a plasma, at about 5,000°K, the degree of ionization for 1 μg ml^{-1} of an element approaches 100 percent and ion populations of 10^{16} ions m^{-3} are exceeded. The ions, relatively free of multiple charged and molecular species, are injected into the axis of a quadrupole type of radio-frequency mass spectrometer. The predicted sensitivity of the system still remains to be realised; however, detection limits of 0.004 μg ml^{-1} have been obtained for Ag and 0.0003 μg ml^{-1} for Co. Apart from assay for the total mass of an element present in a solution, the technique also provides information on isotopic abundances. Future developments in this field are awaited with some interest. Provided that the problems of constructing suitable differential pumping systems can be mastered, an aerosol, at atmospheric pressure generated in a plasma source, can be introduced into the source region of a double focussing mass spectrometer, then multielement analysis and isotopic abundance measurements in conjunction with isotope dilution analysis becomes a possibility, thus overcoming the major problem of conventional thermal source types of excitation.

One of the problems of mass spectrometry, especially spark source instruments, has been the lack of application from those interested in trace elements in biological materials; with few exceptions, emphasis has been directed to their use for the assay of trace contaminants in ultrapure materials such as those required by the electronic and metallurgical industries. The future of plasma optical emission instruments is indeed very bright, but if such sources are combined with a mass spectrometer, perhaps the requirements of a "black box" instrument for versatile element analysis will be reached which will include a facility for biological tracer studies. Before concluding this section on mass spectrometry, one unique technique of analysis directly related to pollution by lead is worth mentioning. The technique, described by Ault, Senechal, and Erlebach (12), Rabinowitz and Wetherill (13), Chow, Snyder, and Earl (14) and Hamilton and Clifton (15), is based on the fact that the isotopic composition of lead can be defined in relation to geographical source; hence on the basis of isotopic composition the entry of one type of natural lead from one area into another can be monitored by isotopic assay. In the United Kingdom for instance, virtually all the lead used by industry is derived from one isotopic type of lead not found naturally in the country, hence the proportion of this industrial lead can be determined in body tissues. Apart from the obvious interest in the uptake by man of lead in the vicinity of smelters, in some areas it is practical to determine the extent to which lead from the car has penetrated body tissues. Such investigations could provide an answer to the question: How much lead in the human body is derived from the car and at what rate does such lead enter and accumulate in body tissues? Magi, Facchetti, and Garibaldi (16) have described a large experiment to be carried out in Italy over a period of three years in which lead present in petrol, with a well-defined isotopic composition and quite different to any local natural lead, will be used to determine the extent to which car lead enters man.

REFERENCES

1. Beynon, J. H.: *Mass Spectrometry and its Applications to Organic Chemistry.* New York, Elsevier, 1960.
2. McDowell, C. A.: *Mass Spectrometry.* New York, McGraw, 1963.
3. Reed, R. I.: *Mass Spectrometry.* New York, Academic Press, 1965.
4. White, F. A.: *Mass Spectrometry in Science and Technology.* New York, Wiley, 1968.
5. Ahearn, A. J.: *Mass Spectrometric Analysis of Solids.* New York, American Elsevier, 1966.
6. Watson, J. T.: *An Introduction to Mass Spectrometry, Biomedical, Environmental and Forensic Applications.* New York, Raven, 1976.
7. Hamilton, E. I.: *Applied Geochronology.* New York, Academic Press, 1965.
8. Tera, F. and Wasserburg, G. J.: Precise isotopic analysis of lead in picomole and subpicomole quantities. *Anal Chem,* 47:2214, 1975.
9. Hamilton, E. I. and Minski, M. J.: Spark source mass spectrometric sensitivity factors for elements in different matrixes. *Int J Mass Spec,* 10:77, 1972/73.
10. Inghram, M. G. and Hayden, R. J. A.: *A Handbook on Mass Spectroscopy.* Nat Acad Sci, Nat Res Council Pub, 311:37, 1974.
11. Gray, A. L.: Plasma sampling mass spectrometry for trace analysis of solutions. *Anal Chem,* 47:600, 1975.
12. Ault, W. U., Senechal, R. G., and Erlebach, W. E.: Isotopic composition as a natural tracer of lead in the environment. *Environ Sci Tech,* 4:305, 1970.
13. Rabinowitz, M. B. and Wetherill, G. W.: Identifying sources of lead contamination by stable isotope techniques. *Environ Sci Tech,* 6:705, 1972.
14. Chow, T. J., Snyder, C. B., and Earl, J. L.: Isotope ratios of lead as pollutant source indicators. In *Isotope Ratios as Pollutant Source and Behaviour Indicators.* Vienna, IAEA, 1975.
15. Hamilton, E. I. and Clifton, R. J.: Isotopic abundances of lead in estuarine sediments. Estuarine and Coastal Marine Sci. In press, 1978.
16. Magi, F., Facchetti, S., and Garibaldi, P.: Essences additionnées de plomb isotopiquement différence. In *Isotope Ratios as Pollutant Source and Behaviour Indicators.* Vienna, IAEA, 1975.

Surface Analysis

So far most of the discussion of element analysis has been concerned with assay for the total mass of an element present in a homogenized sample; such information is of obvious value but usually tells us very little about the micro and biochemical distribution of the elements in man. It is possible to use special chemical techniques to further our knowledge in this field, for instance a specific chemical separation can be performed which will isolate a defined compound and then the total, or extractable mass of an element present in such materials can be determined by bulk assay. In the ideal situation we need to determine both the mass and distribution of elements in biological materials without destruction or damage to biological structures; with advances in technology we are now entering an era in which some investigations of this type can be made; of particular interest is the chemistry of biological surfaces as it is on such materials that important biochemical functions are performed.

For the purposes of this very brief introduction to surface analysis, descriptions of methods and techniques will be concerned with the true surface and horizons just below the surface.

Scanning Electron Microscope (SEM)

The method involves bombarding the surface of a sample, under vacuum, with a focussed beam of electrons produced from a heated filament. The electron backscatter produces a visual image of the sample; electrons which penetrate to depth in the sample become thermalised and do not escape. As the electron beam is moved across a surface, topographic features analagous to those obtained with a light microscope are observed as illustrated in Figure 13. The number of electrons escaping from a surface is related to topographic features; for example, as the beam passes across a sharp edge more electrons are produced and the electron density of the surface picture is increased. By using X-ray fluorescence, or crystal or energy dispersive spectrometers as a part of the instrument or an attachment, element analysis can be performed on the scattered radiation and element distribution maps obtained as illustrated in Figure 14.

Electron Microprobe Analysers (EMA)

This technique, based upon the electron microscope, uses a focussed beam of electrons to excite the characteristic X-rays in an area as small as 1 μ in diameter. Element distributions may be determined by point analyses and line and area scanning. The application of this technique has been described by Birks (1) and Castaing (2), and applications to biological analysis have been reviewed at a recent symposium (3); the basic design of an electron microprobe system is illustrated in Figure 101.

Although analyses can be performed on very small volumes of material, equivalent to about 10^{-14} g, sensitivity of detection for most elements is between 100 and 1,000 ppm. Quite different approaches have to be adopted in microanalysis, and advantage can be taken of high concentrations of elements in very small volumes; tissues of particular interest are the lungs, kidney, liver, and bone surfaces. Williams et al. (4) describe the identification of silicon in retinal vessels following open heart surgery in which a silicone antifoam agent was used. For the unwary, some manufacturers' literature on microanalysis by these techniques can be misleading; pictures are often reproduced showing the levels or distribution of elements, such as osmium and lead, which are not present in the sample but have been added in histological preparations or are products from the source of an instrument; several other elements also enter samples as contami-

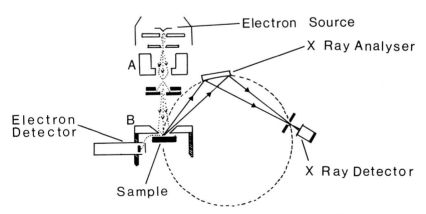

Figure 101: Schematic diagram of an electron microprobe analyser. A = condenser lens; B = objective lens. The X-ray analyser crystal is bent to the diameter of the Rowland circle and ground to the radius. The electron detector is for secondary electrons.

nants or during fixing and staining of materials although most problems can be overcome by use of frozen sections.

Modern electron microscopes have resolution capabilities near the length of a chemical bond but this is difficult to achieve because of damage during preparation of specimens and problems related to contrast. Further, although individual atoms of heavy elements can now be seen (especially following labelling techniques), atom distributions can become disturbed as a result of irradiation. Electron microscopy offers a quantitative approach to molecular resolution and it is therefore of some interest to note the work of Dorignac, Maclachlan, and Jouffrey (5) who have observed the triad of mercury and bromine atoms in the merbromine molecule ($C_{20}H_{10}Br_2HgO_6$) and are now investigating molecules with smaller spacings. With advances in technology very compact and versatile instruments are now commercially available such as the Jeol-100C TEMSCAN which incorporates transmission (TEM), scanning (SEM), and scanning transmission (STEM), together with X-ray analysis.

Ion-Microprobe Analyser (IMA)

The considerable success of electron microprobe X-ray analysis initiated by Castaing was followed by his work, in conjunction with Slodzian (6), in the development of an instrument which is essentially a secondary ion-emission microscope (SIM). A beam of energetic ions is focused upon a surface, secondary atoms are ejected of which a small fraction become ionized. A sector magnetic analyser, through the poles of which the secondary beam passes, acts as a mass analyser. Apart from the original papers, the ion microprobe has been described by Liebl (7) and a recent review of the method has been presented by Morrison and Slodzian (8). The ion optics of the system provide a capability of projecting the origin of the ions from the sample in the final mass resolved beam. The method essentially deals with ion sputtering from surfaces, but by holding the beam in one position, in-depth analysis can be performed from less than 1,000 Å to several microns. Apart from point source and line profile analysis, distribution pictures of the elements can also be obtained by mass dispersion. A magnification of ×1,000 is obtained when the image of the sample is displayed upon a fluorescent screen and provides optimum enlargement as the spatial resolution of the instrument is 1 μm for a field of view of a about 300 μm. The ion microscope permits rapid examination of element distributions; for example, a twenty second exposure can provide ten distribution pictures for ten elements when present at a concentration of greater than 1 percent, while consuming only 200 Å in depth of material; in biological studies the instrument permits correlations between morphological microfeatures and chemical compositions. For biological materials matrix effects arise because of the high abundance of elements such as H, C, O, N, and Ca, also as a result of the production of polyatomic ions which are present between mass 12 and 80. Biologically important elements such as Na, P, Cl, Mg, Mn, Fe, Cu, and Pb can be studied; sulphur does not appear to be detectable. Apart from inorganic components, the organic structure of cells can be identified and localised. The future of the ion microscope holds much promise but at present, for biological materials at least, interest centres upon hard tissue such as bone and teeth; it has yet to be applied to identification of elements in lung debris, calcareous plaques, and dense bodies associated with the kidney. Today there are two commercially available instruments, while ion probes can be added as an extra item to double focussing mass spectrometers.

Laser Probe Analysers

Laser beams may be used in several instruments, for example emission, fluorescence, and mass spectrometers, to excite the emission of elements from small volumes of biological material. Using a Q-switched ruby laser generating a plasma at $10^{5}°K$, material may be volatilised from micro areas with a diameter of between 5 and 25 μm. Using a laser coupled to a time-of-flight mass spectrometer (described by Fenner and Daly, 9), Hamilton (10) examined the distribution of several elements in samples of bone, bovine retina, liver, and liver nuclei; an example for an output from bone for a single laser shot is illustrated in

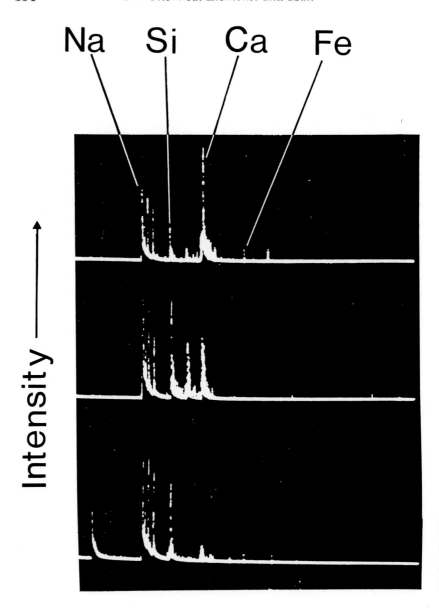

Figure 102: Single shot laser output spectra for rat bone.

Figure 102 for a mass range of 1 to 250; however the instrument used in this study had a resolution of only 30, at mass 30, and was insufficient to resolve higher masses. The technique appears to be valid for elements present at concentrations greater than 1 percent and it is worth noting that lead was detected in nuclei of bovine liver cells. The technique is destructive and the site of impact of each laser shot can be identified by a crater from which a small mass of the sample was volatilised.

Ion Scattering Spectrometer (ISS)

This instrument is capable of analysing for levels of elements and isotopes in the first atomic layer of a surface and is applicable to insulating, semiconducting, and metallic surfaces. The sample is placed in a scattering chamber under ultra high vacuum where it is bombarded with noble gas ions from an ion gun. A fraction of the bombarding ions leave the surface of the sample after only a single binary elastic collision with surface atoms and are then scattered through 90° in an electrostatic energy analyser and finally counted by an ion detector. From knowledge of mass and energy of the ion beam, and scattering phenomena, the mass of surface ions can be determined. Instruments have been described by Goff and Smith (11) and Karasek (12) and are commercially available. By selection of the appropriate ion probe and measurement of the scattered ion energy ratio, the mass of surface atoms is uniquely determined; the height of such peaks, recorded on a strip chart recorder, is a relative measure of the quantity of each constituent present. At present the method is mainly applied to the analysis of corrosion films on surfaces of metals and with depth, by successive removal of surfaces by ion bombardment at a rate of 3 to 150 monolayers/h; the method has an application in biology, for example the examination of frozen cut surfaces, but techniques and applications remain to be developed.

Electron Spectroscopy (ESCA)

Electron spectroscopy for chemical analysis utilises high energy X-rays to cause the emission of inner shell electrons whose binding energy unambiguously defines a particular atom. This instrument provides information for the valence state, bonding, and structural arrangement of materials. At present, work is focussed on the analysis of metal surfaces although the method has been recently applied to biochemical problems such as iron-protein binding studies, assay for N, F, Se, and Hg, the chemical form of sulphur compounds, and copper binding to AMP: it is quite surprising that so few applications of ESCA have been made for biochemical systems.

Auger Spectroscopy (AES)

Auger spectroscopy is based upon the bombardment of surfaces with electrons (2-3,000 ev) or X-ray photons and the ejection from surfaces of Auger electrons which have energy characteristics of the atoms from which they originate. The instrument operates under a high vacuum of 10^{-7} to 10^{-10} torr and concerns secondary emission processes as distinct from ESCA, which is concerned with the measurement of primary photoelectrons. Using Auger spectroscopy, element impurities can be determined in 0.1 percent of an atomic layer; also the chemical state of chemical compounds can be examined through measurement of chemical shifts. Chemical analysis can be made for all the elements of the Periodic Table except H and He. The spectra of biologically important elements, such as C, O, S, P, and Cl tend to be simple; however, as Auger peaks tend to be rather wide, for example about 10 ev, peak overlap occurs but can often be overcome by selecting interference-free peaks. Sample size is only limited by the dimensions of the source, and like ESCA, most studies have been concerned with metal surfaces and semiconductors as the response from insulators tends to be poor. AES can also be used in the scanning mode (SAM) and measures elements in a few atom layers compared with microprobe techniques which are concerned with a layer a few thousand atom layers thick and is an ideal instrument to study levels of impurities on surfaces. The application of AES to surface analysis has been described by Kowalczyk et al. (13).

Surface Composition Determined by Analysis of Impact Radiation (SCANIIR)

This method, described by White, Simms, and Tolk (14) and Tolk, Tsong, and White (15) is based upon the emission of visible, ultraviolet, and infrared radiation when surfaces are bombarded with a beam of low energy or neutral atoms or neutral molecules. The technique is largely nondestructive because of the use of low energy bombardment of less than 4 kev and about thirty minutes are required to obtain a complete spectral scan; analysis is concerned with the top few monolayers of a surface and for homogeneous materials detection sensitivities of 1 part in 10^8 can be achieved. White et al. (14) described the instrument and an application for the analysis of dried blood and plasma; for a 0.1 microlitre sample of dried plasma, radiation was observed from Na, Ca, K, Mg, C, and H; other common elements such as Fe and Cu were not detected.

Alpha Particle Backscatter Analyser (APBSA)

This method of analysis has been described by Patterson, Turkevich, and Franzgrote (16). Monoenergetic alpha particles from a very pure radioactive source are collimated through a system of simple slits and impinge upon the surface of a sample; the whole system is maintained under vacuum. The energy of the backscattered alpha particles is characteristic of elements present in the scattering material which are detected using surface barrier semiconductors. By utilising the low atmospheric pressure existing on the moon, an alpha particle source was used for direct irradiation of surface materials and analyses performed for major, minor, and some trace elements. Because of the short range of alpha particles, the technique is essentially one of surface analysis. Because of the problem of collimating alpha beams and the loss of energy, point source analysis using plated alpha sources is not possible but the technique is applicable to the analysis of fairly homogeneous surface materials.

In considering surface analysis for general areas and point

sources the problem of preparing adequate standards has not yet been discussed; in many applications standards are not required, and the observation of changes in the strength of a particular signal as a sample is scanned horizontally, or by depth profiling, provides valuable information. However, when accurate measurements are required it is possible to use (a) standards of metals, glass, or powders in which the elements of interest are homogenously distributed, (b) ion exchange resin beads to which a known mass of an element is attached, or (c) elements dissolved in gelatine. However, it is difficult to prepare adequate standards when analysing elements contained in very small volumes, although semiquantitative or qualitative data can usually be obtained. In any surface analysis, extreme precautions are required to reduce the deposition of contaminants, particularly if static electric charge is present. In many of the techniques it is possibly to sputter away monolayers of surfaces, but this technique can only be used for those elements which are homogenously distributed throughout the volume of material to be analysed.

Considerable advances are needed for the analysis of elements on surfaces of biological materials; of those which exist at present, microscope techniques, such as the electron and ion probes, provide a unique opportunity for observing the distribution of elements in relation to biological structures. Although detection limits for many elements are poor, many major and minor elements may be readily studied which are pertinent to the study of many diseases. With some exceptions, most techniques are concerned with the alkaline earths and metals together with a few other elements such as Fe, Cu, and Pb, and it is quite unrealistic to expect to attain sensitivities comparable to those obtained by atomic absorption, neutron activation, or mass spectrometric methods. As we probe deeper into microdistributions of elements by instrumental techniques, analytical problems and cost of basic equipment rise sharply but we should not forget the use of simple histochemical staining techniques for determining the cellular distribution of several elements.

Nevertheless, as advances are made in instrumentation, new

items become commercially available which permit a linking together of the requirements of chemistry and biology. An example of such an instrument is the molecular microprobe optical laser examiner (MOLE), produced by Jobin Yvon, France, which provides a microprobe, microscope, and Raman* microspectrophotometer for investigating molecular properties of materials and has been described by Delhaye and Dhamelincourt (17). The instrument permits a reconstruction of a micrographic image of isolated Raman spectrum characteristic of a selected compound and has a future in study of biological materials.

This brief survey for the analysis of elements in biological tissues has only dealt with a few methods and techniques. Most problems require the use of several methods, together with the availability of suitable expertise for analysis and interpretation of data. The total costs can be very high and the requirements of expertise very demanding; therefore, there is a real need to establish national centres where the required equipment can be housed and maintained under suitable environmental conditions; the necessary expertise brought under one roof, if only for limited projects; time allowed for essential development work; and most important, such a structure housed and integrated with medical disciplines for defined and objective study and research. Individuals have their own fields of interest and therefore the success of such multidisciplinary ventures depends upon enlightened and firm management in order that individuals do not become isolated in ivory towers and become detached from the defined objectives of research.

REFERENCES

1. Birks, L. S.: Electron Probe Microanalysis. New York, Wiley, 1963.
2. Castaing, R.: Electron probe microanalysis. *Advan Electron and Electron Physics*, 13:317, 1960.
3. Techniques et Applications de la Microanalyse en Biologiee. *J Microsc Biol Cellulaire*, 22:3, 1976.
4. Williams, I. M., Stephens, J. F., Brunckhorst, L. F. and Brodie,

---

* Based upon Raman scattering—examination of slight changes in photon energy due to interactions with vibrational or rotational energy of molecules.

G. N. J.: Identification of silicon in retinal vessels by electron probe X-ray microanalysis. *Histochem Cytochem*, 23:149, 1975.
5. Dorignac, D., Maclachlan, M. E. C., and Jouffrey, B.: Quantitative approach to molecular resolution electron microscopy. *Nature Lond*, 264:533, 1976.
6. Castaing, R. and Slodzian, G.: Microanalyse par emission ionique secondaire. *J Microsc*, 1:395, 1962.
7. Liebl, H.: Ion Microprobe Analyzer. *J App Phy*, 38:5277, 1967.
8. Morrison, G. H. and Slodzian, G.: Ion microscopy. *Anal Chem*, 47:932A, 1975.
9. Fenner, N. C. and Daly, N. R.: An instrument for mass analysis using a laser. *J Materials Sci*, 3:259, 1968.
10. Hamilton, E. I.: Review of the chemical elements and environmental chemistry—Strategies and tactics. *Sci Total Environ*, 5:1, 1976
11. Goff, R. F. and Smith, D. P.: Surface composition analysis by binary scattering of noble gas ions. *J Vac Sci Technol*, 7:72, 1970.
12. Karasek, F. W.: Surface analysis by ISS and ESCA. *Research and Development*, January 25, 1973.
13. Kowalczyk, S. P., Pollak, R. A., McFeely, F. R., Ley, L., and Shirley, D. A.: $L_{2,3} M_{45} M_{45}$ Auger spectra of metallic copper and zinc, theory and experiment. *Phys Rev B*, 8:2387, 1973.
14. White, C. W., Simms, D. L., and Tolk, N. H.: Surface composition determined by analysis of impact radiation. *Science*, 77:481, 1972.
15. Tolk, N. H., Tsong, I. S. T., and White, C. W.: In situ spectrochemical analysis of solid surfaces by ion beam sputtering. *Anal Chem*, 49:16A, 1977.
16. Patterson, J. H., Turkevich, A. L., and Franzgrote, E.: Chemical analysis of surfaces using alpha particles. *J Geophys Res*, 70:1311, 1965.
17. Delhaye, M. and Dhamelincourt, P.: Raman microprobe with laser excitation. *J Raman Spec*, 3:33, 1975.

Data Retrieval

In trace element analysis an objective is often the retrieval of a number which is a measure of the mass of an element present in a sample. Some types of analysis may be concerned with the rate at which analytical signals from elements change with time, others may require information present in a signal which will identify chemical form, chemical stability, or physical information capable of interpretation of the spatial distribution of elements in samples, for example on surfaces. Many manual and time-consuming techniques exist for achieving such objectives but in recent years the computers, minicomputers, and microprocessors have emerged as essential tools for laboratory data handling. Early problems of interfacing instruments to computors dealt with problems of electronics, especially matching signals from instruments to acceptance requirements of computors. Apart from problems of hardware, those related to development of suitable software were often difficult to overcome. Many of these problems have been solved following developments in the fabrication of integrated circuits and their incorporation into modular packages. Commercially available equipment can be interfaced to a wide variety of instruments and is capable of on-line processing or can be linked to a large computor, supported by versatile software to carry out complex programmes of data retrieval in a form required by defined investigations. Total instrumental analysis can now be achieved based upon some form of computor system and thus a considerable number of processes aimed at verifying data can be accomplished, for example matrix modules can be constructed which carry out tedious calculations involved in determining the presence and extent of spectral overlap, then undertaking any spectral stripping procedures for their removal. By using feedback systems, operating conditions can be optimised during an analysis thus enhancing the quality of data. Finally data can be accumulated on computer files and used in a wide variety of interpretative programmes which are an essential requirement for multielement work. In many applications, developments in signal and data processing have to be accompanied by new items of hardware for automatic loading of samples into source regions of instruments, otherwise many advantages of signal processing become uneconomic particularly for those laboratories concerned with repetitive analyses. The choice of systems has to rest with the analyst and be considered in relation to the nature of the work rather than innovations in electronic technique.

Advances in electronics applied to analytical chemistry are welcomed but a real danger exists in becoming enraptured with the sophistication of electronics and to forget that many of the real problems lie with expertise in basic analytical technique and methodology. Computer systems cannot turn poor analytical data into good data; many of the real problems are not concerned with the final stage of analysis but rather sampling and sample preparation. Total systems of analysis based upon advances in electronics are indispensible in studying the chemical elements and human morbidity. Increasingly, because of the multidisciplinary nature of the work, innovations in electronics provide an essential interface, particularly if developed in central laboratories. For example immediate correlation of computor stored element data with the epidemiology of disease can be used to identify geographical or clinical associations; when linked to system analysis procedures, analytical data can be considered in terms of budget models, metabolic pathways, and perhaps more important as an educational tool in identifying priorities of research.

With increased use of computor-based systems in analytical chemistry, it becomes imperative that acceptable protocols and harmonised methods of analysis are established in order that data from different laboratories can be directly compared. Woodward, Ridgway, and Reilly (1) and Clerc and Ziegler (2) have described an instrumentation-orientated computer which is inexpensive and can optimise data acquisition for the automated laboratory. In recent years, standard commercial instruments can be purchased with built-in data acquisition and automation control systems.

REFERENCE

1. Woodward, W. S., Ridgway, T. H., and Reilley, C. N.: An instrumentation orientated microcomputer. *The Analyst,* 99:838, 1974.
2. Clerc, J. T., and Ziegler, E.: Computor Techniques and Optimization. *Anal Chim Acta,* 95:1, 1977.

Chapter 8

ELEMENTS AND MAN

THE PREVIOUS CHAPTER concerned a general survey of the methods of analysis, and emphasis was placed upon the many problems associated with the final stage of analysis in a chemical laboratory. The analytical quality of such data may be excellent but can be erroneous in terms of the actual levels present in human beings, and quite misleading when attempting to relate levels of elements in tissues with disease because of the lack of information concerning sampling and preparation, also the nature of the chemical form of an element together with dose response relations linking the form of an element and morbidity. Unless acceptable protocols and rigid controls are initiated for all stages of an investigation, the potential value of such work may be diminished or even lost. However, the literature is replete with data for gross levels of the elements in healthy and diseased tissues and it is upon such evidence that cause for concern is often aroused.

First, mention should be made of the few large scale compilations of published data for a large number of elements, for example the Report of the Task Group in Reference Man by ICRP (1) largely based upon the pioneering work of Isobel Tipton and her co-workers, also that of Anspaugh and his co-workers (2) for healthy and diseased states. Then follow several studies describing levels of many elements in various tissues obtained by one laboratory, such as those described by Stitch (3), Tipton and Cook (4), Soman et al. (5), Hamilton, Minski, and Cleary (6), Sumino et al. (7) and Bowen (8).

Literature compilations are very valuable and should provide some guidelines for expected levels but most conceal data of very variable quality, since it is left to the compiler to accept all data or select some, possibly on the basis of the method of analysis used although it does not follow that the best method will always provide the correct answer. Data originating from a single laboratory is usually accompanied with details of the methods used and hence some judgement can often be made concerning quality. The optimum approach consists of a detailed evaluation of all published data and the selection of "best values" based upon clearly defined criteria. This approach is quite acceptable and provided that some general rules can be applied very useful compilations can be obtained; this approach was adopted by Urey (9) in classifying meteorites on the basis of chemical compositions. However, it must be clearly understood that the calculation of mean data from the literature can, at the best, only provide general baseline data by which the relative levels of the elements in human tissues can be estimated. In the absence of a consideration of the influence of geography, diet, age, sex, and, in the instance of diseased tissues, adequate controls, together with information concerning all aspects of the methods used, limitations of baseline data must be recognised. Failure to recognise such limitations could have serious consequences; for example, accepting data obtained from very experienced laboratories, often situated in highly developed countries, as representing normal levels might then be applied to developing countries associated with totally different types of diet and cultures. A further example would be the case of data obtained from one geographical area in which elements are enriched, or depleted, as a consequence of natural or technological processes. Because of the complex nature of cultural and social structures, more attention needs to be paid in establishing regional and national baseline data and perhaps very detailed studies on small communities or even groups of people.

Criteria for establishing essential and nonessential elements for man still remain to be defined; Schwarz (10), Liebscher and Smith (11), and Schroeder (12) have considered several elements but some of the data are obtained from animal experiments which may not be relevant to man. A few papers are available which consider the total role of selected elements in man in relation to biological importance; Halsted, Smith, and Irwin (13) have presented a conspectus of research on zinc requirements for man; copper homeostasis in the mammalian system has been described by Evans (14), and the role of manganese in health and disease by Cotzias (15). One fundamental paper on the metabolic character of the elements as influenced by their position in the Periodic Table has been presented by Durbin (16) for some seventy elements; although the work was carried out on the rat by radiotracer techniques, it was concluded that the properties which determine biological behaviour of the elements are oxidation state in relation to pH, solubility of compounds, rate of incorporation into organic compounds, and the tendency to associate with specific proteins.

In 1966 I was concerned with one study aimed at providing some preliminary baseline data for the United Kingdom. I have selected this Radiological Protection Survey (RPS) 1971, described by Hamilton, Minski, and Cleary (6) and subsequently updated slightly, in order to illustrate problems associated with such exercises. The prime objective was to obtain information for the levels of a large number of elements present in human tissues relevant to the problems of radiological protection. Accurate analysis for a single element can take between three and six months in order to perfect techniques, and quite clearly it is not possible to analyse for a large number of elements and maintain acceptable accuracy for all the elements; compromises thus have to be made. For many elements in normal individuals the cost of obtaining accurate data is too high and often of very little scientific value; for purposes of obtaining baseline data, it is questionable whether the vast increase in effort is worthwhile simply because of biological variability influenced by diet, exposure, the manner in which individuals respond to elements, and circadian rhythms in relation to time of sampling to mention a few. In many investigations, it is more relevant to aim for high precision of assay in order that small differences in levels may be identified, even though the data may not be highly accurate. The extremes of the problem range from the need to obtain data for large populations and also for individuals of a population for which chemical imbalances may be responsible for disease, which is especially pertinent to developing countries.

In selecting the appropriate methods the following factors were considered:

1. The need to use a very sensitive method suitable for minor and trace element assay requiring the minimum of sample preparation and for which major problems of sample matrix could be overcome or accommodated in the exercise. The instrument of choice was the spark source mass spectrometer, in spite of it rarely being used for the analysis of biological materials. A more commonly used method, that of optical emission spectrometry, is less sensitive and a series of analyses are required in order to obtain adequate cover of the elements. Today the instrument of choice would possibly be the inductively coupled plasma emission spectrometer.
2. In order to overcome matrix interferences arising because of the presence of intense signals emitted by major elements they, together with the minor and some trace elements, were determined by crystal dispersion X-ray fluorescence analysis. The method requires very simple techniques for sample preparation and matrix problems can be identified.
3. A third method was required in order to analyse for some elements difficult to determine by the other methods, also to provide a very accurate method of analysis free from matrix problems. The method of choice was activation analysis.

The first stage of the analytical exercise involved the preparation of adequate tissue standards which were cross checked, whenever practical, by all three methods. The basic techniques used were standard addition analysis and use of the few available reference materials such as Bowen Kale, NBS Orchard Leaves, various IAEA standards, and some of the USGS rock standards.

The first two and one-half years of the project were spent in developing techniques of sampling, sample preparation, and standardisation, with the emphasis being placed upon obtaining high precision of assay and acceptable accuracy within the framework of the overall objectives. The magnitude of organisation of such a project is quite different to that required for the assay of a single, or a few, elements: far bolder de-

TABLE LIX*

DETECTION LIMITS FOR ASSAY OF HUMAN TISSUES BY SPARK SOURCE (MS702) MASS SPECTROGRAPHY† AND X-RAY FLUORESCENCE‡ ANALYSIS (CRYSTAL)

Data in µg/g wet weight

Element	Soft Tissues MS702	XRF	Element	Soft Tissues MS702	XRF
U	0.002		Rh	0.0001	
Th	0.002		Ru	0.0008	
Bi	0.002		Mo	0.003	1.1
Pb	0.003		Nb	0.0007	
Tl	0.004		Zr	0.001	
Hg	0.007		Y	0.0007	
Au	0.003		Sr	0.0008	
Pt	0.008		Rb	0.001	0.44
Ir	0.0004		Br	0.002	0.35
Os	0.001		Se	0.002	
Re	0.0004		As	0.0006	
W	0.0009		Ge	0.007	
Ta	0.0003		Ga	0.0009	
Hf	0.0009		Zn	0.002	0.45
Lu	0.002		Cu	0.002	0.26
Yb	0.007		Ni	0.002	
Tm	0.002		Co	0.0005	
Er	0.007		Fe	0.0005	1.1
Ho	0.002		Mn	0.0004	0.13
Dy	0.009		Cr	0.0005	
Tb	0.002		V	0.0004	
Gd	0.01		Ti	0.007	0.3
Eu	0.004		Sc	0.0004	
Sm	0.008		Ca	0.0003	1.5
Nd	0.007		K	0.0003	2.0
Pr	0.002		Cl	0.0004	6.0
Ce	0.002		S	0.0003	5.0
La	0.002		P	0.0003	8.9
Ba	0.002		Si	0.0002	11.3
Cs	0.001		Al	0.0002	
I	0.001		Mg	0.0002	
Te	0.003		Na	0.0002	
Sb	0.002		F	0.0002	
Sn	0.004		B	0.0001	
In	0.0001		Be	0.00008	
Cd	0.007	1.5	Li	0.00006	
Ag	0.002				

* From Hamilton, E. I., Minski, M. J., and Cleary, J. J.: The concentration and distribution of some stable elements in healthy human tissues for the United Kingdom. *Sci Total Environ*, 1:361, 1973.
  † Electrodes 90% ashed tissue 10% carbon.
  ‡ Freeze dried/compressed tablets (see p. 324).

TABLE LX*

THE CONCENTRATION OF ELEMENTS IN BLOOD FROM THE UNITED KINGDOM†

Data in µg/g wet weight

Element	U.K. Master Mix±	R.P.S. Values Mean Values for Individual Bloods§	No. of Samples	Anspaugh et al. Values‖		
				Mean	No. of Samples	Range
U	ND	0.0008 ± 0.0001	21	ND		
Th	ND¶	0.002 ± 0.0004	30	ND		
Pb	0.3 ± 0.1	0.3 ± 0.03	103	0.6	70	
Tl	0.0005 ± 0.0001	ND		ND		
Hg	~(0.006)	ND		0.005	14	0.005-0.005
Au	ND	<0.0003		0.00004	6	0.00003-0.00006
Ba	0.08 ± 0.03	0.1 ± 0.06	103	<0.4		
Cs	0.01 ± 0.006	0.005 ± 0.0006	103	ND		
I	0.06 ± 0.04	0.04 ± 0.003	103	0.07	117	0.02-0.1
Sb	0.002 ± 0.0006	0.005 ± 0.0009	75	0.005	5	0.003-0.009
Sn	0.005 ± 0.002	0.009 ± 0.002	102			
Ag	0.01 ± 0.005	0.008 ± 0.0008	93			
Nb	0.004 ± 0.0005	0.005 ± 0.0007	85			
Mo	0.001	ND	8	0.2	1	
Zr	0.01 ± 0.006	0.02 ± 0.008	98	ND		
Y	0.007 ± 0.003	0.005 ± 0.0006	98	ND		
Sr	0.02 ± 0.005	0.02 ± 0.002	102	0.01	6	0.007-0.03
Rb	2.8 ± 0.1	2.7 ± 0.04	165	1.5	6	1.0-2.0
Se	0.08 ± 0.04	0.06 ± 0.005	98	0.2	251	0.2-0.2

Element	Value ± SD	Value ± SD	n	Value ± SD	n	Range
Br	4.8 ± 0.1	4.6 ± 0.06	163	3.9	16	1.6-12.6
As	<0.01	<0.01	103	0.3	11	0.09-0.7
Zn	7.0 ± 0.2	6.7 ± 0.05	169	7.4	182	4.0-14.8
Cu	1.2 ± 0.02	1.2 ± 0.02	168	0.9	269	0.3-11.5
Fe	498.0 ± 3.9	490.5 ± 2.4	168	390.0	2	
Mn	0.09 ± 0.05	0.05 ± 0.009	102	0.02	42	0.005-0.02
Cr	0.07 ± 0.05	0.03 ± 0.004	102	0.03	155	0.01-0.06
Sc	0.008 ± 0.003	ND				
Ca	60.4 ± 0.3	61.5 ± 0.7	167	60.0	6	40-80
K	1,900 ± 20	1,860 ± 7	169	2,796 ± 403	104	
Cl	3,020 ± 77	3,000 ± 20	170	2,731	7	2,370-3,050
S	1,770 ± 14	1,770 ± 10	167	ND		
P	335.0 ± 10	327.9 ± 1.6	169	400	6	300-550
Al	0.2 ± 0.06	0.4 ± 0.6	94	ND		
Mg	48.8 ± 16.1	46.4 ± 6.6	103	45.1	60	34-54
Na	1,770 ± 100	2,030 ± 20	6	1,700	10	1,500-2,200
F	0.07 ± 0.02	0.2 ± 0.05	100	0.2	37	0.04-0.4
Li	0.002 ± 0.0004	0.006 ± 0.002	101	3.0	6	1.9-4.4

* From Hamilton, E. I., Minski, M. J., and Cleary, J. J.: The concentration and distribution of some stable elements in healthy human tissues for the United Kingdom. *Sci Total Environ*, 1:350, 1973.

† Although as indicated in Figure 103 samples of blood were collected from many different geographical localities in England and Scotland, for practical reasons sample collection was biased towards urban populations. Insufficient numbers of samples from different areas precludes a more detailed consideration of regional differences. However one element, namely lead, has been considered in more detail and it is estimated that the concentration of this element in a rural population is 0.18 ± 0.08 μg gm wet wt.

‡ The United Kingdom Master Mix data are the mean for four separate analyses.

§ Mean values for individual samples of blood are representative for a number of regional areas of the United Kingdom.

‖ Data from Anspaugh, L. R. et al. (2).

¶ ND = Not determined.

TABLE LXI*

THE CONCENTRATION OF TWELVE ELEMENTS IN THE BLOOD OF CHILDREN IN A NORMAL CONTROL GROUP COMPARED WITH PATIENTS SUSPECTED OF LEAD INTOXICATION

Metal	Control Group of Children			Patients		
	Mean Concentration†	Observed Range	No. of Tests	Mean Concentration†	Observed Range	No. of Tests
Iron	389.0	290-525	91	373.4	200-520	206
Zinc	510.0	319-1,040	84	617.9	210-2,284	192
Copper	97.9	55-147	84	102.2	44-162	183
Lead	11.7	1-32	44	32.3	1-140	203
Manganese	1.2	0-5.7	90	1.8	0-15.0	206
Cadmium	0.49	0-1.9	89	0.57	0-7.9	204
Strontium	2.8	0-8.9	77	3.0	0-27.4	182
Chromium	3.5	0-9.7	50	4.1	0-15.0	85
Cobalt	0.46	0-2.1	67	0.32	0-5.0	167
Nickel	2.3	0-9.3	69	1.37	0-8.4	167
Lithium	0.38	0-2.1	69	0.36	0-3.0	146
Bismuth	0.67	0-6.5	48	0.93	0-6.5	131

* From Delves, H. T., Clayton, B. E., and Bicknell, J.: Concentration of trace metals in the blood of children. Br J Preront Soc Med, 27:100, 1973.

† Concentrations are µg/100 ml for all metals except iron, in which the units are µg/ml.

cisions are required as it is quite pointless devoting a large amount of time and energy to perfecting some details and ignoring others which may have a more important bearing on the final value of the data. This Ministry of Health and Medical Research Council sponsored programme was terminated in 1971 following government reorganisation but some progress was made in the study of blood, several tissues of the body, together with various environmental materials.

As an aid in the comparison of published data, it would be very useful if some agreement could be reached for the manner in which data are presented; in this study soft tissues are presented as wet weight and bone as ashed weight; data for wet, dried, and ashed weights are presented in Table LII. Practical detection limits for assay of elements in various tissues determined by spark source mass spectrometry and X-ray fluorescence analysis are given in Table LIX.

Elements and Man

Figure 103: The geographical distribution of the main sources from which human blood was sampled in the United Kingdom (excluding N. Ireland). Altogether some 2500 samples were collected and prepared for analysis by freeze drying 10 mls of whole blood. This collection may serve as reference material to examine any overall changes in the composition of UK blood in the future.

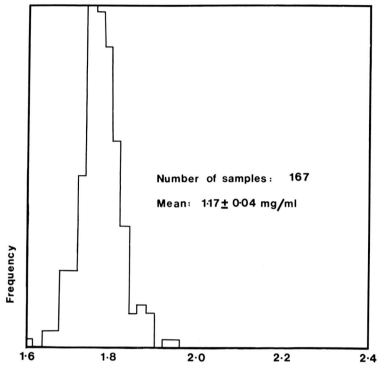

Figure 104: The sulphur content of human blood (mg/ml).

Blood

The geographical distribution for the localities from which samples of whole blood were sampled is given in Figure 103. One of the major problems concerned sampling blood without the introduction of contaminants; while the problems can be reduced by extracting blood from small open "cuts" this is not practical for large scale surveys. Instead normal techniques of sampling were used but the first 10 ml of blood were discarded. In Table LX data are provided for the mean levels of elements present in individual samples, the mean of a master mix containing one gram aliquots from 2,500 samples of blood from United Kingdom residents, and comparative data taken from the literature. Variability in the concentration of sulphur,

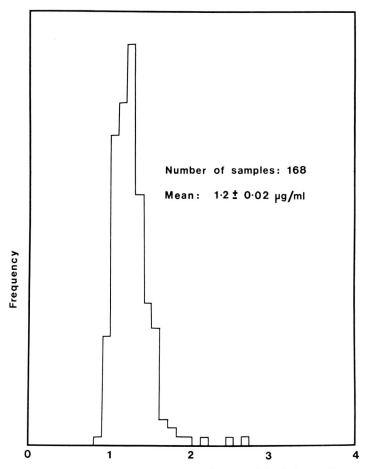

Figure 105: The copper content of human blood (μg/ml).

copper, and manganese in blood are illustrated in Figures 104, 105, and 106. With the exception of hydrogen, carbon, nitrogen, oxygen, and the rare gases, all elements were sought and when not recorded were below the limits of detection. Although levels of elements in children's blood were not studied in this investigation, data are presented in Table LXI and Figure 107 obtained by Delves, Clayton, and Bicknell (17) for estimating normal levels when considering individuals suspected of having been subjected to lead poisoning. The partition of elements

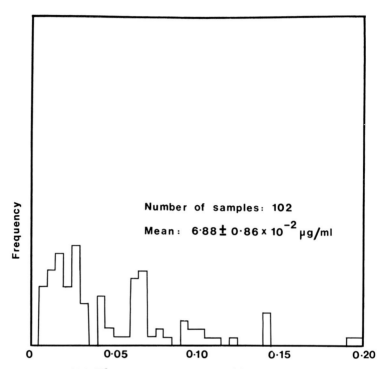

Figure 106: The manganese content of human blood (μg/ml).

between erthyrocytes and plasma has been described by Kasperek et al. (18) using neutron activation and Flint, Lawson, and Standil (19) by X-ray fluorescence analysis.

Considerable problems arise in obtaining a homogeneous sample of dried blood but undoubtedly the most serious source of error is from contamination, which is enhanced if blood is separated into its constituent parts. Examination of literature values for Mn levels in sera illustrates this point, levels between 0.06 and 1.7 μg/100 ml have been reported; in 1966 Cotzias, Miller, and Edwards (20) obtained a value of 0.0587 ± 0.0183 μg/100 ml; in 1976 Versieck et al. (21) a value of 0.057 ± 0.0013 μg/100 ml; while in 1953 Hagenfeldt, Plantin, and Diczfalusy (22) reported values between 0.18 and 0.32 μg/100 ml. The general consensus of opinion is that values greater than 0.08 μg Mn/100 ml are possibly contaminated; great care

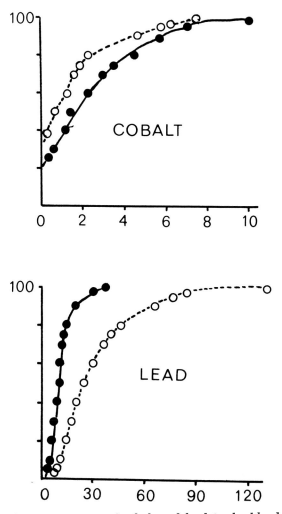

Figure 107: The concentration of cobalt and lead in the blood of children. Open circles represent patients considered to be subject to lead poisoning. Closed circles represent the control group. Concentration, µg/100 mls. From Delves, H. T., Clayton, B. E., and Bicknell, J.: Concentration of trace metals in blood of children. *Br J Preront Soc Med*, 27:100, 1973.

must be taken when making such statements because of real differences attributable to geography, also unique values for levels of elements in blood and tissues may not exist and the exclusion of abnormal data may not be justified. Some elements,

TABLE LXII*

THE CONCENTRATION OF SOME ELEMENTS IN WHOLE HUMAN BLOOD DETERMINED BY NEUTRON ACTIVATIONS ANALYSIS

Element	Average Concentration in μg/g Blood	Range in μg/g Blood	Number of Determinations
Mn	0.015	0.005-0.02	10
Cu	1	0.5-2.5	10
Zn	6	4-8	10
Na	1,700	1,500-2,200	10
P	400	300-550	6
Sr	0.01	0.007-0.03	6
Fe	390		2
Ca	60	40-80	6
Au	0.00004	0.00003-0.00006	6
Sb	0.005	0.003-0.009	5
Cd	0.005	0.002-0.008	6
Cs	0.003	0.002-0.004	4
Rb	1.5	1-2	6

* From Wester, P. O.: Unpublished data.

such as Al, are rarely determined; Galenkamp, Leeuwen, and Das (23) obtained a value of between 0.18 and 0.20 μg/ml determined by neutron activation, which is similar to that found in the RPS survey and provides some indication that the samples have not been contaminated with any appreciable amounts of environmental dust containing mineral debris. Finally a comparison of the RPS data with neutron activation results on individuals from Sweden, obtained by Brune et al. (24), and given in Table LXII, exhibit many similarities.

Attempts have been made to correlate United Kingdom data from different geographical regions in order to determine the extent to which environmental factors influence levels of elements in blood. The number of analyses are too few to draw any firm conclusions, but levels of lead tend to be higher in urban areas while elevated levels of tin were found for a tin miner who had worked underground for thirty years.

Brain

Data from the RPS survey for brain (whole brain, frontal lobes, and basal ganglia) are given in Table LXIII and compared with data from the literature. Because of the complex structure of the brain, data obtained from whole brain are of limited value in the diagnosis of disease; data are required to describe levels of elements in well-defined parts of the brain associated with defined biological functions. Henke, Möllmann, and Alfes (25), see Table LXIV, have determined Cu, Se, Ag, Zn, Rb, Au, La, and Ba in thirteen different parts of the brain and observe that concentrations of elements increase in extrapyramidal motor activity centres rather than in areas of white matter. Ule, Völkl, and Berlet (26) determined Cu, Zn, Ca, and Mg in thirteen areas of the brain, compared with levels of iron during the fourth to eighth decade of life; levels of elements in infant brain tended to be less than those found for adults and, for both, regional patterns were noted for Fe and Cu; for example Cu was enriched in the substantia nigra, higher than the semiovale and lower than the frontal cortex. Schicha et al. (27) describe levels of some elements in surgically removed samples of human brain; Co, Se, and Zn were enriched in the cerebral cortex compared with the medulla possibly because of the presence of ganglion cells. In comparing in vivo with autopsy data, both Rb and Co were lower possibly because of postmortal diffusion. Höck et al. (28) determined Co, Fe, Rb, Se, Zn, Cr, Ag, Cs, and Sb in sixty tissues from thirteen samples of human brain: greater differences were observed for variability of the nonessential compared with the essential elements; Fe and Rb were higher in basal ganglia than cortex. Schicha et al. (29) have observed that the iron content in the brain of the newborn is only 20 percent that of an adult and 60 percent in the case of a four-year-old child; adult iron levels were not reached until an age of seven to nine years. Today attention is being paid to changes in levels of elements from birth to maturity, with particular reference to the unusual exposure to which children are subjected.

The hemispheres of the brain control different sides of the body, usually one hemisphere is dominant over the other and any increase in levels of the elements in one half could be

TABLE LXIII*

THE CONCENTRATION OF SOME ELEMENTS IN HUMAN BRAIN FROM URBAN AREAS OF THE UNITED KINGDOM

Element	Whole Brain	No. of Samples	R.P.S. Values Frontal Lobe‡	Basal Ganglia‡	No. of Samples	Anspaugh et al. Values† Mean Whole Brain	No. of Samples	Range
Bi	0.01	10	0.04 ± 0.03	<0.008 ±	2	<0.03		
Pb	0.3 ± 0.1	10	0.2 ± 0.06	0.1 ± 0.03	2	4.8	14	0.01-19.1
Tl	<0.001	4	ND	ND		ND		
Ba	0.006 ± 0.0003	10	0.008 ± 0.002	0.05 ± 0.04	2	0.03	18	0.01-0.1
I	0.02 ± 0.002	10	0.02 ± 0.15	0.03 ± 0.006	2	ND		
Sb	0.007 ± 0.0004	10	0.01 ± 0.002	0.02 ± 0.003	2	ND		
Sn	0.06 ± 0.01	10	0.03 ± 0.07	0.04 ± 0.02	2	<0.1		
Cd	0.3 ± 0.04	5	0.04 ± 0.004	0.07 ± 0.01	2	<0.8		
Ag	0.004 ± 0.002	10	0.003 ± 0.001	0.004 ± 0.0002	2	0.03	1	
Zr	0.02 ± 0.001	10	0.01 ± 0.006	0.03 ± 0.002	2	2.6	4	0.3-6.4
Y	0.004 ± 0.0001	10	0.007 ± 0.003	0.02 ± 0.0009	2	ND		
Sr	0.08 ± 0.01	10	0.2 ± 0.2	0.07 ± 0.03	2	0.04	145	0.008-0.07
Rb	4.0 ± 1.1	2	3.2 ± 0.1	5.8 ± 0.1	2	ND		
Br	1.7 ± 0.05	2	3.1 ± 0.1	3.1 ± 0.1	2	ND		

Element		n			n		n	Range
Se	0.09 ± 0.02	10	0.05 ± 0.001	0.05 ± 0.02	2	0.2	4	0.2-0.3
As	0.10 ± 0.01	8	ND	ND	2	ND		
Ga	0.0006 ± 0.00003	10	0.001 ± 0.0001	0.002 ± 0.0001	2	ND		
Zn	12.5 ± 0.05	10	13.1 ± 0.1	13.2 ± 0.1	2	13.7	134	5.2-18.6
Cu	5.6 ± 0.2	2	4.7 ± 0.1	6.2 ± 0.1	2	6.2	137	2.2-8.5
Fe	56.6 ± 7.0	2	43.8 ± 0.5	58.1 ± 0.7	2	57.6	108	37.3-81.4
Mn	0.2 ± 0.03	10	0.2 ± 0.03	0.3 ± 0.05	2	0.3	127	0.2-0.5
Cr	0.01 ± 0.001	10	0.03 ± 0.005	0.09 ± 0.06	2	0.03	1	
V	0.03 ± 0.008	10	0.02 ± 0.005	0.09 ± 0.04	2	<0.02		
Ti	0.8 ± 0.05	2	1.0 ± 0.01	0.8 ± 0.01	2	<0.08		
Ca	57.0 ± 0.5	2	62.6 ± 11.1	58.0 ± 5.1	2	100	108	55.9-162.7
K	2,500 ± 600	2	2,450 ± 10	3,150 ± 100	2	3,220	102	2,203-4,237
Cl	1,400 ± 200	2	2,135 ± 10	1,780 ± 54	2	ND		
S	1,500 ± 10	2	1,295 ± 15	1,440 ± 50	2	ND		
P	2,800 ± 10	2	2,585 ± 41	3,096 ± 158	2	3,390	102	2,712-4,407
Si	23.0 ± 4.4	2	13.8 ± 0.4	24.2 ± 4.4	2	ND		
Al	0.5 ± 0.1	10	0.05 ± 0.001	0.07 ± 0.04	2	0.4	127	0.1-0.6
Mg	193.8 ± 38.0	10	198.3 ± 65.9	126.5 ± 7.1	2	155.9	108	98.3-203.4
F	0.03 ± 0.01	10	0.03 ± 0.02	0.01	1	ND		
Li	0.004 ± 0.001	10	0.003 ± 0.001	0.004 ± 0.003	2	ND		

* From Hamilton, E. I., Minski, M. J., and Cleary, J. J.: The concentration and distribution of some stable elements in healthy human tissues for the United Kingdom. *Sci Total Environ*, 1:351, 1973.
† Data from Anspaugh, L. R., et al. (2).
‡ 2 samples each.

TABLE LXIV*

THE CONCENTRATION OF SOME ELEMENTS IN HUMAN BRAIN
DETERMINED BY RADIOACTIVATION ANALYSIS

Data in $\mu g/g$ net weight.

	Cu	Se	Ag	Zn	Fe	Rb	Cs$[10^{-4}]$	Au$[10^{-5}]$	La$[10^{-2}]$	Ba
I. Grosshirn										
1. Substantia corticalis	1.38	0.43	0.081	14.00	24.35	0.66	1.40	12.0	2.84	
2. Substantia alba	1.60	0.61	0.079	8.38	20.05	1.07	6.63	54.0		
3. Corpus callosum	0.56	1.18	0.060	10.71	15.03	1.32	8.27	14.0		
4. Hippocampus	1.78	1.35	0.083	31.62	23.63	1.95	13.7	4.6	4.01	
5. Globus pallidus	3.88	2.85	0.198	14.53	133.10	1.78	9.71	7.2		
6. Caput nuclei caudati	2.94	0.48	0.144	23.03	91.93	2.18	9.13	9.9		
II. Zwischenhirn										
1. Thalamus	1.92	0.37	0.097	15.77		4.25	7.49	8.4		
2. Corpus mamillare	2.80	2.48	0.256	18.41	39.23	1.62	10.5	5.1		3.60
III. Mittelhirn										
1. Nucleus ruber tegmenti	3.12	1.82	0.076	11.09	63.96	1.21	9.13	5.7		
2. Substantia nigra	5.23	2.65	0.308	15.22	97.64	1.41	9.17	17.0		
3. Lamina tecti	2.75	2.34	0.175	17.86	22.45	0.93	7.84	13.0		1.53
IV. Kleinhirn										
1. Nucleus dentatus	3.67	5.21	0.319	13.11	30.52	1.13	8.11	8.0		
2. Vermis cerebelli	1.79	2.11	0.150	10.95	19.52			7.5		

* From Henke, G., Möllmann, H., and Alfes, H.: Vergleichend e Untersuchungen über die Konzentration einiger Spurenelemente in menschlichen Hirnarealen durch Neutronenaktivierunganalyse. Z Neurol, 199:283, 1971. Courtesy of Springer-Verlag, Heidelberg.

of clinical interest; Greiner, Chan, and Nicolson (30) determined Ca, Cu, Mg, and Zn in identical areas of cerebral hemisphere from normal subjects and noted no differences, except for Cu which was higher in the right side. Hu and Friede (31), using atomic absorption spectrophotometry, describe an enrichment of Zn in resistent sector and Ammon's horn; elsewhere Zn levels were rather similar but white matter showed lower levels than grey matter.

Many elements exhibit toxic effects through neurological symptoms, and it is to be hoped that attention will become focussed upon levels of elements present at sites of neurological damage. One element, lead, is considered to be associated with neurophysiological changes. Zaworski and Oyasu (32) have determined a concentration of 0.50 to 0.68 ppm Pb wet weight in 191 normal brains; a sharp increase was found in cerebral lead concentrations during the first two decades of life. Whether or not this is a normal process related to dietary intake is unknown, but Duggan and Williams (33) have described the increased risk of lead intake by children through exposure to street level dirt and thumb sucking. In seeking relations between neurological disease, multielement studies on small histologically identified parts of the brain are very worthwhile; Gooddy, Williams, and Nicholas (34) and Gooddy, Hamilton, and Williams (35) have discussed the role of methods such as spark source mass spectrometry in such investigations, if only as a preliminary to more detailed studies.

Liver and Kidney

Data from the RPS survey for liver and kidney are given in Tables LXV and LXVII. A comparison of the RPS data with that obtained by Hoste (36) for liver by neutron activation for Belgian subjects is given in Table LXVI. The chemical elements in liver tend to show a homogeneous distribution compared with kidney for which significant differences between cortex and medulla exist. Both organs are intimately associated with dietary intake of the elements and the removal of elements from the body, and both contain the highest level of many elements found in the human

TABLE LXV*

THE CONCENTRATION OF SOME ELEMENTS IN HUMAN LIVER FROM AN URBAN AREA OF THE UNITED KINGDOM
(μg/g wet wt.)

Element	R.P.S. Values Liver	No of Samples	Anspaugh et al. Values† Mean	No. of Samples	Range
U	0.0008 ± 0.0001	21	ND		
Bi	~0.004	11	0.09	1	
Pb	2.3 ± 0.6	11	1.5	393	0.2-3.2
Tl	~0.009	4	ND		
La	0.08 ± 0.03	11	0.0003	1	
Ba	0.01 ± 0.003	11	0.04	19	0.01-0.2
I	0.2 ± 0.06	11	ND		
Sb	0.01 ± 0.002	11	0.01	2	0.008-0.01
Sn	0.4 ± 0.08	11	<0.4		
Cd	2.0 ± 0.4	11	0.1	2	0.1-4.4
Ag	0.006 ± 0.002	11	0.04	1	
Mo	0.4 ± 0.2	11	1.0	149	0.4-1.8
Nb	0.04 ± 0.009	11	3.7	3	1.4-9.2
Zr	0.03 ± 0.005	11	6.3	3	1.4-9.2
Y	0.01 ± 0.002	11	ND		
Sr	0.1 ± 0.03	11	0.03	165	0.009-0.09
Rb	7.0 ± 1.0	9	4.7	2	3.8-5.6
Br	4.0 ± 0.3	9	2.5	2	2.0-3.0
Se	0.3 ± 0.1	11	ND		
As	0.0046 ± 0.0030	6	0.007	2	0.005-0.008
Ga	0.0007 ± 0.0001	11	ND		
Zn	56.7 ± 7.3	9	52.4	199	11.0-228
Cu	7.8 ± 1.4	9	9.8	190	3.0-58.0
Fe	206.9 ± 39	9	192.4	211	24-493.2
Mn	0.5 ± 0.08	9	1.7	155	0.7-2.9
Cr	0.008 ± 0.006	11	0.04	124	0.003-0.13
V	0.04 ± 0.01	11	1.9	1	
Ti	0.4 ± 0.1	9	0.2	1	
Ca	54.3 ± 6.7	9	62.8	135	28.6-144.0
K	2,400 ± 100	9	2,597	116	1,818-3,247
Cl	14.30 ± 49	9	ND		
S	2,000 ± 100	9	ND		
P	2,020 ± 120	9	2,570	120	940-3,247
Si	14.9 ± 4	9	ND		
Al	2.6 ± 1.3	11	0.8	155	0.2-1.6
Mg	172 ± 30	11	166.7	133	107.8-233.8
F	0.06 ± 0.01	11	ND		
Li	0.007 ± 0.003	11	ND		

* From Hamilton, E. I., Minski, M. J., and Cleary, J. J.: The concentration and distribution of some stable elements in healthy human tissues for the United Kingdom. Sci Total Environ, 1:353, 1973.

† Data from Anspaugh, L. R., et al. (2).

TABLE LXVI

THE MEAN CONCENTRATION AND RANGES OF TRACE AND MAJOR ELEMENTS OF FIVE HUMAN LIVERS DETERMINED BY NEUTRON ACTIVATION*

	Mean Value μg/g Wet Weight	Range Between Livers μg/g Wet Weight	Coefficient of Variation and 90% Confidence Limits
As	0.0065	0.002-0.012	14.4 (12.1-18.0)
Br	2.06	0.7-9.1	12.8 (10.8-16.0)
Cd	2.61	0.8-7.1	9.8 (8.2-12.3)
Cl	838	740-936	8.1 (6.8-10.1)
Co	0.034	0.023-0.039	12.1 (10.2-15.1)
Cr	0.0054	0.002-0.010	21.5 (18.1-26.9)
Cs	0.012	0.006-0.025	7.7 (6.5-9.6)
Cu	5.98	3.9-7.7	9.0 (7.6-11.3)
Fe	205	21-450	6.7 (5.6-8.4)
Hg	0.077	0.055-0.108	15.2 (12.8-19)
K	3,267	1,013-3,543	3.5 (2.9-4.4)
La	0.020	0.003-0.043	11.1 (9.3-13.9)
Mg	171	148-183	6.3 (5.3-7.9)
Mn	1.41	1.07-2.12	4.9 (4.1-6.1)
Mo	0.371	0.16-0.72	25.7 (21.6-32.1)
Na	564	330-730	11.7 (9.8-14.6)
Rb	4.86	2.9-6.3	8.1 (6.8-10.1)
Sb	0.011	0.003-0.020	19.3 (16.2-24.1)
Se	0.261	0.22-0.33	10.8 (9.1-13.5)
Sc	—	0.0001-0.0004	—
Sn	—	0.08-0.32	—
V	—	< 0.007-0.019	—
Zn	59.0	53-66	7.5 (6.3-9.4)

* J. J. Hoste: Unpublished data.

body. Some elements such as lead and cadmium increase in the kidney with age because of incorporation into some proteins with long turnover rates, as discreet inclusions of material, or accumulation in cells as a result of filtration processes.

Lung

Interest in the lung centres around estimating effects arising from exposure to different atmospheres particularly those of the urban and rural types; the RPS survey data presented in Table LXVIII are for subjects residing in an urban environment, particularly exposed to effluent from cars and the combustion of fossil fuels rather than those derived from industry. Persigehl

TABLE LXVII*

THE CONCENTRATION OF SOME ELEMENTS IN HUMAN KIDNEY FROM AN URBAN AREA OF THE UNITED KINGDOM

(μg/g wet wt.)

Element	Whole Kidney‡	R.P.S. Values Cortex	R.P.S. Values Medulla	No. of Samples	Anspaugh et al. Values† Mean Whole Kidney	Anspaugh et al. Values† No. of Samples	Anspaugh et al. Values† Range
Bi	0.4 ± 0.1	0.4 ± 0.04	0.5 ± 0.01	8	0.5	1	
Pb	1.4 ± 0.2	1.3 ± 0.2	0.7 ± 0.2	8	1.1	388	0.3-2.4
Tl	< 0.003	ND	ND	6	ND		
Ba	0.01 ± 0.001	0.01 ± 0.002	0.01 ± 0.002	8	0.09	18	0.02-0.5
Cs	0.009 ± 0.005	0.003 ± 0.0001	0.01 ± 0.001	8	ND		
I	0.04 ± 0.01	0.03 ± 0.005	0.03 ± 0.01	8	ND		
Sb	0.006 ± 0.001	0.005 ± 0.0004	0.008 ± 0.0004	8	ND		
Sn	0.2 ± 0.04	0.2 ± 0.07	0.3 ± 0.08	8	< 0.3		
Cd	13.9 ± 0.7	22.5 ± 4	7.1 ± 2.1	8	35.1	145	16.5-56.0
Ag	0.002 ± 0.0002	0.001 ± 0.0002	0.002 ± 0.0002	8	< 0.003		
Nb	0.01 ± 0.004	0.01 ± 0.003	0.02 ± 0.007	8	4.7	2	0.7-8.8
Zr	0.02 ± 0.002	0.02 ± 0.002	0.01 ± 0.002	8	2.7	4	0.7-5.4
Y	0.006 ± 0.001	0.005 ± 0.008	0.007 ± 0.003	8	ND		
Sr	0.1 ± 0.02	0.2 ± 0.003	0.1 ± 0.03	8	0.07	161	0.01-0.3
Rb	ND	5.2 ± 0.5	3.3 ± 0.1	8	ND		
Br	ND	8.2 ± 1.0	5.2 ± 0.9	8	ND		

Elements and Man

Element				n		n	Range
Se	0.1 ± 0.02	0.2 ± 0.02	0.1 ± 0.03	8	ND		
As	0.3 ± 0.05	0.3 ± 0.07	0.4 ± 0.09	8	ND		
Ga	0.0009 ± 0.0003	0.0007 ± 0.0002	0.001 ± 0.0005	8	ND		
Zn	37.4 ± 5.9	48.1 ± 4.9	18.4 ± 1.5	8	53.9	146	21.0-81.3
Cu	2.1 ± 0.4	2.8 ± 0.2	1.4 ± 1.1	8	2.9	149	1.3-3.8
Fe	ND	97.8 ± 10.2	88.2 ± 9.2	8	75.6	124	29.0-120.9
Mn	1.3 ± 0.5	0.76 ± 0.04	0.3 ± 0.02	8	1.0	143	0.5-1.6
Cr	0.03 ± 0.005	0.02 ± 0.004	0.04 ± 0.01	8	0.2	128	0.01-1.8
Ti	ND	0.6 ± 0.1	0.3 ± 0.1		0.05	1	
Ca	ND	172.0 ± 30.0	171.0 ± 19.0	8	107.7	121	57.1-175.8
K	ND	2,400 ± 200	2,400 ± 100	8	1,868	114	1,429-2,308
Cl	ND	308 ± 72	348 ± 77	8	ND		
S	1,480 ± 310	1,451 ± 111	852 ± 35	8	ND		
P	ND	1,700 ± 100	1,600 ± 100	8	1,648	116	1,319-1,978
Si	11.2 ± 1.2	37.0 ± 7.0	3.2 ± 0.5	8	ND		
Al	0.4 ± 0.1	0.4 ± 0.3	0.3 ± 0.1	8	0.5	141	0.2-0.9
Mg	205.3 ± 34.7	289.0 ± 58.6	143.7 ± 17.3	8	131.9	121	96.7-186.8
F	0.01 ± 0.002	0.01 ± 0.003	0.02 ± 0.004	8	ND		
Li	0.01 ± 0.003	0.01 ± 0.004	0.01 ± 0.004	8	ND		

* From Hamilton, E. I., Minski, M. J., and Cleary, J. J.: The concentration and distribution of some stable elements in healthy human tissues for the United Kingdom. *Sci Total Environ, 1*:353, 1973.
† Data from Anspaugh, L. R. et al. (2).
‡ Measurements for whole kidney are MS702 results only.

TABLE LXVIII*

THE CONCENTRATION OF SOME ELEMENTS IN HUMAN LUNG FROM AN URBAN AREA OF THE UNITED KINGDOM
(μg/g wet wt.)

Element	R.P.S. Values Lung	No. of Samples	Anspaugh et al. Values† Mean	No. of Samples	Range
U	0.001 ± 0.0005	1	ND		
Th	0.01 ± 0.007	11	ND		
Bi	0.01 ± 0.001	11	<0.05		
Pb	0.4 ± 0.05	11	0.7	141	0.2-1.5
Ba	0.03 ± 0.008	11	0.1	14	0.03-0.4
I	0.07 ± 0.03	11	ND		
Sb	0.06 ± 0.005	11	1.0	3	
Sn	0.8 ± 0.2	11	1.2	1	
Cd	0.48 ± 0.1	4	<0.8		
Ag	0.002 ± 0.0001	11	0.005	1	
Nb	0.02 ± 0.001	11	1.6	2	0.6-2.7
Mo	0.12 ± 0.01	8	ND		
Zr	0.06 ± 0.009	11	3.5	4	1.2-7.5
Y	0.02 ± 0.001	11	ND		
Sr	0.2 ± 0.02	11	0.1	155	0.05-0.5
Rb	3.5 ± 0.4	11	ND		
Br	7.5 ± 0.7	11	ND		
Se	0.1 ± 0.02	11	0.2	4	
As	0.02 ± 0.01	4	ND		
Ga	0.005 ± 0.002	11	ND		
Zn	10.0 ± 0.5	11	15.4	141	9.7-19.8
Cu	1.1 ± 0.1	11	1.4	141	1.0-2.0
Fe	293 ± 47	11	318.7	120	120.9-517.4
Mn	0.08 ± 0.01	11	0.3	141	0.09-0.5
Cr	0.5 ± 0.07	11	0.2	141	0.04-0.5
V	0.1 ± 0.02	11	0.1	1	
Ti	3.7 ± 0.9	11	7.7	1	
Ca	123.0 ± 10.3	11	120.9	119	70.3-175.8
K	2,000 ± 110	11	1,978	110	1,538-2,637
Cl	266 ± 48	11	ND		
S	1,241 ± 36	11	ND		
P	1,000 ± 100	11	1,099	111	769.2-1,538
Si	43.3 ± 13.2	11	ND		
Al	18.2 ± 9.7	11	27.5	141	8.6-43.9
Mg	135.0 ± 28.5	11	100.0	119	68.1-142.9
F	0.04 ± 0.009	11	ND		
Li	0.06 ± 0.01	11	ND		

* From Hamilton, E. I., Minski, M. J., and Cleary, J. J.: The concentration and distribution of some stable elements in healthy human tissues for the United Kingdom. *Sci Total Environ*, 1:354, 1973.

† Data from Anspaugh, L. R. et al. (2).

et al. (37) have compared lung burdens for rural and industrial areas of Germany and observe that levels of scandium were enhanced in rural areas by a factor of about 100, those for Cs and Al by a factor 10, while europium was detected in all samples from urban areas but not found in those from rural areas. The fact that levels of Sc and Al do not differ in proportion is of interest; if the Sc is derived from terrestrial mineral debris, natural terrestrial sources have a rather constant Sc/Al ratio, furthermore pure sources of scandium are very rare and it is very difficult to separate these elements from one another. Analyses obtained by neutron activation are particularly suitable for Sc but special methods are required for Al and also Si and hence are seldom reported. This is unfortunate as further information is required concerning the Sc/Al and related Si/Al ratios, as they can be of diagnostic value in determining whether or not samples contain mineral debris which may be natural or the result of contamination. Analytical techniques available today are particularly suitable for determining the chemical composition of particulate matter in the lung, particular interest being directed towards the identification of mineral debris such as the amphibole group of minerals, especially the asbestos types and ferruginous nodules.

Muscle

Very little data exists for levels of elements in muscle, particularly body muscle; analysis can be very difficult because, of all the major tissues of the body, levels of most elements in muscle are the lowest. The RPS data are given in Table LXIX and compared with published values. With present information it is difficult to identify the role of elements in the aeitology of muscular diseases. Webb et al. (38) have described the distribution of trace metals in beef heart tissue at fifteen discreet anatomical sites and showed that many types of tissue, of the beef cardiovascular system, had distinctive trace element compositions.

Bone and Teeth

Data for the abundance of elements in human rib are presented in Table LXX, while further data are provided in

TABLE LXIX*

THE CONCENTRATION OF SOME ELEMENTS IN HUMAN MUSCLE FROM AN URBAN AREA OF THE UNITED KINGDOM
(μg/g wet wt.)

Element	R.P.S. Values Muscle	No. of Samples	Anspaugh et al. Values† Mean	No. of Samples	Range
U	0.0002 ± 0.0001	8			
Bi	0.007	1	<0.02		
Pb	0.05 ± 0.009	6	0.2	1	
Ba	0.02 ± 0.006	6	0.07	2	
I	0.01 ± 0.001	6			
Sb	0.009 ± 0.003	6			
Sn	0.07 ± 0.01	6	0.1	1	
Cd	0.03 ± 0.01	6	<0.9		
Ag	0.002 ± 0.0005	6	0.006	1	
Mo	~0.01	4	<0.05		
Nb	0.03 ± 0.008	6			
Zr	0.02 ± 0.003	6	2.6	1	
Y	0.004 ± 0.002	6			
Sr	0.05 ± 0.02	6	0.03	137	0.004-0.06
Rb	5.0 ± 0.5	6			
Se	0.11 ± 0.01	6	0.4	14	0.3-0.4
As	0.002 ± 0.001	6			
Ga	0.0003 ± 0.00004	6			
Zn	38.7 ± 3.2	6	57.8	137	34.9-81.9
Cu	0.7 ± 0.02	6	1.0	136	0.6-1.4
Fe	30.7 ± 1.0	6	42.2	120	25.3-61.5
Mn	0.04 ± 0.007	6	0.07	133	0.02-0.2
Cr	0.005 ± 0.001	6	0.06	1	
V	0.01 ± 0.003	6	<0.01		
Ti	0.2 ± 0.01	6	0.06	1	
Ca	41.0 ± 5.3	6	42.2	119	19.2-62.7
K	2,850 ± 380	6	3,015	141	2,229-3,735
Cl	720 ± 10	4	740.1	64	694.0-858.0
S	1,108 ± 21.0	6			
P	1,410 ± 350	6	1,795	122	1,325-2,289
Si	4.1 ± 0.9	6			
Al	0.5 ± 0.2	6	0.4	136	0.07-0.9
Mg	231.7 ± 60.2	6	193.2	129	118.1-289.2
F	0.01 ± 0.004	6			
Li	0.005 ± 0.002	6			

*From Hamilton, E. I., Minski, M. J., and Cleary, J. J.: The concentration and distribution of some stable elements in healthy human tissues for the United Kingdom. *Sci Total Environ, 1*:356, 1973.

† Data from Anspaugh, L. R. et al. (2).

Table LXXI for levels of some elements in vertebrae. The distribution of elements in bone needs to be studied in relation to bone type, in particular cortical and trabecula bone and factors associated with changes with increasing age of individuals; some elements such as lead accumulate in bone with age as illustrated in Figure 108. Elements found in bone can be considered in relation to those present in mineralised tissues and those present on endosteal surfaces; very little data exists for elements present in bone marrow and bone lipids. The levels of many trace elements (such as strontium) in bone are influenced by geographical locality (see page 28), while other elements (such as uranium) tend to be rather similar irrespective of geography.

The chemistry of bone is mainly concerned with the alkaline earth-phosphate system; following the release of ^{90}Sr as a result of nuclear tests, considerable interest was expressed in stable strontium in order to predict sites of deposition and retention times of ^{90}Sr in bone. The surface area of bone is large and it is upon these surfaces that important chemical processes take place; sometimes very elegant methods exist for investigating some elements on such surfaces, such as the fission track method (see p. 293) illustrated in Figure 84.

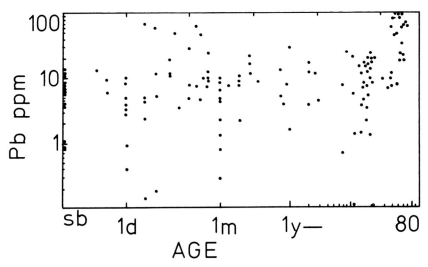

Figure 108: Relations between the concentration of lead in human bone (vertebrae and rib) with age. (n = 149). Sb = still born.

TABLE LXX*

THE CONCENTRATION OF SOME ELEMENTS IN HUMAN BONE†
FROM HARD AND SOFT WATER AREAS OF
THE UNITED KINGDOM

| Element | R.P.S. Values‡ | | Anspaugh et al. Values§ |
	Rib From Hard Water Area (22 Samples)	Rib From Soft Water Area (22 Samples)	General Bone
U	<0.02	<0.02	
Th	<0.04	<0.04	
Bi	<0.2	<0.2	
Pb	29.0 ± 3.9	34.5 ± 2.9	31.0
Tl	<0.6	<0.6	
Hg	<0.7	<0.7	
Au	<0.03	<0.03	
Lu	<0.2	<0.2	
Yb	<0.7	<0.7	
Tm	<0.2	<0.2	
Er	<0.7	<0.7	
Ho	<0.2	<0.2	
Dy	<0.9	<0.9	
Tb	<0.2	<0.2	
Gd	<1.0	<1.0	
Eu	<0.4	<0.4	
Sm	<0.8	<0.8	
Nd	<0.7	<0.7	
Pr	<0.2	<0.2	
La	<0.2	<0.2	
Ba	18.0 ± 2.8	19.3 ± 2.0	7.1
Cs	0.1 ± 0.02	0.09 ± 0.01	
Sb	1.3 ± 0.2	1.7 ± 0.3	
Sn	4.1 ± 0.6	3.7 ± 0.6	18.0
Cd	3.6 ± 0.5	4.8 ± 0.9	
Ag	1.1 ± 0.2	1.1 ± 0.2	
Nb	<0.07	<0.07	
Zr	<0.1	<0.1	
Y	0.07	0.07	
Sr	155.9 ± 14.6	138.7 ± 9.0	149.0
Zn	217.4 ± 35.4	164.6 ± 17.8	181.0
Al	73.4 ± 16	60 ± 10	50.0
B	10.2 ± 5	6.2 ± 2.1	0.9

* From Hamilton, E. I., Minski, M. J., and Cleary, J. J.: Problems concerning multi-element assay in biological materials. *Sci Total Environ*, 1:359, 1973.

† μg element/gm of ash (hydrogen peroxide decomposition and low temperature ashing).

‡ Preliminary data.

§ Data from Anspaugh, L. R. et al. (2).

There is a paucity of multielement data on bone, although a few selected elements have been considered in some detail; Strehlow and Kneip (39) have determined an average skeletal

TABLE LXXI

THE CONCENTRATION* OF SOME ELEMENTS IN THE VERTEBRAE OF THE NEWBORN (n = 20)

Element	Element µg/g Ash
U	<0.01
Th	<0.04
Bi	<0.2
Pb	4.9
Ba	6.4
I	0.3
Sb	<0.1
Sn	1.1
Cd	<2
Ag	<1
Mo	2
Zr	<0.1
Rb	26.2
Br	52.6
Zn	190
Cu	80
Sr	130
Al	32
B	4

* Assay by spark source mass spectrography.

concentration of 120 ppm for lead and 189 ppm for zinc. Very little data exists for levels of elements in cartilage and changes associated with calcification processes. Spadaro and Becker (40) describe levels of Cu, Fe, and Mn in collagenous material and Rb, Si, Sr, and V in the mineralised fraction of bone where these elements appear to be present in interstitial areas.

It is often very difficult to obtain samples of fresh bone, particularly when it is necessary to follow changes in levels of elements with age; most samples come from the very young or very old. An alternative to bone is the use of unerupted teeth, or teeth from individuals who have not had any dental fillings. The presence of elements in dental amalgams such as Au, Ag, Zn, and Sn can sometimes be responsible for elevated levels in soft tissues of the body as a result of transport through solution processes and diffusion; Halse (41) has described the penetration of Zn along dental tabules adjacent to amalgam fillings. Although teeth have their own particular chemical characteristics, they can provide some information for the type of elements likely to accumulate in mineralised materials. Using a laser probe, Goldman, Ruben, and Sherman (42) have illustrated enhanced levels of Mg, Al, Ca, P, Fe, and Cu in supragingival calculus, compared with levels in mandibular cortical bone. Losee, Cutress, and Brown (43) determined levels of thirty-nine elements in enamel: for twenty-eight samples of bicuspid teeth sampled in the United States for the age range twelve to twenty years, general levels observed were as follows:

greater than 10 ppm   F Mg S Cl K Zn Sr
between 1 and 10 ppm   B Al Cr Fe Mo Ba Pb
between 0.1 and 1.0 ppm Li Mn Cu Se Br Rb Nb Ag Cd Sn Sb
less than 0.1 ppm   Ti V Y Zr I Cs Ce Pr Nd Bi

Many elements in teeth are not homogeneously distributed; for example, Brudervold et al. (44) noted high levels of zinc (430-2100 ppm Zn) on the surface of teeth and a decrease in depth; the distribution of zinc is similar to that observed for fluorine and like fluorine pretreatment of teeth with zinc increases the resistance of the hydroxyapatite to acid attack.

The problem of dental caries has attracted considerable at-

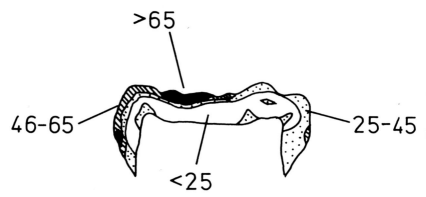

Figure 109A: Fluoride concentration (ppm) in different areas of a premolar from a forty-four-year-old subject.

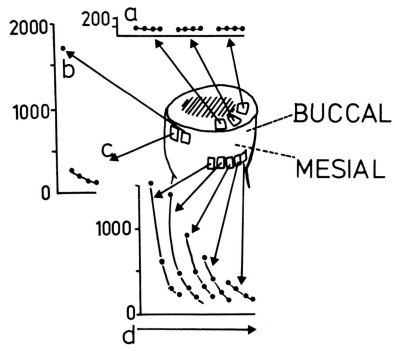

Figure 109B: Each curve represents the gradient in fluoride concentration from surface to the interior of the enamel. Each point on the curves represents the fluoride concentration in a layer of enamel about 20 μm thick. From Weatherell, J. A.: Composition of dental enamel. *Br Med Bull, London, 31*:115, 1975. Reproduced by permission of the Medical Department, The British Council.

tention, in particular the dietary intake of fluoride ions. Daily intake of fluorine in diet is about 4 mg per day and any excess is absorbed and excreted rapidly in urine; the retention of fluorine with time has been described by Spencer, Osis, and Wiatrowski (45). Factors related to the occurrence of dental fluorisis have been described by Nanda et al. (46) who studied 16,563 children and noted that several factors were involved, such as climate, water supply, culture, and work type; in this study approximately half the daily intake of fluorine was obtained from drinking water. Adkins, Barmes, and Scham-

TABLE LXXII*

THE CONCENTRATION OF SOME ELEMENTS IN LYMPH NODES FROM AN URBAN AREA OF THE UNITED KINGDOM

(μg/g wet wt.)

Element	R.P.S. Values Lymph Node	No. of Samples	Anspaugh et al. Values† Mean
U	0.01 ± 0.002	6	
Th	0.2 ± 0.1	6	
Bi	0.02 ± 0.001	6	
Pb	0.4 ± 0.1	6	
Ce	0.4 ± 0.1	6	
La	0.07 ± 0.02	6	
Ba	0.8 ± 0.3	6	0.09
Cs	0.02 ± 0.01	6	
I	0.03 ± 0.01	6	
Sb	0.2 ± 0.1	6	
Sn	1.5 ± 0.6	6	
Cd	0.06 ± 0.02	6	
Ag	0.001 ± 0.0002	6	
Nb	0.06 ± 0.007	6	
Zr	0.3 ± 0.2	6	
Y	0.06 ± 0.02	6	
Sr	0.3 ± 0.08	6	0.2
Rb	5.5 ± 1.1	6	
Br	0.9 ± 0.3	6	
Se	0.05 ± 0.01	6	0.2
As	<0.2	6	
Ga	0.007 ± 0.002	6	
Zn	14.1 ± 2.0	6	
Cu	0.8 ± 0.06	6	
Fe	106.0 ± 20.1	6	
Mn	1.1 ± 0.6	6	
Cr	2.2 ± 0.8	6	0.01
V	0.4 ± 0.2	6	
Ca	316.3 ± 67.2	6	
K	687.4 ± 162.4	6	
Cl	2,149.9 ± 113.0	6	
S	1,010 ± 190	6	
P	1,420 ± 280	6	
Si	489.0 ± 219.0	6	
Al	32.5 ± 18.0	6	
Mg	179.4 ± 54.1	6	
F	0.09 ± 0.03	6	
Li	0.2 ± 0.07	6	

* From Hamilton, E. I., Minski, M. J., and Cleary, J. J.: Problems concerning multi-element assay in biological materials. *Sci Total Environ*, 1:355, 1973.
† Data from Anspaugh, L. R. et al. (2)

schula (47) have been able to identify individuals with dental caries on the basis of levels of Mo, Mg, Cu, Pb, P, and Sr in urine; fluorine levels in urine were not of diagnostic value although they had a considerable influence on urinary output of Mg, P, Sr, Zn, and Cu. Weatherell (48) has described the composition of dental enamel, in particular the distribution of fluorine as illustrated in Figure 109. The cause of dental caries is clearly not only related to availability of fluoride ions although regular brushing of teeth with toothpastes containing fluoride and other ions seems to reduce the number of caries; like so many diseases there are many factors which have to be considered in the development of caries.

Lymph Nodes

Data for the abundance of the chemical elements in lymph nodes are seldom reported; the results of the RPS survey for urban subjects are given in Table LXXII. The different types of lymph nodes and their complex microstructure should benefit from microdistribution investigations as the function of the lymphatic tissue is one of absorbing foreign materials by filtration processes and phagocytic action. Very little can be said concerning the data given in Table LXXII except perhaps to note levels of Ce, Sn, Zr, Cr, and Al, which tend to be associated with particulate matter. Studies are wanting for levels of elements present in lymphatic tissue associated with occupational exposure to defined elements or substances.

Testes and Ovaries

Apart from detailed studies for such elements as zinc and copper related to fertility, little data are available; the results of the RPS survey are presented in Tables LXXIII and LXXIV.

In completing this brief survey of elements determined in human tissues performed in the preliminary phase of the RPS survey, further data are given in Table LXXV for some other elements which have been determined in a limited number of samples.

TABLE LXXIII*

THE CONCENTRATION OF SOME ELEMENTS IN TESTES FROM AN URBAN AREA OF THE UNITED KINGDOM

(μg/g wet wt.)

Element	R.P.S. Values Testes	No. of Samples	Anspaugh et al. Values† Mean	No. of Samples	Range
Bi	<0.002		0.2	1	
Pb	0.1 ± 0.03	5	0.3	1	
Ce	0.006 ± 0.002	5	ND		
La	0.002 ± 0.0001	5	ND		
Ba	0.02 ± 0.009	5	0.03	1	
Cs	<0.001		ND		
I	0.02 ± 0.003	5	ND		
Sb	0.008 ± 0.003	5	ND		
Sn	0.3 ± 0.1	5	0.4	1	
Cd	0.3 ± 0.09	5	0.5	1	
Ag	0.002 ± 0.0004	5	0.005	1	
Nb	0.009 ± 0.004	5	0.6	2	0.4-0.8
Zr	0.01 ± 0.003	5	1.4	4	0.9-2.0
Y	0.004 ± 0.001	5	ND		
Sr	0.09 ± 0.002	5	0.06	70	0.02-0.1
Rb	19.6 ± 6.2	5	ND		
Br	5.1 ± 1.7	5	ND		
Se	0.2 ± 0.04	5	0.5	4	
As	0.2	5	ND		
Ga	0.0009 ± 0.0001	5	ND		
Zn	13.0 ± 2.3	5	16.5	71	9.9-24.2
Cu	0.8 ± 0.2	5	1.0	71	0.6-1.2
Fe	41.2 ± 5.9	5	28.6	68	15.4-49.5
Mn	0.1 ± 0.04	5	0.2	72	0.09-0.2
Cr	0.03 ± 0.01	5	0.02	1	
V	0.2 ± 0.08	5	<0.01		
Ca	135.3 ± 51.5	5	105.5	68	70.3-153.8
K	1,540 ± 580	5	2,088	67	1,319-2,637
S	1,310 ± 320	5	1,260	5	
P	1,670 ± 530	5	1,319	65	769.2-1,758
Si	3.1 ± 1.6	5	ND		
Al	0.4 ± 0.2	5	0.5	72	0.2-0.9
Mg	127.7 ± 45.5	5	120.9	68	85.7-164.8
F	0.02 ± 0.01	5	ND		
Li	0.003 ± 0.001	5	ND		

* From Hamilton, E. I., Minski, M. J., and Cleary, J. J.: The concentration and distribution of some stable elements in healthy human tissues for the United Kingdom. *Sci Total Environ*, 1:357, 1973.

† Data from Anspaugh, L. R. et al. (2).

TABLE LXXIV*

THE CONCENTRATION OF SOME ELEMENTS IN OVARIES FROM AN URBAN AREA OF THE UNITED KINGDOM

(μg/g wet wt.)

Element	R.P.S. Values ppm Wet Wt.	No. of Samples	Anspaugh et al. Values† Mean	No. of Samples	Range
Pb	0.09 ± 0.03	6			
Ba	0.02 ± 0.01	6	0.02	1	
Cs	0.009 ± 0.002	6			
I	0.07 ± 0.03	6			
Sn	0.32 ± 0.19	6			
Cd	0.10 ± 0.03	6	0.8	1	
Ag	0.002 ± 0.0005	6			
Zr	0.03 ± 0.009	6			
Y	0.01 ± 0.005	6			
Sr	0.14 ± 0.06	6	0.1	17	0.04-0.3
Rb	5.0 ± 0.90	6			
Br	3.3 ± 1.1	6			
Se	0.09 ± 0.03	6	0.5	4	
Ga	0.002 ± 0.0005	6			
Zn	12.5 ± 4.8	6	18.0	16	8.4-27
Cu	1.2 ± 0.3	6	1.3	16	0.9-1.8
Fe	68.1 ± 22.0	6	59.0	16	22-94
Mn	0.7 ± 0.3	6	0.2	16	0.09-0.3
Cr	0.06 ± 0.02	6	0.5	16	0.007-1.5
Ca	344.1 ± 161.7	6	150	16	46-350
K	1,350 ± 170	6	1,400	16	1,100-1,700
Cl	464.1 ± 138.9	6			
S	2,100 ± 310	6	2,430	1	
P	$1.0 \pm 0.25 \times 10^3$	6	900	16	700-1,200
Si	7.4 ± 2.1	6			
Al	0.4 ± 0.1	6	0.6	16	0.1-1.5
Mg	96.7 ± 12	6	94.0	16	70.0-120
F	0.01 ± 0.003	6			
Li	0.002 ± 0.0004	6			

* From Hamilton, E. I., Minski, M. J., and Cleary, J. J.: The concentration and distribution of some stable elements in healthy human tissues for the United Kingdom. *Sci Total Environ,* 1:358, 1973.

† Data from Anspaugh, L. R. et al. (2).

Tissues of the Newborn

Of all the human tissues available for study, those from fetal deaths, the placenta at full term, and newborn deaths can be most readily obtained; therefore, it is surprising to note that very few comprehensive studies have been made on

TABLE LXXV*

THE MEAN CONCENTRATION OF SOME ELEMENTS PRESENT AT TRACE LEVELS IN SOME HUMAN TISSUES FROM THE UNITED KINGDOM†

(ppm wet wt.)

Element	Blood	Liver	Kidney	Muscle	Lung	Lymph Node	Brain	Testes
Bi	< 0.003							
Er	< 0.009				0.002			
Ho	< 0.009				0.001			
Dy	< 0.1				0.002	0.008		
Tb	< 0.009					0.006		
Gd	< 0.1				0.02	0.02		
Eu	< 0.005				0.001	0.002		
Sm	< 0.009				0.003	0.001		
Nd	< 0.008	0.03			0.006	0.05		0.005
Ce	< 0.003	0.08	0.003		0.05			0.001
La	< 0.003		0.003		0.01		0.0008	
Mo		0.9	0.4				< 0.1	
Ge	0.2	0.04	9.0	0.03	0.09	0.009	0.1	< 0.5
Ni	0.03	0.2	6.0	0.2	0.2	0.3	0.4	0.4
V	0.02		2.0					
B	0.4	0.2	0.6	0.1	0.6	0.6	0.06	0.09

* From Hamilton, E. I., Minski, M. J., and Cleary, J. J.: The concentration and distribution of some stable elements in healthy human tissues for the United Kingdom. *Sci Total Environ, 1*:360, 1973.

† The data are considered of a preliminary nature because of problems related to interference of the selected element mass spectral lines by various unknown spectral lines.

such tissues. Baglan et al. (49) have determined levels of Hg, Pb, Cd, Se, Rb, Zn, Fe, and Co in the placenta, and in maternal and fetal blood and hair. In general, placental levels exceeded both maternal and fetal levels for blood and maternal blood levels exceeded those of the fetus; of all the tissues examined, the highest concentration of mercury was found in hair which, in adults, is a recognised route for excretion. The data presented (Baglan et al., 49) for Pb, Cd, and Zn are several orders of magnitude less than those obtained in an earlier study by Dawson et al. (50).

The possibility of using the placenta for monitoring environmental levels of elements has been described by Thieme et al. (51) with reference to air and water. Widdowson et al. (52)

have described the accumulation of copper and zinc in liver before birth; for example, the level of copper in fetal liver was 10 times that of maternal liver, while full term level was 1.5 times maternal level; no accumulation was observed for Mn, Cr, and Co. Chaube, Nishimura, and Swinyard (53) observed an increase in zinc content of fetal livers between the thirty-first and thirty-fifth day of gestation, and cadmium was present in all tissues at a concentration of between 0.032 and 0.071 ppm wet weight. Henke, Sachs, and Bohn (54) also detected cadmium in all livers from juveniles and observed that an increase by a factor of about 200 occurred in the first three years of life.

Few attempts have been made to determine element imbalance during normal and abnormal pregnancies. A study by Thieme et al. (55) established a reliable curve for serum copper levels during a normal pregnancy and following abortions and severe preeclampsia at the seventh month, the curve showed a significant decrease in maternal serum copper levels and this element appears to hold promise for investigating placental insufficiencies. Panteliadis, Boenigk, and Janke (56) have described levels of twenty-five elements in fifteen newborn infants and thirty-four sucklings; significant differences were noted for some elements in whole blood between adults and the newborn but not for serum; levels of some elements also differed in urine. Hopefully, more attention will be directed to studies of human material at the onset of life and to trace changes in childhood in order to identify those elements which are normal and essential, as distinct from those which are not required for normal development. A possibility exists of identifying elements which enter because of environmental exposure compared with those from diet, for example lead.

Hair

Human hair is easy to obtain and suitable cleansing techniques to remove surface contaminants have been described by Bate and Dyer (57) and Perkins and Jervis (58). Coleman et al. (59) have described levels of elements in hair for the United

Kingdom for purposes of forensic investigations. Chattopadhyay and Jervis (60) have described the value of hair as an indicator of multielement exposure of population groups. Hair registers changes in the excretion of many elements, and numer-

TABLE LXXVI*

THE CONCENTRATION OF SOME TRACE ELEMENTS IN HUMAN HAIR AND DUST

A. TRACE METAL CONCENTRATIONS IN HUMAN HAIR IN CHILDREN (AGES 0-15) AND ADULTS (AGES $\geq$ 16)

	No. pbs.	Geometric Mean	Min μg/g	Max	± 1 Geom. SO
Children					
Barium	267	0.762	0.054	9.29	0.271-2.135
Boron	265	0.881	0.030	22.0	0.263-2.956
Cadmium	281	0.88	0.14	6.90	0.42-1.85
Chromium	261	0.56	0.076	4.80	0.26-1.22
Copper	279	12.11	1.01	144.0	4.97-29.48
Iron	282	20.83	2.70	152.00	10.52-41.24
Lead	284	13.47	2.12	100.0	6.07-29.91
Lithium	277	0.044	0.009	0.300	0.022-0.088
Manganese	267	0.56	0.05	12.0	0.22-1.45
Mercury	280	0.672	0.048	11.30	0.236-1.914
Nickel	265	0.51	0.036	11.0	0.20-1.30
Selenium	260	0.320	0.025	1.65	0.158-0.645
Silver	266	0.205	0.007	6.20	0.065-0.637
Tin	265	0.561	0.034	8.30	0.231-1.361
Vanadium	267	0.250	0.010	2.90	0.085-0.738
Zinc	224	90.50	10.50	450.0	46.21-177.24
Adults					
Barium	197	1.41	0.121	29.00	0.419-4.77
Boron	197	0.981	0.037	25.00	0.297-3.23
Cadmium	201	0.76	0.08	8.73	0.33-1.74
Chromium	192	0.52	0.06	5.30	0.26-1.46
Copper	204	18.25	2.22	184.00	7.28-45.75
Iron	202	22.30	3.60	177.00	10.46-47.53
Lead	207	12.21	1.96	155.00	5.31-29.07
Lithium	206	0.055	0.009	0.228	0.025-0.083
Manganese	197	0.95	0.07	11.0	0.34-2.67
Mercury	203	0.774	0.050	14.00	0.276-2.17
Nickel	194	0.74	0.045	11.0	0.27-2.07
Selenium	188	0.303	0.025	1.58	0.140-0.653
Silver	198	0.165	0.007	4.30	0.049-0.550
Tin	191	0.785	0.048	12.00	0.283-2.17
Vanadium	193	0.182	0.009	2.20	0.056-0.588
Zinc	167	108.54	20.10	313.0	64.02-170.68

B. MEAN TRACE-METAL CONCENTRATIONS IN DUSTFALL, HOUSEDUST, AND SCALP-HAIR, BY COMMUNITY, FOR SCALP-HAIR METALS WITH A SIGNIFICANT ENVIRONMENT EXPOSURE EFFECT

	Mean Dustfall, mg/m^2 Per Month	Geometric Mean for Housedust	Scalp Hair Children's Geometric Mean Metal, $\mu g/g$	Adults' Geometric Mean
Chromium				
Riverhead	0.05	31.7	0.52	...
Queens	0.16	42.3	0.45	...
Bronx	0.37	45.0	0.80	...
Lead				
Riverhead	2.01	278.9	11.74	10.39
Queens	10.12	629.0	14.07	14.50
Bronx	16.37	766.3	17.13	12.88
Nickel				
Riverhead	0.21	14.6	0.50	...
Queens	0.38	26.1	0.40	...
Bronx	0.85	42.0	0.70	...
Barium				
Riverhead	...	65.2	0.64	1.11
Queens	...	137.6	0.63	1.36
Bronx	...	312.4	1.26	2.34
Mercury				
Riverhead	...	1.9	0.48	0.66
Queens	...	5.9	1.00	1.01
Bronx	...	3.6	0.73	0.67
Tin				
Riverhead	...	6.4	0.54	...
Queens	...	6.2	0.44	...
Bronx	...	7.8	0.83	...
Vanadium				
Riverhead	...	3.3	0.20	0.12
Queens	...	18.8	0.24	0.20
Bronx	...	40.2	0.40	0.35

* From Creason, J. P., Hinners, T. A., and Bumgarner, J. E.: Trace elements in hair related to human exposure. *Clin Chem,* 21(4):603, 1975.

ous studies have been made describing changes in levels of elements along hair with growth, for example Gangadharan, Lakshmi, and Das (61) have investigated Na, Cl, Mn, Co, Sn, Zn, As, Se, Ag, I, and Au. Baumslag et al. (62) have shown a gradual increase in levels of Zn, Cu, Fe, and Pb in hair from infants and children and suggest the possibility of placental transfer of at least lead. Henke and Fitzek (63) describe the variability in levels of thallium in hair with length: while the mean level was about 4 ppm Tl, at specific sites along the hair levels of above 30 ppm were recorded. Although very little data exist for the levels of thallium in human materials, this element is very toxic and in nature occurs in association with lead; geochemical pathways also show similarities with potassium.

In pollution studies, Oleru (64) has used levels of elements in hair as a means of distinguishing occupational from non-occupational exposure. Skerfving (65) has used hair and blood to investigate exposure to mercury for citizens of Sweden who consume fish; levels of mercury in hair were 250 times times greater than those in blood, and axillary head hair contained less mercury than occipital hair. Nord et al. (66) describe levels of 11.0 ppm Hg in male hair and 18.9 ppm Hg for women for residents of Los Alamos, New Mexico, United States, and 25 ppm Hg for residents of Pasadena, California, United States; as the normal value for nonoccupational exposure is about 10 ppm Hg, higher values may reflect differences in environmental exposure of the two populations to mercury. Creason et al. (67) have illustrated the value of hair in relation to environmental exposure; data are given in Table LXXVI for three regions of New York.

In spite of the problems of contamination arising because of the use of shampoos and various cosmetics, hair does provide a useful record of historic changes in elements excreted by this pathway; detailed examination of changes along the length of a hair could provide useful information in relation to various diseases and also the influence of some drugs on metabolic processes.

Skin

Very few investigations have been made for human skin apart from studies of electrolytes. In a series of papers, Molokhia and Portnoy (68, 69, 70) have investigated levels of Cu, Mn, and Zn and have emphasised the problem of regional differences for skin obtained from different parts of the body. They also note the need, because of the higher water content of the dermis, to separate the dermis from the epidermis, techniques for which have been described by Baumberger, Suntzeff, and Cowdry (71).

Gaseous compounds which pass through the skin may be detected by gas mass spectroscopy as described by Adamczyk, Boerboom, and Kistemaker (72). Apart from studies of common gases, there is some interest in analysing air extracted from the stomach in relation to the production of alkyl-metal compounds derived from the presence of microoganisms in the gastrointestinal tract.

Radioelements

The unique and characteristic properties of ionizing radiation permit the detection for very small levels for both static and dynamic processes. A general description of the exposure to man by ionizing radiation has been given by Morgan (73). Only very brief comments will be made here concerning the value of radioactive measurements in describing the distribution of elements in man. Direct measurements of naturally occurring radioelements such as the uranium and thorium series and ^{40}K may be made, if present in sufficient quantities, on living man using highly collimated detectors or whole body counters. The chemical toxicity for the naturally occurring radioelements is usually unimportant compared with damage arising from ionizing radiation; the natural uranium series contain extremely toxic daughter products such as ^{210}Po; Stannard and Casarett (74) consider ^{210}Po to be five times more toxic than ^{226}Ra in shortening the life span of the rat and mouse. Polonium 210 enters the human body through diet and inhalation and smokers tend to have higher body burdens than nonsmokers. The concentration of ^{210}Po and its parent ^{210}Pb in soft human tissues has been described by Blanchard (75), skeletal levels by Holtzman (76), while Hursh (77) has described the natural ^{210}Pb content of man.

The radon daughter decay pair of radionuclides, $^{210}Pb-^{210}Po$, illustrate a source of entry through diet; people such as the

TABLE LXXVII*

THE CONCENTRATION OF THORIUM IN URINE,
BONE, AND BLOOD DETERMINED BY
NEUTRON ACTIVATION ANALYSIS

A. THE CONCENTRATION OF ^{232}Th IN URINE

Th compound	Type of work	Period exposed	^{232}Th(ng/ml)
No exposure 5 samples+			< 0.001
Monazite sand	Transport	‡	0.25
Mixed: mainly nitrate	Processing	‡	0.25
	Packer	1 yr	< 0.001
	Maintenance	5 yr	< 0.001
	Maintenance	9 yr	0.093
	Yard Labourer	8 yr	< 0.001
Sulphation (sulphate, hydrate, chloride)	Chemical processing	7 yr	< 0.001
		7 yr	1.1
		7 yr	< 0.001
Thorium nitrate (gas mantle industry)	Open evaporation of solutions	6 months	0.20
		2 yr	0.84
		7 yr	1.83
	Process work	15 yr	< 0.001
		7 yr	2.24
		6 months	0.58
		10 months	0.26
		21 yr	0.33
		19 yr	0.18
	General work	11 months	< 0.001
		3 months	0.12

B. ^{232}Th IN RIB ASH OF INDIVIDUALS NOT OCCUPATIONALLY EXPOSED TO THORIUM

No.	Age	Sex	^{232}Th(ng/g ash wt.)
1	34	F	5.8
2	36	M	163.8
3	39	M	17.7
4	47	F	7.3
5	49	M	6.8
6	49	M	50.9
7	50	F	85.0
8	53	F	2.4
9	54	M	1.1
10	56	F	1.0
11	58	F	8.5
12	58	M	22.5
13	59	M	0.8

Mean 28.72 ± 13.14

C. ^{232}Th IN BLOOD OF INDIVIDUALS NOT OCCUPATIONALLY EXPOSED TO THORIUM

No.	Age	Sex	^{232}Th(ng/ml)	No.	Age	Sex	^{232}Th(ng/ml)
948	22	F	8.3	163	37	M	0.2
224	22	F	7.3	982	37	F	2.0
1111	23	F	3.0	147	42	F	6.4
975	24	F	6.1	979	44	M	2.6
1560	29	M	0.1	674	46	F	2.4
314	30	M	1.1	891	47	F	0.1
974	30	M	1.4	947	49	M	1.2
980	30	F	3.3	161	53	F	2.0
120	31	F	0.8	140	55	M	1.5
879	33	M	0.1	976	58	F	1.3
873	33	F	2.0	530	58	M	1.4
858	34	M	0.2	673	59	F	6.5
1571	34	M	0.3	136	63	F	2.9
509	35	F	1.9	900	63	F	1.4
950	35	F	1.3	157	66	F	3.5

Mean 2.42 ± 0.42

* From Clifton, R.J., Farrow, M. and Hamilton, E.I.: Measurements of ^{232}Th in normal and industrially exposed humans. *Ann Occup Hyg, 14*:303, 1971.
+ Ages 25-35 yr.
‡ Period of exposure unknown.

Lapps or Eskimos living in high latitudes use reindeer and caribou as a source of food; a major source of food for such animals is moss and lichen which accumulate aerial debris on surfaces including many natural and man-made radionuclides. Lead 210 is also a very useful tracer for stable lead and relations between radioactive and stable lead have been described by Jaworowski (78).

Unlike uranium, thorium is very insoluble in body fluids and is mainly found in bone in spite of low transport across the gut; Lucas, Edgington, and Markun (79) have described the concentration of thorium in rib bone which varies from less than 0.1 to 72 ng/g ash weight; levels in bone increase with age and exposure. Clifton, Farrow, and Hamilton (80) estimated that the thorium content of soft tissue was less than 3 ng/g and relationships between occupational exposure and urine levels are provided in Table LXXVII.

As a result of nuclear testing and generation of nuclear power, a large number of radionuclides are present in man's environment. In those instances where the radionuclides and stable elements become intimately mixed, such as transport through the system soil→plant→man, assay for radionuclides after chemical processing, or in vivo whole body counting, can provide valuable information concerning metabolic pathways; for example the globally distributed radionuclide ^{137}Cs can be used as a tracer for potassium.

REFERENCES

1. Report of the Task Group on Reference Man. ICRP Pub 23, Oxford, Pergamon, 1976.
2. Anspaugh, L. R., Robison, W. L., Martin, W. H., and Lowe, O. A.: Compilation of published information on elemental concentrations in human organs in both normal and diseased states, Parts I, II, III. Berkeley, Lawrence Livermore Laboratory, University of California, UCRL-51013, 1971.
3. Stitch, S. R.: Trace elements in human tissues. Part I. A semi-quantitative spectrographic survey. *Biochem J*, 67:97, 1957.
4. Tipton, I. H. and Cook, M. J.: Trace elements in human tissue. Part II. Adult subjects from the United States. *Health Phys*, 9:103, 1963.
5. Soman, S. D., Joseph, K. T., Raut, S. J., Mulay, C. D., Parameshwaran, M., and Panday, V. K.: Studies on major and trace element content in human tissues. *Health Phys*, 19:641, 1970.

6. Hamilton, E. I., Minski, M. J., and Cleary, J. J.: The concentration and distribution of some stable elements in healthy human tissues from the United Kingdom. *Sci Total Environ, 1*:341, 1972/73.
7. Sumino, K. S., Hayakawa, K., Shibata, T., and Kitamura, S.: Heavy metals in normal Japanese tissues. *Arch Environ Health, 30*:487, 1975.
8. Bowen, H. J. M.: *Trace Elements in Biochemistry.* London, Academic Press, 1966.
9. Urey, H. C.: The composition of the stone meteorites and origin of the meteorites. *Geochim Cosmochim Acta, 4*:36, 1953.
10. Schwarz, K.: Elements newly identified as essential for animals. In *Nuclear Activation Techniques in the Life Sciences, 1972.* Vienna, IAEA, 1972.
11. Liebscher, K. and Smith, H.: Essential and nonessential trace elements. *Arch Environ Health, 17*:881, 1968.
12. Schroeder, H. A.: Environmental metals. In McKee, W. D. (Ed.): *Environmental Problems in Medicine.* Springfield, Thomas, 1974.
13. Halsted, J. A., Smith, J. C., and Irwin, M. I.: A conspectus of research on zinc requirements of man. *J Nutri, 104*:345, 1974.
14. Evans, G. W.: Copper homeostasis in the mammalian system. *Physiol Rev, 53*:535, 1973.
15. Cotzias, G. C.; Manganese in health and disease. *Physiol Rev, 38*:503, 1958.
16. Durbin, P. W.: Metabolic characteristics within a chemical family. *Health Physics, 2*:225, 1960.
17. Delves, H. T., Clayton, B. E., and Bicknell, J.: Concentration of trace metals in the blood of children. *Br J of Prev Soc Med, 27*:100, 1973.
18. Kasperek, K., Schicha, J., Siller, V., Feinendegen, L. E., and Höck, A.: Trace-element concentrations in human serum: Diagnostic implications. In *Nuclear Activation Techniques in the Life Sciences, 1972.* Vienna, IAEA, 1972.
19. Flint, R. W., Lawson, C. D., and Standil, S.: The application of trace element analysis by X-ray fluorscence to human blood serum. *J Lab Clin Med, 85(1)*:155, 1975.
20. Cotzias, G. C., Miller, S. T., and Edwards, J.: Neutron analysis: The stability of manganese concentration in human blood and serum. *J Lab Clin Med, 67*:836, 1966.
21. Versieck, J., Barbier, F., Speecke, A., and Hoste, J.: Normal manganese concentrations in human serum. *Acta Endocrinol, 76*:783, 1974.
22. Hagenfeldt, K., Plantin, L. O., and Diczfalusy, E.: Trace elements in the human endometrium. 2. Zinc, copper, manganese levels in the endometrium, cervical mucus and plasma. *Acta Endocrinol (Kbh), 72*:115, 1973.

23. Galenkamp, H., van Leeuwen, J., and Das, H. A.: The determination of aluminium in blood by activation. Pethen, The Netherland Reactor Centre, Rept RCN-208, 1974.
24. Brune, D., Frykberg, B., Samsahl, K., and Wester, P. O.: Determination of elements in normal and leukemic human whole blood by neutron activation analysis. Stockholm, Sweden, Aktiebolaget Atomenergi, Report AE-60, 1961.
25. Henke, G., Möllmann, H., and Alfes, H.: Vergleichends Untersuchungen über die Konzentration einiger Spurenelemente in menschlichen Hirnarealen durch Neutronenaktiverungsanalyse. Z Neurol, 199:283, 1971.
26. Ule, G., Völkl, A., and Berlet, H.: Spurenelemente in menschlichen Gehirn. II. Cu, Zn, Ca, Mg. Z Neurol, 206:117, 1974.
27. Schicha, H., Müller, W., Kasperek, K., and Schröder, R.: Neutron activation analysis of trace element content in surgically removed sample of human brain. Beitr Pathol, 146:366, 1972.
28. Höck, A., Demmel, U., Schicha, H., Kasperek, K., and Feinendegen, L. E.: Trace element concentrations in human brain. Brain, 98:49, 1975.
29. Schicha, H., Kasperek, K., Feinendegen, L. E., Siller, V., and Klein, H. J.: The iron content of human brain and its correlation with age. Beitr Pathol Path, 142:268, 1971.
30. Greiner, A. C., Chan, S. C., and Nicolson, G.: Human brain contents of Ca, Cu, Mg, and Zn in some neurological pathologies. Chemica Chim Acta, 64:211, 1975.
31. Hu, K. H. and Friede, R. L.: Topographic determination of zinc in human brain by atomic absorption spectrometry. J Neurochem, 15:677, 1968.
32. Zaworski, R. E. and Oyasu, R.: Lead concentration in human brain tissue. Arch Environ Health, 27:383, 1973.
33. Duggan, M. J. and Williams, S.: Lead-in-dust in city streets. J Sci Total Environ, 7:98, 1977.
34. Gooddy, W., Williams, T. R., and Nicholas, D.: Spark source mass spectrometry in the investigation of neurological disease. I. Multielement analysis in blood and cerebrospinal fluid. Brain, 97:327, 1974.
35. Gooddy, W., Hamilton, E. I., and Williams, T. R.: Spark source mass spectrometry in the investigation of neurological disease. II. Element levels in brain, cerebrospinal fluid and blood: Some observations on their abundance and significance. Brain, 98:65, 1975.
36. Hoste, J.: Private communication.
37. Persigehl, M., Schicha, H., Kasperek, K., and Feinendegen, L. E.: Behaviour of trace element concentration in human organs independence of age and environment. Intern Conf Modern Trends in Activation Analysis, Munich, 1:395, 1976.

38. Webb, J., Kirk, K. A., Niedermeier, W., Griggs, J. H., Turner, M. E., and James, T. N.: Use of pattern recognition to classify beef cardiovascular tissues on the basis of their trace metal compositions. *Bioinorganic Chem, 5*:261, 1976.
39. Strehlow, C. D. and Kneip, T. J.: The distribution of lead and zinc in the human skeleton. *Am Ind Hyg Assoc Ann J, 30*:372, 1969.
40. Spadaro, J. A. and Becker, R. O.: The distribution of trace metal ions in bone and tendon. *Calcif Tissue Res, 6*:49, 1970.
41. Halse, A.: Metals in dental tubules beneath amalgam filling in human teeth. *Arch Oral Biol, 20*:87, 1975.
42. Goldman, H. M., Ruben, M. P., and Sherman, D.: The application of laser spectroscopy for the qualitative and quantitative analyses of the inorganic components of calcified tissues. *Oral Surgery, 17*:102, 1964.
43. Losee, F. L., Cutress, T. W., and Brown, R.: Natural elements of the Periodic Table in human dental enamel. *Caries Res, 2*:123, 1974.
44. Brudervold, F., Steadman, L. T., Spinelli, M. A., Amdur, B. H., and Grøn, P.: A study of zinc in human teeth. *Arch Oral Biol, 8*:135, 1963.
45. Spencer, H., Osis, D., and Wiatrowski, E.: Retention of fluoride with time in man. *Clin Chem, 21*:613, 1975.
46. Nanda, R. S., Zipkin, I., Doyle, J., and Horowitz, H. S.: Factors affecting the prevalence of dental fluorosis in Lucknow, India. *Arch Oral Biol, 19*:781, 1974.
47. Adkins, B. L., Barmes, D. E., and Schamschula, R. G.: Etiology of caries in Papua, New Guinea. *Bull WHO, 50:*495, 1974.
48. Weatherell, J. A.: Composition of dental enamel. *Br Med Bull London 31*:115, 1975.
49. Baglan, R. J., Brill, A. B., Schulert, A., Wilson, D., Larsen, K., Dyer, N., Mansour, M., Schaffner, W., Hoffman, L., and Davies, J.: Utility of placental tissue as an indicator of trace element exposure to adult and fetus. *Environ Res, 8*:64, 1974.
50. Dawson, E. B., Croft, H. A., Clark, R. A., and McGanity, W. J.: Study of seasonal variation in nine cations of normal term placenta. *Am J Obstet Gynecol, 102*:354, 1968.
51. Thieme, R., Schramel, P., Klose, B. J., and Waidl, E.: Spurenelemente in der menschlichen Plazenta. *Geburtshilfe Frauenheilkd, 35*:349, 1975.
52. Widdowson, E. M., Chan, H., Harrison, G. E., and Milner, R. D. G.: Accumulation of Cu, Zn, Mn, Cr, and Co in the human liver before birth. *Biol Neonate, 20*:360, 1972.
53. Chaube, S., Nishimura, H., and Swinyard, C. A.: Zinc, cadmium in normal human embryos and fetuses. *Arch Environ Health, 26*:237, 1973.

54. Henke, G., Sachs, H. W., and Bohn, G.: Cadmium determination in the liver and kidneys of children and juveniles by means of neutron activation analysis. *Arch Toxikol,* 26:8, 1970.
55. Thieme, R., Klose, B. J., Schramel, P., and Samsahl, K.: Das verhalten von essentiellen Spurenelementen in Verlauf der normalen und pathologischen Schwangershaft. *Geburtshilfe Frauenheilkd,* 33:652, 1973.
56. Panteliadis, C., Boenigk, H-E, and Janke, W.: Ueber das Spektrogramm der Spurenelemente bei Neugeborenen und Sauglingen. *Infusionstherapie,* 2:1, 1975.
57. Bate, L. C. and Dyer, F. F.: Trace elements in human hair. *Nucleonics,* 23:74, 1965.
58. Perkins, A. K. and Jervis, R. E.: Trace elements in human head hair. *J Forensic Sci,* 11:50, 1966.
59. Coleman, R. F., Cripps, F. H., Stimson, A., and Scott, H. D.: The trace element content of human hair in England and Wales and the application to forensic science. *Atomics,* 123:12, 1967.
60. Chattopadhyay, A. and Jervis, R. E.: Hair as an indicator of multielement exposure of population groups. In Hemphill, D. D. (Ed.): *Trace Substances in Environmental Health, VII.* Columbia, U of Mo, 1974.
61. Gangadharan, S., Lakshima, V. V., and Das, M. S.: Growth of hair and the trace element profile as studied by sectional analysis. *J Radioanal Chem,* 15:287, 1973.
62. Baumslag, N., Yeager, D., Levin, L., and Petering, H. G.: Trace metal content of maternal and neonate hair. *Arch Environ Health,* 29:186, 1974.
63. Henke, G. and Fitzek, A.: Nachweis von Thalliumvergiftungen durch Neutronenaktivierungsanalyse. *Arch Toxikol,* 27:266, 1971.
64. Oleru, U. G.: Epidemiological implications of environmental cadmium. *Am Ind Hyg Assoc J,* 36:229, 1975.
65. Skerfving, S.: Methylmercury exposure, mercury levels in blood and hair, and health status in Swedes consuming contaminated fish. *Toxicology,* 2:3, 1974.
66. Nord, P. J., Kadaba, M. P., and Sorensen, J. R. J.: Mercury in human hair. *Arch Environ Health,* 27:40, 1973.
67. Creason, J. P., Hinners, T. A., Bumgarner, J. E., and Pinkerton, C.: Trace elements in hair, as related to exposure in metropolitan New York. *Clin Chem,* 21:603, 1975.
68. Molokhia, M. and Portnoy, B.: Neutron activation analysis of trace elements in skin. I. Copper in normal skin. *Br J Dermatol,* 81:110, 1969.
69. Molokhia, M. and Portnoy, B.: Neutron activation analysis of trace elements in skin. III. Zinc in normal skin. *Br J Dermatol,* 81:759, 1969.

70. Molokhia, M. and Portnoy, B.: Neutron activation analysis of trace elements in skin. IV. Regional variations in copper, manganese and zinc in normal skin. *Br J Dermatol, 82*:254, 1970.
71. Baumberger, J. P., Suntzeff, V., and Cowdrey, E. V.: Methods for the separation of epidermis from dermis and some physiologic and chemical properties of isolated epidermis. *J Natl Cancer Inst, 2:* 413, 1942.
72. Adamczyk, B., Boerboom, A. J. H., and Kistemaker, J.: A mass spectrometer for continuous analysis of gaseous compounds excreted by human skin. *J Appl Physiol, 21:*1903, 1966.
73. Morgan, K. Z.: Ionizing radiation exposure. In McKee, W. D. (Ed.): *Environmental Protection Problems in Medicine.* Springfield, Thomas, 1974.
74. Stannard, J. N. and Casarett, G. W.: Metabolism and biological effects of an alpha particle emitter, polonium-210. *Rad Res Suppl 5,* 1974.
75. Blanchard, R. L.: Concentrations of ^{201}Po and ^{210}Pb in human soft tissues. *Health Physics 13:*625, 1967.
76. Holtzman, R. B.: Measurements of the natural contents of RaD (Pb^{210}) and RaF (Po^{210}) in human bone—estimates of whole—body burdens. *Health Physics, 9:*385, 1963.
77. Hursh, J. B.: Natural lead-210 content of man. *Science, 132:*1666, 1960.
78. Jaworowski, Z.: *Stable and Radioactive Lead in Environment and Human Body,* NEIC-RR-29, Review Rept. Warsaw, Poland, Nuclear Energy Information Center, 1967.
79. Lucas, H. F., Edgington, D. N., and Markun, F.: Natural thorium in human bone. Argonne Nat Lab, Radiol Phys Div, Rept ANL 7615: 53, 1969.
80. Clifton, R. J., Farrow, M., and Hamilton, E. I.: Measurements of ^{232}Th in normal and industrially exposed humans. *Ann Occup Hyg, 14:*303, 1971.

Chapter 9

DISEASE

MY MAIN INTEREST in the role of the chemical elements and disease concerns general populations not occupationally exposed to harmful materials. Relations between disease and the elements are best illustrated in the discipline of toxicology which provides many clear examples of cause and effect. With recent advances in toxicity testing, at all levels of concern, we can expect some harmful materials to be identified but because of the multifactorial nature of many modern diseases simple cause-effect and dose-response relationships seem doubtful. In spite of the many ethical problems, I personally expect more advances in the prevention of disease to come from studies of epidemiological surveys, but do not foresee that even if an area of concern is identified simple solutions will be forthcoming, or if specific elements are identified that obvious solutions will be acceptable to man. Because of the complex etiology of modern diseases (in fact, how modern these diseases are is a matter for debate) and the apparent lack of simple dose-response relations, there is a great danger of rendering investigations so complex that instead of focussing attention upon causes they will become directed to a study of interesting details, which may or may not be important; the use of such terms as synergistic, antagonistic, additive, and multiplicity illustrate the complex issues which confront man today when seeking the eradication or control of disease. The eradication of some major diseases identified with contamination of man's environment by the discovery of some form of drug or medicine seems unlikely, and increasingly, decisions appear to rest with politicians and the will of nations to accept when essential a radical change in lifestyles.

In the absence of disease the World Health Organization defines health as "a state of complete physical, mental and social well-being, and not merely the absence of disease or infirmity." A healthy individual can enjoy life to the full and within the limits of his own genetic structure and personality can contribute to society, rather than depend upon society for support. With increase in age individuals have to rely more and more on society for assistance and the care of such people has to be an integral part of our culture. Man does not possess a set number of diseases, they change with time as a result of cultural development; today it is generally accepted that some of the major diseases of concern arise from our technological achievements and associated social changes in ways of life, which can result in both somatic and psychological illness of which mental stress is significant. Apart from factors associated with time, disease is also related to geography; a review of medical geography through the ages has been described by Howe (1). We can therefore study the etiology of disease in relation to geography, and the significance of the elements is likely to be more pronounced for those individuals who live distant from highly sophisticated societies; unfortunately such people are becoming extremely rare and any simple relationship with environmental conditions is becoming rapidly eroded by cultural advances. For many of these people the common causes of disease are hunger and infection associated with poor personal hygiene and inadequate sanitation. When great differences in types of culture exist between people, they cannot be bridged in the space of a few years. The tragic consequences of bringing together people of different standards of life are to be found in fringe areas of technically advanced communities and in instances where people have become engulfed by the advance of modern society and become isolated from past traditions. By nature man is a curious animal, and evolutionary processes result in him having a strong desire for survival and to put off the eventual termination of material life as long as possible. Eventual death results from the failure to overcome an insult to the body because of the lack of resources to overcome a challenge. Modern medicine can provide temporary support in order to overcome some ailments associated with ageing, but eventually they also fail; death is inevitable.

In historic times many diseases related to environmental factors would tend to be local and soon overcome by processes of adaptation. Today matters are very different, especially for industrialised nations in which a vast number of substances are dispersed into man's environment; the potential hazards to health for many single substances are unknown, and even less is known concerning substances which interact with one another to produce new materials. Today the trend seems to be one of calculating the number of individuals at risk and to estimate total costs and benefits to society as a whole in maintaining an acceptable incidence of disease for large populations and those occupationally exposed to hazards. Very great improvements in medicine and health care are now taking place, but if we have to deal with multicause phenomena new techniques are still required. As far as environmental and geographically associated diseases are concerned, a case can be made for the use of large scale system modelling in order to identify possible causes and also to provide an educational tool for strategic planning. When this approach is attempted it is often found that the correct type of information required for a model does not exist; for example, adequate historical medical records of diseases are lacking and there is compartmentalisation of recorded incidences of disease in terms of nations, countries, or counties rather than geographical limits containing areas of similar environmental chemistry.

When we come to consider problems of the Third World, quite simple solutions are sometimes available and can result in quite dramatic improvements in health. For example, malnutrition may not be caused by lack of important items of diet but rather the presence of gastrointestinal infections or failure to maintain the efficacy of available vaccines because of poor refrigeration; some diseases such as schistosomiasis, filariasis, and leprosy cannot be treated by vaccines while others such as tuberculosis, rare in developed countries, are rampant in the Third World but are amenable to treatment. Many states of poor health and disease are associated with survival on marginal diets, with little capacity to overcome transient episodes of reduced health and through which element imbalances can occur. Most of medicine is concerned with organic systems and there is a tendency to neglect inorganic constituents, but yet important constituents of the body such as the enzymes are intimately associated with the chemical elements.

The role of elements and disease may be obscure but the following examples illustrate the urgent need for proper multidisciplinary investigations with particular attention being directed to forging sound links between the presence of elements and biological function.

Air

Clinical interests in the chemistry of air is concerned with the lungs, whose function is to transfer air to and from the alveoli and permit the exchange of oxygen for carbon dioxide in the blood. Air is transferred through the bronchial tree, which is lined by the bronchial epithelium covered by a double layer of mucus about 15 μm thick. The upper part of the system contains cilia projecting into the bronchial tubes whose function is to transport mucus and entrained foreign bodies which are removed to the gut. The airways of the pulmonary system terminate in the respiratory bronchioles consisting of alveolar ducts from which protrude the alveolar sacs. The diameter of the ducts and sacs varies from about 300 to 600 μm and the alveoli are covered by the alveolocapillary membrane which is between 0.4 and 2.5 μm thick and covered by a secretory substance 0.1 to 0.3 μm thick. Particulate matter present in the alveoli is removed by phagocytes which may originate from the epithelial cells, leucocytes, capillary epithelium, or connective tissue. When considering environmentally induced diseases of the lung, the delicate biological structure of surfaces must be considered, as well as the extent to which the presence of unusual gases, particulate matter, and their interactions can impair biological processes, for example the effects arising from partial loss of the mucus layer and subsequent effects upon the ciliary escalator.

Ignoring catastrophic episodes of air pollution and occupational exposure to chemicals, it is apparent that urban air is more harmful to health than rural air. Considerable attention has been focussed upon the effects of gases and their chemical products such as the oxides of nitrogen, sulphur, and carbon monoxide. In the air these gases can interact in very complex processes to give rise to some toxic substances; for example, industrial smog produces peroxyacetyl nitrate which irritates the eyes. Although many organic compounds have been determined in air the levels, or presence of many others, are unknown; the tendency is to consider that such materials relate to the general urban factor associated with lung disease. Far more general information is available for the levels of the elements in air but their significance in contributing to an urban disease factor is unknown. Very little data are available describing chemical changes which take place in the lung following inhalation of the elements, in dissolved or solid phases, and impaction onto mucus surfaces. It seems reasonable to assume that microchanges in pH on surfaces could be important. Some types of biochemical change must take place in the deep lung following exposure to air-borne pollutants; sites at which changes could take place are many, for example impairment of fibrous proteins responsible for the elastic properties of the lung, loss of activity by proteolytic enzymes, and the release of enzymes as a result of cell damage. The overall effect could be the disruption of the fibrous network, and because the lung is under continuous mechanical stress, natural repair may be difficult. In element-lung investigations attention is often directed towards the total mass of an element which enters the body through the lung compared with that entering through diet; for almost all elements dietary intake far exceeds that from inhalation and the tendency is to ignore intake through the lung as a significant pathway. Although the total respiratory system is very efficient in excluding and removing particulate matter, some reaches the deep lung where residence times can be very long and some elements accumulate with age thus allowing long periods of time for chemical reactions to take place.

Analytical techniques are now available to study levels of elements in lung tissue and also to investigate the composition of deposited particulate matter. Until far more information is available for elements present in the lung, from different environments, it is not possible to comment further on harmful effects except to note that the levels of many elements in the human lung tend to be similar irrespective of geography. Levels of some elements in the body associated with mineral debris do increase for some environments, but there seems to be little evidence that any very large differences exist. Therefore it is necessary to consider the significance of quite small differences for adequately sampled populations.

Water

Mortality from cardiovascular disease is one of the major causes of death and is particularly high for developed countries. More disturbing, younger age groups are becoming increasingly affected, thus the trend towards an increase in the expected life span of man is reversed. The death rate from arteriosclerotic and degenerative heart disease for the age group 45-54 years for the period 1955-1967 has increased in many countries; for example, as reported by WHO (2), Netherlands 51%; Finland 41%; Hungary 32%; United Kingdom 29%; compared with Canada 9%; United States of America 1%; while for Switzerland there has been a 3% decrease. With a few exceptions the general consensus of opinion is that hard water areas usually experience lower cardiovascular mortality rates than areas supplied with soft water; there appears to be a water factor as yet unidentified. Unlike lung disease, very much attention has been directed towards the role of the chemical elements, for example Zn, Cr, Mn, Co, As, V, Ni, Cd, Se, Si, F, and Cu. Associations between cardiovascular disease, water quality, and the chemical elements have been described by Kobayashi (3), Morris, Crawford, and Heady (4), Dingle et al. (5), Masironi (6), Schroeder (7), Elwood et al. (8), and Pansar et al. (9). Shacklette, Sauer, and Miesch (10) have investigated the significance of the chemical elements in soils and plants from areas in the United States which have contrasting rates of disease of the heart, and they conclude that, if geochemical differences between the areas do have a causal relationship to death from cardiovascular diseases, the cause would appear to be one of deficiency rather than an excess of some elements. Other diseases such as malformations of the central nervous system have also been correlated with softness of water and have been described by Lowe, Roberts, and Lloyd (11).

One of the outstanding problems is the time required following exposure to soft water before changes detrimental to health occur. Anderson (12) and Robertson (13) have noted a sudden increase in the death rate from cardiovascular disease for towns in which the water was softened and a decrease for towns where the water has been hardened, which is indeed alarming, but far more evidence is required to identify these findings in terms of a specific feature associated with water quality. There are of course many other possible causes of this type of disease which do not concern trace elements, such as the quality of diet, intake of sugar or lipids which have been discussed by Yudkin (14) and Masironi (15), lack of exercise, and various factors associated with an increase in the affluence of society. Seldom are really large scale epidemiological surveys undertaken incorporating expertise from many disciplines which are essential: cardiovascular diseases provide an excellent example of multifactorial causation.

In considering those studies involving the chemical elements, there is room to doubt the diagnostic value of many findings, although when all are considered the evidence favours an association of the disease with water quality. Some of the problems are as follows:

1. Often very general statements are made concerning hard and soft water and cognizance is not taken that there are many types of hard and soft water.
2. Often very little attention is paid to the selection of those elements considered to influence the disease. Some are

determined simply because of the availability of a suitable method and are carried out in isolation to clinical and epidemiological investigations. Even when good analytical measurements are made, the data can be of limited value because of poor or inadequate sampling techniques.
3. Studies often concern the "naked elements" and no consideration is given to chemical form or association with organic compounds.
4. The significance of organic materials in water is neglected.
5. Many hard and soft water areas are geographically very extensive and impose regional chemical characteristics which affect many materials apart from water; it is for this reason that it is essential to determine whether or not the act of changing only the quality of water influences the incidence rate of the disease.
6. Hard water areas are usually associated with the presence of calcareous rocks, such as limestones or chalk, which were deposited from clear oceanic environments distant from terrigenous debris. Calcareous rocks are permeable to water because of solution processes, and water deposited on the surface tends to drain down until trapped above an impervious layer, such as a clay band. Drinking water is usually pumped up from these subterranean reservoirs and once depleted by drought can take a considerable time to be replenished. Calcareous rocks tend to support thin soils, and large forests are not as common as grasslands; although climate influences the type of vegetation, deep organic litter tends to be absent. The general picture, for which there are many exceptions, is one of rapid permeation of rainfall, often through considerable thicknesses of bedrock, which can act as a form of filter favouring transport of dissolved rather than particulate matter. For some hard waters a further stage of filtration occurs during settlement in natural reservoirs and in tanks at water treatment plants.

In soft water areas mechanical filtration of particulate matter can be quite efficient, for example transport of water through sandstone aquifers but, because of the lower content of dissolved solids there is a lack of chemical processes, combining both coprecipitation and absorption of dissolved and suspended matter. Many soft water areas support dense vegetation and such environments are associated with surface peat deposits, a constant source of complex organic compounds. It is very difficult to accept that the presence of major cations such as Ca and Mg is of importance; it is possible that the filtration processes associated with hard water and the lack of large amounts of organic matter in many areas may have a bearing on cardiovascular diseases.

In order to account for the role, if any, of the chemical elements and cardiovascular disease, there is a clear need for multidisciplinary research; areas for which the quality of water is to be changed should be selected for detailed examination before and after any changes. There is much in favour of considering small, but detailed studies, in order to reduce the number of variables encountered in regional investigations. For example, it is possible to select small islands remote from industry which have distinctly different geochemical environments but supporting people of the same race and culture and subjected to the same type of climatic influences. In attempting such studies, there is often a lack of medical records or such areas are associated with special diseases and unusual diets making extrapolation to technological societies difficult.

Cancer

Cancer is now a major fatal disease and together with the cardiovascular diseases is considered by many to have reached epidemic proportions. Since about 1933, the overall cancer death rate has been increasing at about 1 percent each year and in recent years there have been some suggestions that quite sudden increases have occurred. While some types of cancer have undoubtedly increased and the disease appears to be a feature of advanced societies, Silverberg and Holleb (16) have observed that the overall incidence of cancer has decreased over the past twenty-five years except for some such as cancer of the lung, colon, pancreas, and bladder. The most striking change has been a drop in stomach cancer since about 1935, which has been described by Haenzel (17) for the United States, and supported by Segi and Kurihara (18) for selected sites in twenty-four countries. In common with cardiovascular diseases, causes of cancer are many but if one general component can be identified it is possibly concerned with diet. It is commonly accepted that possibly 90 percent or more of all cancers are caused by carcinogens present in the environment and therefore they should be amenable to identification. Significant improvements have been made in testing for the carcinogenicity of chemicals but many are based upon the results of animal experiments. Epstein (19, 20) has described some of the problems concerned with laboratory animal tests many of which appear to be inadequate or quite unsuitable when determining whether or not a chemical has carcinogenic properties. An alternative approach is through epidemiological surveys which is tantamount to saying that people should be used as a test bed; on grounds of ethics many will find such a suggestion totally unacceptable, but it has to be acknowledged that for many years a vast number of chemicals have entered the natural environment which are carcinogens, or through interactions in the environment give rise to carcinogens, and which cannot be removed. Further, the latent period between exposure and when effects are seen can vary from about five to forty years, therefore it is conceivable that the full impact of man's release of carcinogens into his environment remains to be registered; it is only since the end of World War II that the dramatic advances in technology have taken place. With the vast improvements in laboratory testing, it seems possible to identify harmful materials, and efforts can be made to reduce or prevent their release into man's environment. However, it seems inevitable that, as some have already entered the environment and cannot be removed, except by natural processes, and others avoid detection, the ultimate test will be the incidence of disease for exposed populations or individuals from defined geographical areas.

One feature of cancer is its geographical distribution, described by Doll (21). Doll, Payne, and Waterhouse (22) and Higginson and Maclennan (23) note that in different countries cancer in children shows a low range of incidence compared with cancer in adults. Very high incidences of some types of cancer have very well-defined geographical distributions. For example, cancer of the oesophagus varies considerably in different countries; a very high incidence rate has been observed for central Asia in Kazakhstan, northeast Iran, and extending across south Siberia into Mongolia and North China. In the littoral zone of the Caspian Sea, in Iran, Mahboubi et al. (24) have noted regional patterns reflecting ecological variables such as rainfall, soil types, vegetation, and farming practice. Davies and Griffith (25) have described an association between death from cancer of the stomach in north Wales, United Kingdom, and residence of individuals on a particular soil group who were not engaged in agricultural practice.

Several elements have been shown to be carcinogenic and one, arsenic, is known to cause skin cancer; in Taiwan, How and Yeh (26) and Yeh (27) have described various forms of skin cancer found among people with a high percentage of chronic arsenious intoxication because of the presence of arsenic in drinking water. Milham and Strong (28) have described an increase in respiratory cancer for men working at a copper smelter while Blot and Fraumeni (29) describe an enhanced level of lung cancer for workers in Cu, Pb, and Zn smelting industries and the influence of community air pollution from industrial emissions containing inorganic arsenic.

Several types of cancer are associated with the inhalation of particulate matter, for example, Boyd et al. (30) observed a lung cancer mortality about 70 percent higher than normal for haematite (iron-ore) workers in West Cumberland, United Kingdom; Wagner et al (31) and Gilson (32) describe the etiology of asbestos and cancer, the incidence of which can be very high for exposed workers. It is beyond the scope of this monograph to deal with these matters in detail but suffice it to say that some elements play a role in the etiology of cancer which is a multifactorial type of disease possibly associated with man's technological progress.

In recent years many elements such as lead, cadmium, and mercury have been subjected to detailed study in relation to disease. Beyond those incidences attributable to industrial exposure and accidents, or following the release into defined environments of industrial effluents, the evidence for their effects on general unexposed populations is sparse, although levels of these elements are enhanced in urban compared with rural populations. If a linear dose-response relationship exists between exposure and effects, then many elements could be associated with disease; for some a threshold level may exist, while for others the effect is influenced by complex interactions between elements and compounds. Excessive concentrations of copper in the brain and liver have been described by Osborn and Walshe (33) while Leu et al. (34) note that an accumulation of copper in muscle only occurs when binding sites in the liver become saturated. Crapper, Krishnan, and Dalton (35) observe an increase in aluminium associated with neurofibrillary degeneration in senile and presenile dementia of the Alzheimer type and suggest that in man aluminium may be a neurotoxic factor. The chemical form of the aluminium is unknown but, if present as the oxide, hydroxide, or phosphate, it would provide suitable sites for the absorption and coprecipitation of many elements. Interactions between elements and disease are illustrated in the case of excessive iron storage in organs, for example alcoholics with cirrhosis of the liver and hemochromatosis, which are associated with an increase in several other elements, such as Pb, Mo, Ca, and Cu, and a decrease in Al; Butt et al. (36) have discussed this type of element association and comment that elevated hepatic lead levels cannot always be interpreted as prima facie evidence of lead intoxication.

It is no longer acceptable just to recognise that the presence of a particular element in man will result in disease; it is necessary to identify the responsible biochemical processes. All the chemical elements are toxic to man at some level of exposure; many are essential for our way of life. Increasingly, evidence is presented which shows that many could be harmful to man or components of the natural environment at present observed levels; their removal is, at present, impossible and is only likely to be achieved by a drastic change in our lifestyle which is not likely to occur for large populations but can be accomplished to a degree by individuals. Patterson (37) has described, in a very careful manner, lead in the environment and concludes that it is likely that present day blood lead levels are 100 times greater than would be expected for a natural environment exposure; lead is now globally distributed and levels in man are well documented, but there is little evidence that levels for nonexposed individuals are responsible for any well-defined patterns of disease. The possibility that environmental levels of lead are harmful cannot be ruled out, but historical exposures are likely to have been greater for geographically restricted populations.

The possible hazards presented by the chemical elements and their compounds are very difficult to decipher since the whole advancement of man has been through the exploitation of metals and their use in society. Identification of toxic metals is obtained from those occupationally exposed; by the application of various types of safety factors, acceptable levels for nonoccupationally exposed populations are derived. Those elements which are not covered by this approach but could have toxic properties can be revealed by various types of laboratory toxicity testing, but eventually, the proof of their safety rests with epidemiological investigations through which elements of concern are identified. Thus the appropriate biochemical pathways can be investigated.

REFERENCES

1. Howe, G. M.: *Man, Environment and Disease in Britain.* New York, B & N, 1972.
2. *International Work in Cardiovascular Diseases, 1959-1969.* Geneva, WHO, 1969.
3. Kobayashi, J.: Geographical relationship between chemical nature of river water and death rate from apoplexy. Berichte Ohara Inst Landw Forsch, *11*:12, 157.

4. Morris, J. N., Crawford, M. D., and Heady, J. A.: Hardness of local water-supplies and mortality from cardiovascular disease. Lancet, 2:506, Sept 6, 1961.
5. Dingle, J. H., Paul, O., Sebrell, W. H., Strain, W. H., Wolman, A., and Wilson, J. R.: Water composition and cardiovascular disease. Ill Med J, 125:25, 1964.
6. Masironi, R.: Trace elements and cardiovascular diseases. Bull WHO, 40:305, 1969.
7. Schroeder, H.: Environmental metals: Specific effects on the human body. In McKee, W. D. (Ed.): Environmental Problems in Medicine. Springfield, Thomas, 1976.
8. Elwood, P. C., Bainton, D., Moore, F., Davies, D. F., Wakley, E. J., Langham, M., and Sweetnam, P.: Cardiovascular surveys in areas with different water supplies. Br Med J, 2:362, 1971.
9. Pansar, S., Erämetsä, O., Karvonen, M. J., Ryhänan, Hilska, P., and Nornamo, H.: Coronary heart disease and drinking water. J Chronic Dis, 28:259, 1975.
10. Shacklette, H. T., Sauer, H. I., and Miesch, A. T.: Geochemical environments and cardiovascular mortality rates in Georgia. Statistical studies in field geochemistry. Washington, D.C. Geological Survey, Prof Paper 574-C:1, 1970.
11. Lowe, C. R., Roberts, C. J., and Lloyd, S.: Malformations of central nervous system and softness of local water supplies. Br Med J, 2: 357, 1971.
12. Anderson, T. W., Le Riche, W. H., and MacKay, J. S.: Sudden death and ischemic heart disease. Correlation with hardness of local water supply. N Engl J Med, 280:805, 1969.
13. Robertson, J. S.: Mortality and hardness of water. Lancet, 2:348, 1968.
14. Yudkin, J.: Sugar and disease. Nature (Lond), 239:197, 1972.
15. Masironi, R.: Dietary factors and coronary heart disease. Bull WHO, 42:103, 1970.
16. Silverberg, E. and Holleb, A. I.: Cancer Statistics 1975—25 Year Cancer Survey. Geneva, WHO, 1975.
17. Haenszel, W.: Variation in incidence of and mortality from stomach cancer, with particular reference to the United States. J Natl Cancer Inst, 21:213, 1958.
18. Segi, M. and Kurihara, M.: Cancer mortality for selected sites in 24 countries. No. 5 (1964-1965) Department of Public Health, Tohoku University School of Medicine, Sendai, Japan, 1969.
19. Epstein, S. S.: The carcinogenecity of Dieldrin. Sci Total Environ, 4:1, 1975.
20. Epstein, S. S.: Carcinogenicity of heptachlor and chlordane. Sci Total Environ, 6:103, 1976.
21. Doll, R. (Ed.): Methods of Geographical Pathology. Report of the study group convened by The Council for International Organisation of Medical Services, Springfield, Thomas, 1959.

22. Doll, R., Payne, P., and Waterhouse, J. (Eds.): *UICC Cancer Incidence in Five Continents. A Technical Report.* Berlin, Springer. 1966.
23. Higginson, J. and Maclennan, R.: The world pattern of cancer incidence. In Raven, R. O. (Ed.): *Modern Trends in Oncology, Part I.* London, Butterworths, 1972.
24. Mahboubi, E., Kmet, J., Cook, P., Day, N. E., Ghadirian, P., and Salmasizadeh, S.: Oesophageal cancer studies in the Caspian Littoral of Iran. *Br J Cancer, 28:*197, 1973.
25. Davies, R. I. and Griffith, G. W.: Cancer and soils in the county of Anglesey. A revised method of correlation. *Br J Cancer,* 8:594, 1954.
26. How, S. W. and Yeh, S.: Studies on ecemic chronic arsenical poisoning. 2. Histopathology of arsenical cancer, with special reference to Bowen's disease. *Rept Inst Pathology, National Taiwan U, 14:*75, 1963.
27. Yeh, S.: Relative incidence of skin cancer in Chinese Taiwan: With special reference to arsenical cancer. *Natl Cancer Inst Monograph* No. 10, 1962. (See Conf on Biol of Cutaneous Cancer, Philadelphia, Pennsylvania, April 6-11, 1962.)
28. Milham, S. and Strong, T.: Human arsenic exposure in relation to a copper smelter. *Environ Res, 7:*176, 1974.
29. Blot, W. H. and Fraumeni, J. R.: Arsenical air pollution and lung cancer. *Lancet, 2:*142, July 26, 1975.
30. Boyd, J. T., Doll, R., Faulds, J. S., and Leiper, J.: Cancer of the lung in iron ore (haematite) miners. *Br J Ind Med, 27:*97, 1970.
31. Wagner, J. C., Gilson, J. C., Berry, G., and Timbrell, V.: Epidemiology of asbestos cancers. *Br Med Bull, 27:*71, 1971.
32. Gilson, J. C.: Asbestos cancer: Past and future hazards. *Epidem Prev Med, 66:*395, 1973.
33. Osborn, S. B. and Walshe, J. M.: Studies with radioactive copper (^{64}Cu and ^{67}Cu) in relation to the natural history of Wilson's disease. *Lancet, 1:*346, 1967.
34. Leu, M. L., Strickland, G. T., Beckner, W. M., Chen, T. S. M., Wang, C. C., and Yeh, S. J.: Muscle copper, zinc and manganese levels in Wilson's disease: Studies with the use of neutron activation analysis. *J Lab Clin Med, 76:*432, 1970.
35. Crapper, D. R., Krishnan, S. S., and Dalton, A. J.: Brain aluminium distribution in Alzheimer's disease and experimental neurofibrillary degeneration. *Science, 180:*511, 1973.
36. Butt, E. M., Nusbaum, R. E., Gilmour, T. C., and DiDio, S. L.: Trace metal patterns in disease states: Hemochromatosis, and refractory anemia. *Am J Clin Pathol, 26:*225, 1956.
37. Patterson, C. C.: Contaminated and natural lead environments of man. *Arch Environ Health, 11:*344, 1965.

Chapter 10

BIOCHEMICAL CONSIDERATIONS

THE EVOLUTIONARY PATHWAY linking the chemical elements and man starts with the cosmogenic production of elements, their participation in the biosphere, and finally entry into man through diet. Elements of the biosphere exhibit a strong correlation with those found in the oceans which are derived from the primordial leaching of the protocrust. Throughout the biosphere living cells are involved in a vast number of chemical processes and many elements are utilised to perform specific functions, for example iron in electron transfer processes and calcium in structural materials. In natural systems most chemical processes have been subjected to a wide range of exposures to the elements, indeed the most dramatic occurred very early in biological evolution. Survival in the biosphere requires a constant supply of essential elements, the absence of toxic levels of elements, and some ability to overcome excesses or deficiencies of the elements. If we consider man in terms of biological evolution, an outstanding feature is the development of the brain which can be vulnerable to element exposure for levels for which no evolutionary adaptive experience is available; it is worth noting that hazards arising from many elements and compounds common to our present environments are first registered by neurological disorders.

In order to try and understand chemical effects on man, assay for gross tissue levels is a starting point and most of this monograph has been concerned with such matters. However, the real problem lies in attempting to understand the distribution of the elements within tissues, cells, and parts of cells, ultimately focussing attention upon mechanisms of molecular structure in biochemical compounds; pathways for investigation of elements in the natural environment through to incorporation in biochemical compounds is illustrated in Table LXXVIII.

TABLE LXXVIII

THE PATHWAY FOR DETERMINING TOTAL ELEMENT DISTRIBUTIONS IN MAN

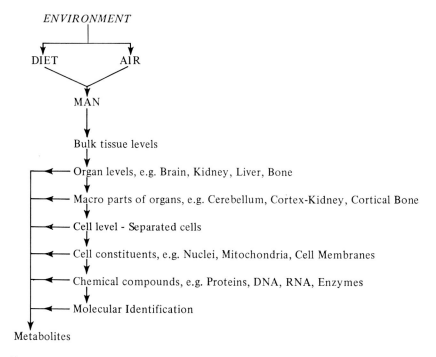

In a manner similar to the gap which exists between extrapolation of animal data to man, the gap existing between gross levels of elements in tissue and those present in parts of some cells is difficult to bridge. For some tissues, such as muscle and cortical bone, bulk data may suffice, but in others some elements become enriched in small volumes, are associated with surfaces, or are essential components of metal-enzyme systems. Today, instrumentation is available to investigate all elements, directly or indirectly, in relation to biological function. Simple empirical rules describing similarities in element characteristics can be obtained from their classification in the

Periodic Table and hence biochemically controlled element anomalies can be identified. This is an interesting problem and poses such questions as "To what extent, and under what conditions, can iron be substituted by manganese in oxygen transfer processes? To what extent can calcium be replaced by strontium and potassium by rubidium?" Through the Periodic Table, which has a universal application, the chemical properties of the elements can be rationalised at all levels of investigation, for example through properties of ionization energy, electronegativity, polarising power, and binding energy.

In furthering our understanding of element behaviour in man, bulk assay should be extended to examine element distributions in tissues, cell distributions, and levels in isolated chemical compounds; methods range from those which are very simple to those requiring considerable expertise and use of sophisticated instrumentation.

The gross distribution of the chemical elements in tissues can be determined by simple histological staining techniques in which reagents are used to produce a specific colour by reaction with elements in particular chemical forms. Ancillary techniques, such as irradiation by ultraviolet light, can destroy chemically bound elements which can then react with specific reagents thus permitting a consideration of chemical form. Timm (1) has described very elegant techniques for staining various elements and compounds in biological tissues; rapidly frozen sections of tissues are exposed to dihydroxy-dinapthyl-disulphide which will identify sulfhydryl and disulphide groups of proteins. Exposure to hydrogen sulphide deposits insoluble sulphides of many elements, which if then exposed to silver results in a catalytic reduction of silver to produce molecular silver; a useful reducing solution is hydroquinone and a silver salt. Using this technique Fjerdingstad, Danscher, and Fjerndingstad (2) have described a tenfold increase of lead in the hippocampus of the rat; while Danscher et al. (3) have used the Timm method to examine the distribution of various metals (Cu, Zn, and Pb) in the rat as illustrated in Figures 110 and 111. Brun and Brunk (4) have used a modified sulfide-

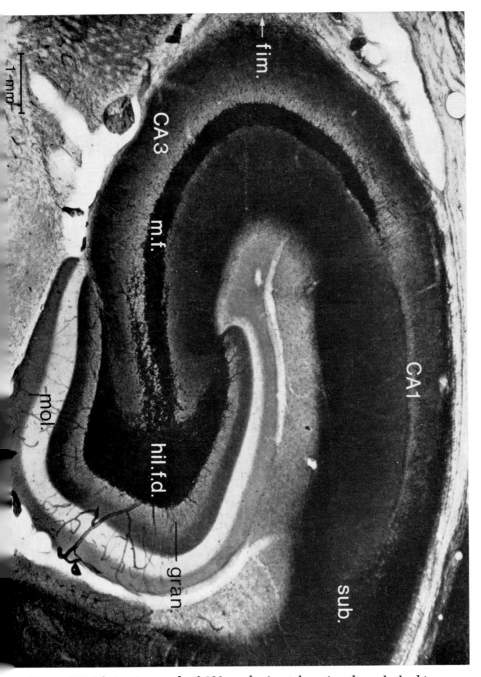

Figure 110: Photomicrograph of 100 μm horizontal section through the hippocampus of adult rat stained with the Timm's method for heavy metals. Abbreviations: CA, cornu amonis; fim., fimbria; gran., granular layer of area dentata; hil.f.d., hilus fasciae dentatae; m.f., layer of mossy fibers; mol., molecular layer; sub., subiculum. From Fjerdingstad, E. J., Danscher, G. and Fjerdingstad, E.: Hippocampus: selective concentration of lead in the normal rat brain. *Brain Res*, 80:350, 1974.

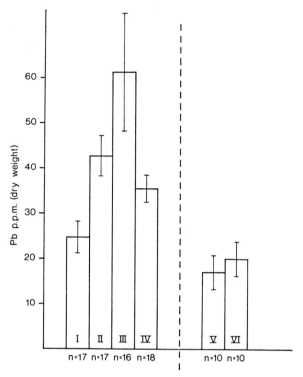

Figure 111: Mean content of lead ± S.E.M. in the subdivisions of the hippocampus: I, II granule and molecular layers of fascia dentata; III, IV hilus fasciae dentatae; V undivided dorsal; and VI ventral hippocampus. The difference between means of the groups I-IV is highly significant (one-way analysis of variance, $F = 4.59$, $P < 0.01$). From Danscher, G., Fjerdingstad, E. J., Fjerdingstad, E. and Fredens, K.: Heavy metal content in subdivisions of the rat hippocampus (zinc, lead and copper). Brain Res, 112:442, 1976.

silver method to demonstrate heavy metals (Zn, Cu, Hg, and Pb) in hepatic parenchymal cells, Kupffer cells, neurons, and glial cells; under pathologic conditions, such as Wilson's disease and experimentally induced intoxications, lysosomes were shown to take up Cu, Hg, and Pb. Once sulphide deposition has been achieved element analysis of precipitated material can be undertaken by electron or ion-microprobe examination. A considerable number of element-staining techniques are avail-

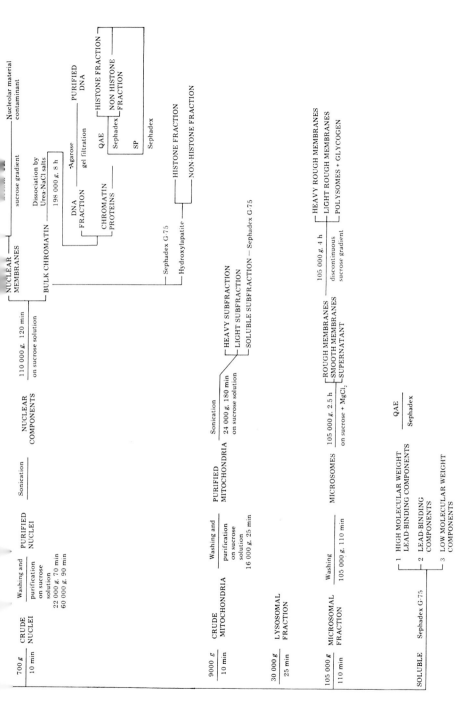

Figure 112: Flow diagram for the fractionation of rat liver homogenate into its intracellular components. From Sabbioni, E. and Marafante, E.: Identification of lead-binding components in rat liver: in vivo study. *Chem-Biol Interactions, 15*: 1, 1976.

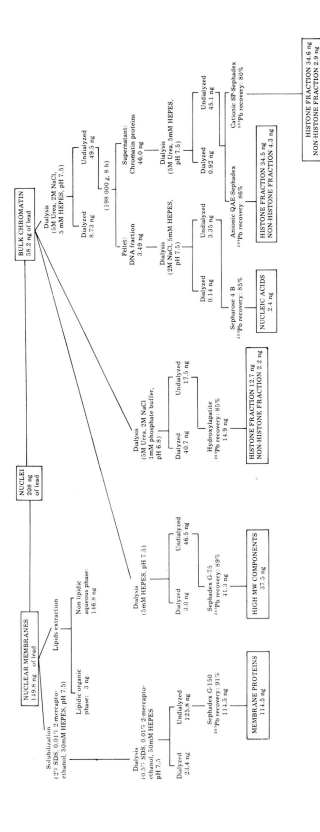

Figure 113: Flow diagram and results of experiments to identify the nuclear lead-binding components (Pb-BC) from rat liver. (NB Undialysed material refers to residual material in the bag after dialysis.) From Sabbioni, E. and Marafante, E.: Identification of lead-binding components in rat liver: in vivo study. *Chem-Biol Interactions*, 15:1, 1976.

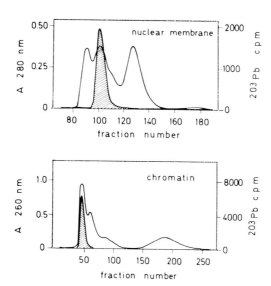

Figure 114: Nuclear membranes: Ultraviolet (UV) and ^{203}Pb elution profiles, from Sephadex G-150, of the non-delipidated ^{203}Pb-labelled nuclear membranes solubilized in SDS. The column (5 × 100 cm) was equilibrated with 0.5% SDS, 0.01% 2-mercaptoethanol, 50 mM HEPES, pH 7.5. Bulk chromatin: UV and ^{203}Pb elution profiles, from Sephadex G-75, of undissociated ^{203}Pb-labelled bulk chromatin. The column (5 × 100 cm) was equilibrated with 5 mM HEPES, pH 7.5. From Sabbioni, E. and Marafante, E.: Identification of lead-binding components in rat liver: in vivo study. *Chem-Biol Interactions,* 15:1, 1976.

able for experiment and hopefully more use will be made of these methods in conjunction with analytical chemistry.

In some tissues various elements can be concentrated by normal biochemical processes and identified by microscopy. Choie and Richter (5) observed that within one to six days following a single dose of 0.05 mg Pb/g body weight in the rat characteristic intranuclear inclusions developed in the epithelium of proximal tubules of the kidney. The presence of these inclusions appears to provide an indication of acute lead poisoning and they may be derived from a soluble protein-lead complex. Hsu et al. (6) have also described this type of inclusion in the liver, kidney, humerus, and femur of the pig. Murakami and Hirosawa (7) describe inclusion in the kidney

TABLE LXXIX*

THE DISTRIBUTION OF LEAD IN TISSUES OF THE RAT, INTRACELLULAR DISTRIBUTIONS, AND THOSE FOR MITOCHONDRIA AND MICROSOMAL FRACTIONS

A. DISTRIBUTION OF LEAD IN TISSUES OF RATS AND INTRACELLULAR DISTRIBUTION IN LIVER AND KIDNEY

(12 rats in each experiment. Results are mean of 3 experiments with standard deviation. Animals were killed 24 h after an I.V. injection of 18 μg of $^{203}Pb^{2+}$/rat.)

Tissue	% of the Dose Administered	ng of Lead	ng of Lead g of Tissue
Kidney	8.6 ± 0.45	1,556 ± 81.4	740.9
Liver	4.06 ± 0.22	735 ± 39.8	75
Lung	0.11 ± 0.016	19.9 ± 2.89	17
Spleen	0.075 ± 0.006	13.6 ± 1.08	39
Trachea	0.032 ± 0.003	5.79 ± 0.54	5.8
Testicles	0.03 ± 0.002	5.43 ± 0.36	2.7
Heart	0.016 ± 0.001	2.89 ± 0.18	2.1
Brain	0.0067 ± 0.0003	1.21 ± 0.05	0.7
Total blood	—	—	66.6
Serum	—	—	5.4

Fraction	Liver		Kidney	
	% total homogenate	ng of lead	% total homogenate	ng of lead
Nuclei	28.3 ± 4.3	208 ± 31.6	34.2 ± 3.2	532 ± 49.8
Mitochondria	9.5 ± 1.1	69.8 ± 8.08	26.1 ± 2.1	406 ± 32.7
Lysosomes	10.8 ± 1.3	79.4 ± 9.55	5.3 ± 0.9	82.5 ± 14
Microsomes	10.9 ± 1.2	80.1 ± 8.82	5.0 ± 0.7	77.8 ± 10.9
Soluble	39.2 ± 3.1	288.1 ± 25.7	32.1 ± 4.6	499 ± 71.5

B. DISTRIBUTION OF LEAD INTO LIVER INTRACELLULAR MEMBRANES AFTER EXTRACTION OF LIPIDS

(Results are mean of 3 experiments with standard deviation.)

	Aqueous Phase		Organic Phase	
	% of Lead	ng of Lead	% of Lead	ng of Lead
Nuclear membranes	97 ± 2.3	145.3 ± 3.44	1.3 ± 0.5	1.94 ± 0.75
Mitochondrial inner membranes†	97.6 ± 3.5	45.3 ± 1.62	1.6 ± 0.9	0.75 ± 0.42
Mitochondrial outer membranes‡	97.5 ± 2.7	4.3 ± 0.12	0.9 ± 0.3	0.04 ± 0.013
Microsomal rough membranes	98.5 ± 2.4	49.4 ± 1.2	2.1 ± 1	1.05 ± 0.5
Microsomal smooth membranes	99 ± 1.9	23.2 ± 0.44	1.2 ± 0.4	0.28 ± 0.093

C. DISTRIBUTION OF LEAD INTO PURIFIED LIVER MITOCHONDRIAL AND MICROSOMAL SUBFRACTIONS
(Results are mean of 3 experiments.)

Fraction	% of Lead in Total Cell Fraction	ng of Lead in the Subfraction
Heavy subfraction	66.5 ± 4.5	46.4 ± 3.14
Soluble subfraction	22.4 ± 1.9	15.6 ± 1.33
Light subfraction	8.1 ± 0.9	4.4 ± 0.63
Heavy rough membranes	52.3 ± 4.98	41.8 ± 3.91
Smooth membranes	29.2 ± 3.4	23.4 ± 2.72
Light rough membranes	8.3 ± 0.88	6.64 ± 0.7
Supernatant	7.8 ± 2.6	6.24 ± 2.08
Free polysomes	2.14 ± 0.75	1.71 ± 0.6

* From Sabbioni, E. and Marafante, E.: Identification of lead-binding components in rat liver: in vive study. *Chem-Biol Interactions*, 15:1, 1976.

† Correspond to isolated mitochondrial heavy subfraction.

‡ Correspond to isolated mitochondrial light subfraction.

of the rat following administration of ^{210}Pb and have identified inclusions in mitochondria and cytoplasm, half of which were found to be associated with cytoplasmic membranes; hence lead deposits are involved with both membrane and cytoplasm.

Sabbioni and Marafante (8) employed very exacting techniques to fractionate rat liver homogenate into its intracellular components, as illustrated in Figures 112 and 113; they identified nuclear lead binding components from rat liver, illustrated in Figure 114 and Table LXXIX, following in vivo experiments using ^{203}Pb and radioactively labelled precursors such as (^{14}C) arginine and (^{3}H) trytophan. Significant quantities of lead were found to be associated with the endoplasmic reticulum and on rough surface of microsomes; no significant levels were present in free polysome subfractions or in lipids of the endoplasmic reticulum.

The pathogenesis of lead inclusions is not known but they are acid fast, indicating a sulfhydryl-containing protein, presumably of the nonhistone type without DNA or RNA. Walton (9) has described granules containing lead and calcium in the mitochondria of rat kidney as illustrated in Figures 115, 116, and 117 and

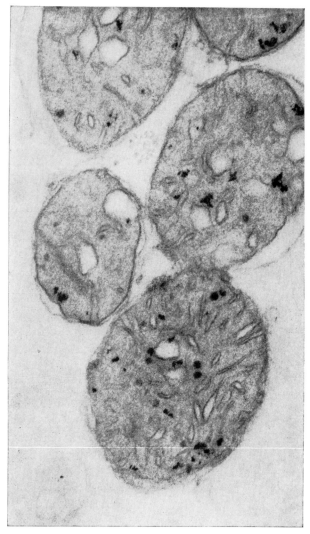

Figure 115: An illustration of lead granules in isolated mitochondria of rat kidney. Mag. ×78,816. Photograph supplied by J. R. Walton.

suggests that effects of lead uptake on the mitochondria are twofold: one preventing ATP synthesis and interfering with the maintenance of ionic concentration gradients across membranes, and the second related to ATP-forming complexes with the -SH groups of mitochondrial enzymes, such as those involved

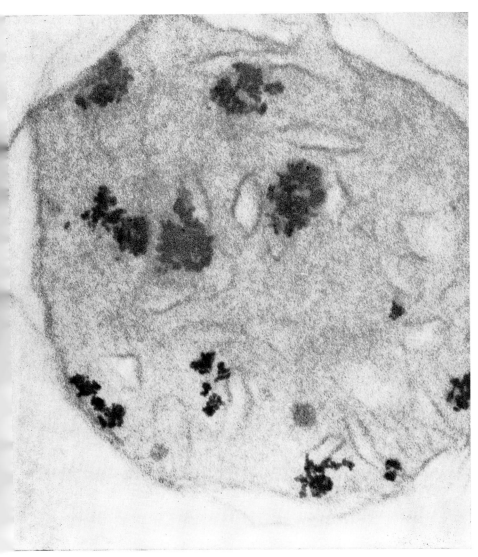

Figure 116: Granules containing lead in isolated mitochondria of rat kidney. Mag. ×198,400, illustrating the high resolution obtained by electron microscopy. Photograph supplied by J. R. Walton.

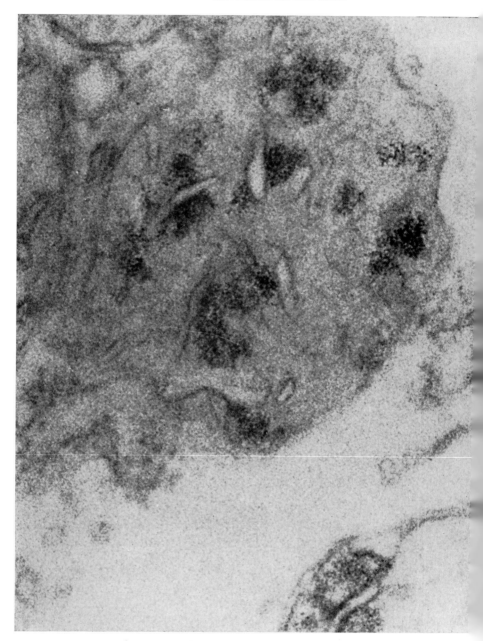

Figure 117: Calcium granules in rat kidney mitochondria. Mag. ×173,750. Photograph supplied by J. R. Walton.

in cytochrome synthesis and by reaction with available anions such as phosphates. Continued accumulation of lead granules would ultimately result in a breakdown of metabolic processes and structural damage to cells.

Walton (10) suggests that cell transport of lead ions may take place through the same mechanisms as used for calcium ion translocation which could be important in the rates of deposition and remobilisation of lead in bone. The sequestering of lead is not restricted to mammalian systems: Malone, Koeppe, and Miller (11) have described the precipitation of lead in the corn plant (*Zea mays L.*); apart from any adverse affects on physiology, productivity, and efficiency of the plant, deposited lead is made available to man in corn based foods, although the G.I. tract transfer of such forms of lead is not known. Many elements, for example Ca, Sr, Mn, Ba, and Mg, can form electron-dense granules, a study of which may help an understanding of possible toxic effects at the cell level.

The next level of investigation concerns a determination of elements present in separated biochemical compounds such as proteins and enzymes. Hamilton and Minski (12) determined the abundance of elements present in commercially available DNA and RNA by spark source mass spectrometry; as illustrated in Table LXXX, the results are tainted by those elements which were lost during commercial processing and those added as a result of contamination. It seems reasonable to assume that the presence of an element indicates some form of attachment to DNA and RNA; levels of Sr and Ba will be related to Ca, other elements such as Pb, Br, Zn, Cu, Ni, Fe, and Mn may serve some functional purpose. The presence of Ti may indicate contamination from preparative materials; levels of Si and Al are low considering that much of the separation processes would have been carried out in materials made of glass.

Data given in Table LXXXI were obtained by Wester (13) using neutron activation analysis and therefore as great care was taken to prevent pre-irradiation contamination of the samples the data are more likely to represent true levels of the

TABLE LXXX*

THE ABUNDANCE OF SOME ELEMENTS IN DNA AND RNA
(ppm dry wt) FOR INDUSTRIALLY PREPARED MATERIALS

Element	DNA (Mg Salt) Salmon Sperm	DNA Herring Sperm	DNA (Na Salt) Calf Thymus	DNA Calf Thymus	RNA Yeast	RNA Yeast
Pb	3.8	99.8	6.2	0.5	10.8	370
La	0.08	0.06	0.06	0.5	0.5	0.5
Ba	6.2	4.9	1.5	5.0	17.6	109
Cs	0.02	0.07	0.004	0.02	0.02	0.008
I	2.3	0.6	0.1	0.03	0.1	0.4
Sb	0.1	0.1	0.04	0.06	0.04	0.1
Sn	0.2	6.1	0.4	0.3	0.7	7.6
Cd	0.2	0.6	0.3	0.2	0.3	11.4
Ag	0.05	0.04	0.02	0.003	0.02	0.2
Mo	0.6	0.2	0.5	1.5	0.03	0.2
Zr	0.06		0.1	0.03	0.6	0.1
Y	1.1		0.8	0.3	0.6	
Sr	89.2	2.1	1.1	3.5	129.6	25.0
Br	2.3	3.9	1.8	1.0	22.1	4.3
Se	0.2	0.03	0.05	0.008	0.2	0.05
As	0.9	0.4	0.2	0.2	2.6	0.8
Ge	0.7	5.3	2.7	2.9	3.3	0.3
Ga	0.04	0.06	0.02	0.01	0.07	0.02
Zn	146.0	11.4	36.0	1.9	1,410	205
Cu	0.8	6.4	4.0	1.1	14.3	69.0
Ni	1.6	2.1	12.9	2.1	7.7	3.7
Fe	157.3	9.7	92.2	9.9	3,040	44.3
Mn	2.2	1.0	1.1	1.7	379.5	183.9
Cr	0.7	0.6	1.8	1.0	6.5	10.5
V	0.2	0.05	0.04	48.7	0.4	5.2
Ti	7.4	5.4	7.7	5.6	14.2	12.5
Ca			Major†			
K			Major			
S			Major			
P			Major			
Si	65.8	154	24.3	426	191	462
Al	0.7	2.5	1.6	0.1	0.9	0.2
Mg			Major			
Na			Major			
F	0.2	2.5	2.6	1.4	0.2	0.9
B	0.6	0.7	0.2	1.2	0.8	0.8
Be	0.007	<0.001	<0.001	<0.001	<0.001	<0.001
Li		0.2	0.02	0.1	0.06	0.1

* From Hamilton, E. I. and Minski, M. J.: Comments upon the organic constituents present in DNA and RNA. *Sci Total Environ,* 1:105, 1972.

† Major corresponds to a major constituent with concentration > 100 ppm.

elements, although it should be noted that often the range of observed levels can be quite considerable. Further data obtained by Sabbioni and Girardi (14) for calf thymus DNA, determined by neutron activation analysis, are given in Table LXXXII for which levels of Cu, Zn, and Mn are of interest.

Sabbioni and Girardi (14) have also determined the levels of several elements in microsamples of enzymes, as illustrated in Table LXXXIII: calf intestine alkaline phosphatase, cow milk xanthine oxidase, and calf thymus deoxynucleotidyl transferase; levels of As, Au, Cu, Fe, Hg, Mo, Sb, and Zn clearly indicate the involvement of many elements in proteins. As all separation and chemical processes were carried out under very carefully controlled conditions, including the use of metal-free animal cages, data such as this permit a realistic examination of the extent to which the chemical elements participate in protein synthesis. Using neutron activation methods, Sabbioni

TABLE LXXXI*

A. TRACE ELEMENTS IN RNA (μg/g) AND SACROTUBULAR FRACTION OF BEEF HEART TISSUE (μg/g)

Element	RNA (Beef Heart Tissue) Present Study (Range)	Sarcotubular Fraction of Beef Heart Tissue (Range)	RNA (Beef Heart Tissue)/Whole Beef Heart Tissue
Ag	0.12-0.45		66
Au	0.00012-0.0028		12
Ba	42;56	0.33-0.63	
Br	0.80-1.8	0.60-3.5	0.12
Cd	0.12-0.31		47
Ce	0.46-4.4		75
Co	0.11-3.5	0.071-0.12	57
Cr	3.7-190		5,600
Cs	0.04;0.14	0.0049-0.046	
Cu	86-280	60.2-66.2	9.5
Fe	1,000-1,800	3,700-4,600	6.7
Hg	0.54-1.3	1.1-1.9	4.9
La	0.011-0.16	0.020-0.11	36
Mo	1.1-2.6	0.86-1.6	6.1
Rb	12-27	0.70-0.88	1.4
Sc	0.00039-0.0036		25
Se	0.31-0.61	0.43-0.82	1.5
W	0.39-0.84	0.16-1.2	110
Zn	210-3,100	54-80	17

(Table continues on next page)

B. TRACE ELEMENTS IN RNA FROM BEEF HEART TISSUE, AND DATA ILLUSTRATING MAXIMAL POSSIBLE CONTAMINATION FROM REAGENT SOLUTIONS

Element	Amounts Found (Range)	Maximal Possible Contamination From Reagent Solution
Ag	0.0025-0.012	0.00017
As	0.0021-0.062	0.00080
Au	0.0000039-0.00015	0.000000043
Ba	0.75-1.4	0.0023
Br	0.017-0.066	0.0089
Cd	0.0041-0.0095	0.00063
Ce	0.018-0.05	0.00074
Co	0.0022-0.12	0.00024
Cr	0.049-6.0	0.0010
Cs	0.00053-0.0029	0.0097
Cu	3.7-5.1	0.0054
Fe	17.5-62.5	0.045
Hg	0.0077-0.046	0.00047
La	0.00024-0.0083	0.000010
Mo	0.023-0.13	0.0030
Rb	0.38-0.93	0.26
Sb	0.011-0.11	0.020
Sc	0.000022-0.00019	0.0000021
Se	0.0064-0.032	0.0015
Sm	0.0022-0.022	0.0010
W	0.0081-0.031	0.000085
Zn	6.8-102	1.82

* From Wester, P. O.: Trace elements in RNA from beef heart tissue. *Sci Total Environ*, 1:100-101, 1972.
Method of assay, neutron activation.

and Marafante (8) have considered the binding of metals to rat-liver-binding protein and have shown that Cu, Zn, Ag, Cd, Sn, and Hg are incorporated into cadmium-binding protein while no incorporation of Be, V, Cr, Mn, Co, Ni, Se, As, Mo, Te, Ir, Au, Tl, Pb, and Bi was observed following the injection of single labelled metals to two groups of rats, one set being treated with cadmium and the other cadmium-free. By using ^{35}S as an indication of protein synthesis, the biosynthesis of cadmium de-novo was clearly established, as illustrated in Figure 118. The realistic nature of this type of work is illustrated by the use of long term, low level exposure of rats to many elements. Experiments carried out over a period

TABLE LXXXII*

METAL CONTENT OF CALF THYMUS DNA AS DETERMINED BY NEUTRON ACTIVATION ANALYSIS

Element Determined	Metal Content (μg/g)	
	DNA Sample	Blank (Reagent Solution)
Ag	0.2	0.015
As	0.008	0.0002
Au	0.0004	0.000008
Ba	8	0.16
Br	0.06	0.0005
Cd	0.05	0.0011
Co	0.06	0.0048
Cr	1.2	0.07
Cs	0.03	0.0016
Cu	211	0.003
Eu	0.0001	0.000008
Fe	11.6	0.13
Ga	< 0.002	—
Hf	< 0.0003	—
Hg	0.3	0.014
Ir	< 0.00004	—
In	< 0.002	—
La	0.0008	0.00001
Mn	16	0.005
Mo	0.013	0.0011
Os	< 0.0006	—
Pd	< 0.0004	—
Pt	< 0.05	—
Rb	1	0.03
Ru	< 0.008	—
Sb	0.06	0.0003
Sc	0.003	0.0005
Se	0.1	0.0003
Sn	< 0.95	—
Sr	2.2	0.22
Ta	< 0.003	—
Th	0.008	0.0002
U	< 0.007	—
Zn	1,450	0.09
Zr	< 6.5	—
W	< 0.00032	—

* From Sabbioni, E. and Girardi, F.: Metallobiochemistry of heavy metal pollution: Nuclear and radiochemical technique for long-term low level exposure (LLE). *Sci Total Environ,* 7:145, 1977.

of one year using high specific activity radionuclides showed that, in the case of cadmium, accumulation occurred in the kidney, liver, intestine, and pancreas; the intracellular distribu-

TABLE LXXXIII*

HEAVY METAL CONTENT IN MICROSAMPLES† OF ENZYMES BY NEUTRON ACTIVATION ANALYSIS

	Calf Intestine Alkaline Phosphatase (MW = 140,000)			Cow Milk Xanthine Oxidase (MW = 275,000)			Calf Thymus Deoxynucleotidil Transferase (MW = 33,000)		
	Amount of Element Determined (μg/g)		g Atom Mole Protein	Amount of Element Determined (μg/g)		g Atom Mole Protein	Amount of Element Determined (μg/g)		g Atom Mole Protein
	Sample	Blank		Sample	Blank		Sample	Blank	
As	4.8	<0.001	0.009	16.4	0.04	0.06	16	0.05	0.007
Au	7.1	0.002	0.005	5.7	0.003	0.008	2.4	0.002	0.0004
Cd	8	<0.01	0.01	24.5	<0.03	0.06	10.2	<0.03	0.003
Co	1.3	0.005	0.07	6.43	0.008	0.03	1.4	0.4	0.0008
Cs	8.5	0.0004	0.009	3.9	0.004	0.008	28.2	1.7	0.007
Cu	362.8	0.04	0.8	13.8	0.003	0.06	38.4	8.2	0.02
Fe	320	0.3	0.8	1,649	0.32	8.1	84.8	0.1	0.05
Hg	28.6	0.009	0.02	36.4	0.2	0.05	—	—	—
Mn	2.76	0.0006	0.007	8	0.02	0.04	5	0.04	0.003
Mo	54.8	<0.02	0.08	684.2	0.07	1.96	—	—	—
P	199.3	0.001	0.9	484.2	0.006	4.3	37.5	0.02	0.04
Sb	17.4	0.07	0.02	17.7	0.009	0.04	14.7	0.09	0.004
Zn	1,892	0.001	4.05	21.4	0.01	0.09	1,803	0.01	0.91

* From Sabbioni, E. and Girardi, F.: Metallobiochemistry of heavy metal pollution: Nuclear and radiochemical techniques for long-term low level exposure (LLE). *Sci Tota Environ*, 7:145, 1977.

† All results are the mean of more determinations ranging from 3 to 4. Samples of protein ranged from 250 to 700 μg.

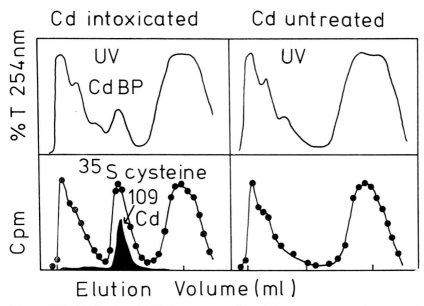

Figure 118: Sephadex G-75 chromatography of the supernatant fraction of livers of ^{109}Cd- and ^{35}S-cysteine treated and control rats. From Sabbioni, E. and Marafante, E.: Identification of lead binding. *Chem Biol Int*, 15:1, 1976.

tion demonstrated that the element was essentially present in soluble cytoplasmic fractions but absent from nuclei, mitochondria, lysosomes, and microsomes. By using radionuclides of high specific activity and long half lives it is possible, for a limited number of elements, to carry out experiments over long periods of time and using levels of elements commonly found in natural diets.

Today very elegant chromatographic techniques and ultracentrifugation techniques exist for tissue, cell, and cell constituent separations. Unfortunately it is often difficult to purify the basic materials used in these separations, although this is not important for neutron activated samples. However, there are available several simple techniques as illustrated in Figures 119-122; because of small size they are amenable to clean-

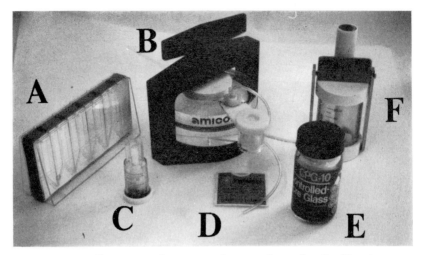

Figure 119: Illustration of some simple items for molecular filtration.
A. The Minicon Concentrator (Amicon Ltd., High Wycombe, Backs, United Kingdom) illustrated consists of five cells; a sample for molecular separation is pipetted into the central channel which is backed by a membrane and an absorbent pad. Depending upon the molecular cut-off value of the membrane, water and membrane-permeable species are extracted and then the concentrate retained in the central channel can be recovered with a pipette.
B. The Amicon Model CECI On-line Column Eluate Concentrator directly concentrates column eluates as high as 20 : 1. The unit can be coupled to the outlet of a chromatographic column, the eluate from which passes through a thin channel path over the surface of a molecular filter paper and emerges in two outlet tubes, one for the ultrafiltrate phase and the other for the concentrate.
C. The Bio-rad Laboratories Bio Fiber® 50 (see Figure 120).
D. Nuclepore Filter (Nuclepore, General Electric) papers composed of polycarbonate which have been irradiated with fission fragments followed by chemical etching to provide an even distribution of pores ranging from 0.5 to 8.0 microns (see page 293) with a pore density of 3×10^7 cm². The precise shape of these holes allows various type of cells to be separated, for example, large cancer cells.
E. Controlled Pore Glass (Electro Nucleonics, Inc.) is for use in gel permeation chromatography providing a range of pore sizes from 75 to 2,000 Å.
F. Magnetically stirred ultrafiltration cells (Amicon Ltd.) are a simple system for concentrating macromolecules, consisting of a cell for operation under pressure and using membrane filters for separations.

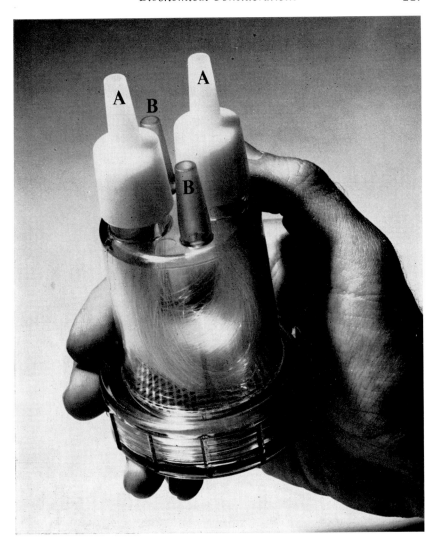

Figure 120: A Bio-rad Bio-Fiber Minibeaker. The system is based upon the use of hollow fibres through which a solution is passed; permeable species pass through the walls of the fibres and macromolecules are retained. In the illustration the sample solution circulates through the ports A and the permeable species are removed through the ports B; the base of the beaker can be fitted with a magnetic stirrer. This system is particularly suitable for protein separations. Photograph supplied by Bio-Rad, United Kingdom.

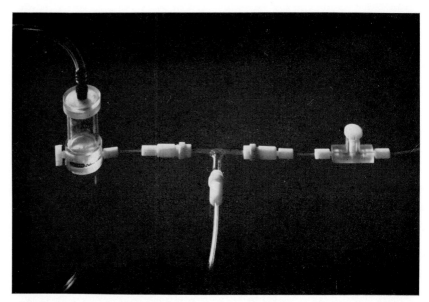

Figure 121: A Bio-rad T-Tube Hollow Fiber System. In the illustration the left hand unit consists of a stirred filtering cell (see Figure 119F), the filtrate from which passes into a bundle of hollow fibres. The concentrate passes along the tube while the ultrafiltrate is removed through the "T" arm. Photograph supplied by Bio-Rad, United Kingdom.

up techniques and are particularly useful for separating proteins and enzymes from large dilute solutions and for desalting liquids.

The living cell is a complex structure composed of membranes, one around the cell, one around the nucleus, and others surrounding various intracellular bodies such as lysosomes, Golgi apparatus, endoplasmic reticulum, and mitochondria. Individual parts of a cell, and the cell as a whole, perform individual functions; collections of similar cells produced by processes of differentiation undertake special body processes concerned with many exchange processes such as those of metabolism and excretion. At a general level of study the chemical elements have clearly defined functions; for example Mg^{+2}, K^+, and PO_4^- are typically found within cells, while Ca^{+2} and Na^+

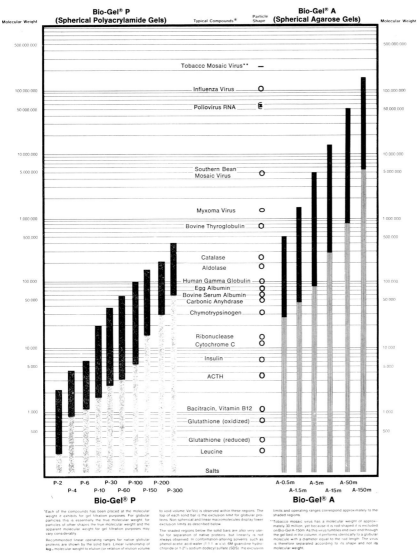

Figure 122: Examples of molecular separations using spherical polyacrylamide and agarose gels. Photograph supplied by Bio-Rad, United Kingdom.

are extracellular; cell permeability is dependent upon the divalent metal ions which are not associated with very strong bonding. Metals such as Fe, Cu, Co, and Mo are associated with enzymes, participate in redox changes, and are strongly bound. In proteins, metal ions may form an integral part of the structure, or are loosely attached in order that they can be involved in electron transfer processes which require the rapid transport of electrons. The stability of metal enzymes in aqueous systems is governed through relationships between complex molecules and the degree of stability can be evaluated in relation to the type of coordination or chelated compounds. In the absence of complexing substances the order of stability for transition metal complexes is as follows:

$$Mn^{+2} < Fe^{+2} < Co^{+2} < Ni^{+2} < Cu^{+2} < Zn^{+2}$$

This sequence is called the Irving-Williams (15) order and is applicable for nearly all complexes irrespective of the nature of coordination ligands or number of ligand molecules involved; the order is based upon ionic radius, ionization potential of metals, and preferred configuration of bonding and ligand type which results in a highly stable complex. The presence of some metal ions can alter the rate of some catalytic reactions; Pedersson (16) studied the hydrolysis of nitro-acetic acid and showed that decarboxylation was retarded by metal ions following the formation of cation complexes with anion with the following order of stability:

$$H^{+1} > Al^{3+} > Cu^{2+} > Be^{2+} > Cd^{2+} > Pb^{2+} >$$
$$Ni^{2+} > Zn^{2+} > Co^{2+} > Mg^{2+} > Ca^{2+} > Ba^{2+}$$

Those metal ions with chalcophilic properties are often involved in interaction with sulphydryl groups, the order of metal displacement being—

$$Ag^{+2} \approx Hg^{+2} > Cu^{+2} > Zn^{+2} > Cd^{+2} > Pb^{+2} >$$
$$Co^{+2} > Ni^{+2} > Mn^{+2}$$

In aquatic environments, Shaw (17) postulated that cation toxicity arises through a combination of metal ions with an essential group, possibly -SH or a key enzyme; data obtained for urease diastase showed the following order of toxicity:

$$Mn^{+2} < Fe^{+2} < Co^{+2} < Ni^{+2} < Cu^{+2} > Zn^{+2}$$

It is now possible to identify both stability and activity of many metal-organic systems, expecially those concerned with enzymatic activity, but interest centres upon identifying sites of exchange and the nature of the reaction processes. Williams (18) proposed that proteins generate a highly constrained site for metal enzymes which endow the site with a specific catalytic significance. In attempting to formulate structured links, the tendency is to ignore the essential requirement of electron transfer, namely speed of transfer; clearly the topographic form of molecular surfaces is important and special configurations such as grooves in enzymes may concentrate those surface properties essential for surface catalytic processes. Today it is difficult to determine how the present configurations and sites were developed as all we can observe is specificity of natural enzymes as a result of organic evolution, such as iron and copper in hemerythrin and haemocyanin, molybdenum in nitrogen fixation, and zinc in the hydrolysis of phosphate ester bonds. Williams (19) considers that electron transfer processes involve metal to metal, metal-free sulphydryl bonding sites, metal to aromatic, and metal to some organic sites; excluded are electron transfer from metals to organic groups such as phenyl rings, alcohol oxygen, or carboxylate, as there is no evidence they are involved although all are present in haemoglobin.

In terms of normal body function, the key role of metal ions is defined but they only function in order for electron transfer processes to take place. Williams (19) emphasises that because of the need for speed of transport one-electron reactions do not require atom transport between reaction centres, therefore at critical initial exchange sites two-electron transfers are likely to be more acceptable if considered as two one-electron transfers. An analogy between electron exchange and transfer processes is provided by simple electric circuits in which capacitor protein units, such as two or more metal centres, for example Fe-S, provide important redox equivalents for which electrons would be supplied at a rate limited by a conformational change to a site of chemical reduction. An operational chain to provide a steady delivery of energy would be a chemical cell, a capacitor isolated from the main energy transfer stream, and a switching mechanism, all linked by some form of conduction pathway. Today, concepts based upon "hard wiring" can be examined in terms of solid state electronic systems in which electron transfer and storage can be considered in terms of electron conduction bands controlled by the presence of element impurities in structures. A direct analogy with semiconductors is not possible because of the time required to carry out electron diffusion processes.

So far we have considered the importance of several minor and trace elements as essential constituents in maintaining health. Since the Industrial Revolution, around 1850, technological elements have pervaded man's environment which are new to processes of natural evolution. Some types of adaptation undoubtedly take place but there exist natural pathways for rejecting excessive amounts of many elements. Nevertheless, because of similarities between elements of the Periodic Table many enter man, and on the basis of simple chemical effects, are likely to influence enzymatic processes. An alternative viewpoint would be to consider that exposure of developing cell systems, in the reduced environment of pre-biotic seas, provided a template for dealing with unwanted elements and compounds, at least to a degree that some protection is obtained. Microorganisms play an important role in element chemistry: they actively participate in many reactions involving the chemical elements and are capable of changing oxidation states and the chemical properties of elements. The work of Wood and his co-workers (20) has clearly illustrated the role of microorganisms in the natural environment, particularly in the alkylation of metals. These mechanisms have an impact on man, for example soluble Hg^{+2} can be reduced to Hg^0 by enzyme processes and can be visualised as a detoxification process. Vitamin B_{12} (cobalt complex), in which the cobalt ion can exist in three oxidation states, can change the oxidation state of inorganic complexes. The polarity of organometallic complexes is very much related to toxicity; for example, if polar, they can diffuse through membranes while nonpolar compounds partition into lipids and lipoproteins where they become enriched. Chance similarity in molecular structure can result in an enrichment of some compounds; Wood (21) describes similarities between the methyl mercury-homocysteine complex and the essential amino acid methionine. Electrophilic attack by metal ions on Co-C (Vitamin B_{12}) bonds yield products which are very toxic; methylcobalamin is excellent for the synthesis of methyl and dimethylmercury. Methylation of mercury by microorganisms active in aquatic sediments resulted in the Minamata disaster. Several other metals can also be transferred in a similar manner, for example methylation of Pt, Au, and

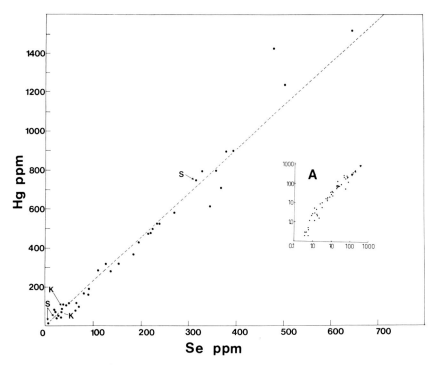

Figure 123: Correlation between mercury and selenium in the liver of seals, spleen (S) and kidney (K). Method of analysis X-ray fluorescence, data in terms of ppm dry weight. The insert A presents data (obtained by Koeman, J. H., Vande Ven, W. S. M., Goeij, J. J. M., Tijoe, P. S., and van Haaften, J. L.: Mercury and selenium in marine mammals. *Sci Total Environ*, 3:279, 1975) for various types of marine mammals. From Hamilton, E. I.: Review of the chemical elements and environmental chemistry—strategies and tactics. *Sci Total Environ*, 5:1, 1976.

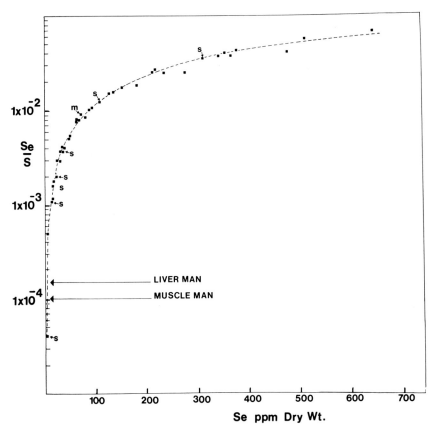

Figure 124: Correlation between the ratio Se/S and total Se (ppm dry wt) for various organs of the seal; s, spleen, m, muscle, remaining data points are for liver. From Hamilton, E. I.: Review of the chemical elements and environmental chemistry—strategies and tactics. *Sci Total Environ,* 5:1, 1976.

Tl, while the metalloids of As, Se, and Te all produce very toxic volatile compounds and are associated with neurological disease in man. In the case of selenium, dimethylselenide is a product which is vented, hence the methyl groups are transferred from the more toxic methylmercury cycle to the less toxic methylselenium cycle.

Natural associations between Hg and Se have clearly been demonstrated in marine mammals by Koeman et al. (22) and

Hamilton (23), who have described an increase in levels of selenium in the liver of seals with increase in levels of mercury as illustrated in Figure 123. Pentreath (24-26) has presented an excellent account describing the accumulation of mercury in fish with particular reference to the importance of dietary intake. There is much to be learned from the study of marine animals which is pertinent to man. In laboratory animals the correlation between Se and Hg has been considered beneficial and is often referred to as an illustration of an animal using Se to detoxify Hg, rather than perhaps emphasising fortuitous similarities of coordination chemistry of benefit to the animal. Man, and most vertebrates, obtain their source of Se from sulphur compounds in which Se is ubiquitously present as a result of isomorphous substitution; problems of disease are usually associated with a deficiency in Se. Hamilton (23) considered the possibility of the effect of Se utilisation on stable sulphur levels in seals, as illustrated in Figure 124; although Se levels varied between $\approx$ 10 and 700 ppm dry weight, those for sulphur were constant, within experimental errors, of 9132 ± 910 ppm dry weight. In the sea there are no discrete sources of Se, apart from those produced by biochemical processes, and most of a seal's supply of Se will be from fish muscle.

Much of the work in identifying toxic effects of various elements and compounds is derived by extrapolation of animal data, particularly that of the rat and mouse. Neither animal has a very long life span while the use of primates in research is expensive. Very little work in relation to toxic effects of the elements on man has been carried out on marine mammals such as the seal, although such data may be relevant to furthering our understanding of human diseases and the toxicity of some elements; for example in the marine environment cadmium is a conservative element (that is, it mixes freely with salt water and does not form insoluble compounds nor is it precipitated by saline processes)—since life evolved from the seas this element has always been able to enter living systems and, in man, is associated with discrete types of protein. On the other hand most of the lead in the sea is in an insoluble

A COMPARISON BETWEEN THE CONCENTRATION (ppm wet wt.) FOR S

Element	Blood Seal (12)†	Blood Man (2,500)	Spleen Seal (12)	Heart Seal (12)	Seal (12)
Pb	0.1 ± 0.1	0.3 ± 0.1	0.08 ± 0.03	0.04 ± 0.01	0.1 ± 0
Ba	0.3 ± 0.1	0.08 ± 0.03	0.02 ± 0.01	0.004 ± 0.001	0.02 ± 0
Cs	0.01 ± 0.01	0.01 ± 0.01	~ 0.0008	0.006 ± 0.006	0.04 ± 0
I	0.1 ± 0.01	0.06 ± 0.04	0.04 ± 0.01	0.002 ± 0.001	0.02 ± 0
Sb	0.002 ± 0.001	0.002 ± 0.001	0.02 ± 0.01	0.03 ± 0.01	0.004 ± 0
Sn	0.02 ± 0.01	0.005 ± 0.002	0.05 ± 0.02	0.03 ± 0.01	0.1 ± 0
Cd	0.02 ± 0.01	—	0.1 ± 0.05	0.03 ± 0.01	0.2 ± 0
Ag	0.005 ± 0.001	0.01 ± 0.005	0.002 ± 0.001	0.0009 ± 0.003	0.002 ± 0
Nb	0.005 ± 0.001	0.004 ± 0.0005	—	—	< 0.0
Zr	0.09 ± 0.03	0.01 ± 0.006	0.01 ± 0.005	0.006 ± 0.002	0.01 ± 0
Y	0.005 ± 0.001	0.007 ± 0.003	0.008 ± 0.002	0.003 ± 0.001	0.02 ± 0
Sr	0.1 ± 0.01	0.02 ± 0.005	0.2 ± 0.05	0.07 ± 0.03	0.2 ± 0
Rb	< 0.5	2.8 ± 0.1	1 ± 0.5	0.7 ± 0.1	1.1 ± 0
Br	16.0 ± 4	4.8 ± 0.1	11 ± 3	7 ± 1	4.0 ± 0
As	0.08 ± 0.03	< 0.01	0.2 ± 0.1	0.1 ± 0.05	0.3 ± 0
Ga	0.003 ± 0.001	—	0.0009 ± 0.0001	0.0003 ± 0.0001	0.02 ± 0
Zn	4.0 ± 1.0	7.0 ± 0.2	20 ± 6	28 ± 3	26 ±
Cu	< 0.1	1.2 ± 0.02	0.2 ± 0.1	1 ± 0.5	< 0.3
Ni	3.4 ± 2.0	—	< 10	< 8	< 10
Co	0.02 ± 0.01	—	< 0.05	< 0.01	0.03 ±
Fe	1,557 ± 172	493.0 ± 3.9	1,166 ± 881	240 ± 115	442 ±
Mn	0.2 ± 0.1	0.09 ± 0.05	0.2 ± 0.1	0.4 ± 0.3	0.4 ±
Cr	0.03 ± 0.01	0.07 ± 0.05	0.004 ± 0.001	0.007	0.01 ±
V	0.02 ± 0.01	—	0.008 ± 0.002	0.02 ± 0.01	0.02 ±
Ti	1.4 ± 0.2	—	1.3 ± 0.2	2 ± 0.3	1.9 ±
Ca	74 ± 23	60.4 ± 0.3	41 ± 8	36 ± 10	30 ±
K	223 ± 79	1,900 ± 20	3,380 ± 400	2,630 ± 420	3,970 ±
Cl	2,044 ± 141	—	1,176 ± 100	720 ± 90	365 ±
S	1,831 ± 123	1,770 ± 14	2,242 ± 120	2,300 ± 173	2,507 ±
P	574 ± 136	335 ± 10	3,325 ± 343	2,042 ± 180	2,380 ±
Si	23 ± 8	—	29 ± 8	22 ± 8	44 ±
Al	0.2 ± 0.2	0.2 ± 0.06	0.5 ± 0.3	0.07 ± 0.04	0.2 ±
Mg	29 ± 7	49 ± 16	193 ± 24	103 ± 73	82 ±
F	0.007 ± 0.002	0.07 ± 0.02	0.02 ± 0.01	0.001 ± 0.0005	0.007 ±
B	2.0 ± 0.4	—	0.5 ± 0.2	0.1 ± 0.05	0.3 ±
Be	—	< 0.01	< 0.005	< 0.03	< 0.02 ±
Li	0.003 ± 0.001	0.002 ± 0.0004	0.0007 ± 0.0003	0.002 ± 0.001	0.003 ±

* From Hamilton, E. I.: Review of the chemical elements and environmental chem strategies and tactics. *Sci Total Environ*, 5:36-37, 1976.
† Number of samples analysed.

Biochemical Considerations

XIV*

MENTS IN SEAL AND HUMAN TISSUES FOR THE UNITED KINGDOM

	Kidney			Liver	
Man (6)	Seal (12)	Man (8)	Seal (12)	Man (11)	
5 ± 0.009	0.4 ± 0.2	1.4 ± 0.2	0.3 ± 0.2	2.3 ± 0.6	
2 ± 0.006	0.02 ± 0.01	0.01 ± 0.001	0.005 ± 0.001	0.01 ± 0.003	
—	0.002 ± 0.001	0.009 ± 0.005	—	—	
± 0.001	0.2 ± 0.1	0.04 ± 0.01	3 ± 3	0.2 ± 0.06	
± 0.003	0.004 ± 0.001	0.006 ± 0.001	0.008 ± 0.003	0.01 ± 0.002	
± 0.01	0.08 ± 0.02	0.2 ± 0.04	0.1 ± 0.05	0.4 ± 0.08	
± 0.01	10 ± 9	14 ± 0.7	0.8 ± 0.7	2.0 ± 0.4	
± 0.0005	0.005 ± 0.001	0.002 ± 0.002	0.1 ± 0.1	0.006 ± 0.002	
± 0.003	< 0.002	0.01 ± 0.004	< 0.03	0.04 ± 0.009	
± 0.003	0.002 ± 0.001	0.02 ± 0.002	0.03 ± 0.01	0.03 ± 0.005	
± 0.002	0.002 ± 0.001	0.006 ± 0.001	0.01 ± 0.005	0.01 ± 0.002	
± 0.02	0.2 ± 0.1	0.1 ± 0.02	0.1 ± 0.05	0.1 ± 0.03	
± 0.5	1 ± 0.2	—	0.8 ± 0.2	7.0 ± 1.0	
—	16 ± 2	—	12 ± 4	4.0 ± 0.3	
± 0.001	0.3 ± 0.1	0.004 ± 0.0030	0.10 ± 0.06	0.0006 ± 0.0030	
± 0.00004	0.001 ± 0.0005	0.0009 ± 0.0003	0.001 ± 0.001	0.0007 ± 0.0001	
± 3	39 ± 14	37 ± 6	62 ± 16	57 ± 7	
± 0.02	2 + 0.5	2 ± 0.4	21 ± 13	0.8 ± 10	
—	< 10	—	9 ± 6	—	
—	0.03 ± 0.01	—	0.02 ± 0.005	—	
± 1	277 ± 63	—	1,318 ± 690	207 ± 39	
± 0.007	2 ± 0.7	1 ± 0.5	8 ± 3	0.5 ± 0.08	
± 0.001	0.01 ± 0.005	0.03 ± 0.005	0.006 ± 0.004	0.008 ± 0.006	
± 0.003	0.05 ± 0.02	—	0.2 ± 0.001	0.2 ± 0.01	
± 0.01	2 ± 0.3	—	2 ± 0.2	0.4 ± 0.1	
± 5	61 ± 6	—	45 ± 12	54 ± 7	
± 380	2,310 ± 340	—	2,950 ± 360	2,400 ± 100	
± 10	1,686 ± 180	—	974 ± 148	1,430 ± 49	
± 21	2,473 ± 128	1,480 ± 310	2,911 ± 401	2,000 ± 100	
± 350	2,718 ± 432	—	3,186 ± 94	2,020 ± 120	
± 0.9	37 ± 20	11.2 ± 1.2	36 ± 12	14.9 ± 4	
± 0.2	0.5 ± 0.2	0.4 ± 0.1	0.3 ± 0.3	3 ± 1	
± 60	79 ± 9	205 ± 35	149 ± 32	172 ± 30	
± 0.004	0.01 ± 0.005	0.01 ± 0.002	0.007 ± 0.005	0.06 ± 0.01	
—	0.01 ± 0.01	—	0.2 ± 0.1	—	
—	< 0.04	—	< 3	—	
± 0.002	0.002 ± 0.002	0.001 ± 0.003	0.02 ± 0.005	0.007 ± 0.003	

TABLE LXXXV

	W. Scotland			Farne Islands		
Element	Male	Female	Pup	Male	Female	Pup
Hg	1547	1424	579	31	<9	<9
Pb	0.4	1.5	0.2	0.06	0.10	0.09
Se	647	480	280	<22	<22	<22
V	0.3	0.2	0.2	0.04	0.007	0.005
Zn	69	82	76	40	55	48

The concentration (ppm dry weight) for some elements in seal liver from the Hebrides (W. Scotland) and the Farne Islands (North Sea). The islands of the Hebrides are situated off the west coast of Scotland and although distant from sources of contamination and pollution they receive waters from the contaminated Irish Sea to the south. The Farne Islands are adjacent to areas of high technical effluents but yet, compared with the Hebrides, are relatively uncontaminated. Although mortality, directly attributable to contaminants present in the water and items of diet, for both populations of seals is unknown, nevertheless such groups of mammals provide an ideal opportunity to study biochemical effects for defined elements and their compounds.

form and for man, considering the large amounts of technical lead to which he is exposed, very little enters the body. Finally in the sea an element such as aluminium is almost totally associated with insoluble phases and, although man is exposed to very large amounts of this element, which is a major constituent of rocks and minerals of the earth's crust, it is virtually absent from the human body. Data are presented in Table LXXXIV for the abundance of several elements in the seal compared with those for man, while some selected data are given in Table LXXXV for some elements present in seals from clean and polluted waters. Allowing for characteristic biochemical features of the seal, for example high levels of myoglobin in muscle cells and a 75-fold increase in blood volume in the seal in order to distribute heat in marine environments, both species have much in common; in fact requirements for muscle coordination are possibly more stringent in seals than man. Hence seals could be a very suitable species to study neurological response following metal pollution. With advances in analytical chemistry we are now entering an era in which element interactions can be studied at the cell level of investigation. In relation to effects attributable to elements, the field of neurological response is important. There is a need to further our knowledge of the manner in which elements and their compounds interact with biochemical processes taking place at synaptic junctions, the extent to which they can enter the basic structure of synaptic material, for example tubulin and other proteins, and also to identify any specific interactions taking place at nerve cell contact regions in relation to brain function. Processes taking place at synaptic junctions are relevant to many diseases, for example myasthenia gravis, and today we can apply both direct and indirect methods of analysis to study the crucial area of membrane receptor processes.

An element such as Se provides many illustrations of the importance of a trace element. Hulstaert et al. (27) describe selenium in vitamin E deficiency and the occurrence of myopathy as a symptom of vitamin E deficiency; ducklings have shown that symptoms of Se deficiency can easily be produced by feeding animals with a low Se synthetic diet obtained by using a low Se caesin as a source of Se. This feature is found in the natural environment in relation to the distribution of Se and geography; in fact because of the ubiquitous association of Se and S, selenium-deficient caesin has to be obtained from milk of cows grazing on Se-deficient herbage.

Schwarz and Pathak (28) describe fatal selenium response deficiencies in many animals illustrated by dietary liver necrosis which can be corrected by administering Se in diet. By studying different chemical forms of selenium, it has been shown that the required potency can be obtained, in the absence of toxic effects, by selecting Se in a particular chemical form, for example selenodiacetylalanine with a ip. LD_{50} of 1.1 g Se per kg compared with selenite with a value of 4 mg Se per kg. The possibility therefore exists of selecting other elements in particular chemical forms for therapeutic purposes and thus overcoming their inherent toxicity. Selenium cannot be considered a pollutant but quite clearly is an important element and has recently been reviewed by MBEEP (29).

The scope for element-biochemical research, fundamental and applied and directly related to the health of man, should not ignore a consideration of inorganic in favour of pure organic study. Hopefully, this monograph has identified, in general terms, the role of inorganic studies, especially the central role of analytical chemistry and the essential requirement of true, unfettered, multidisciplinary cooperation.

REFERENCES

1. Timm, F.: Zür Histochemie der Schwermetalle. Das sulfid-silber Verfahren. *Dtsch Z Ges Gericht Med, 46*:706, 1958.
2. Fjerdingstad, E. J., Danscher, G., and Fjerdingstad, E.: Hippocampus: Selective concentration of lead in the normal rat brain. *Brain Res, 80*:350, 1974.
3. Danscher, G., Fjerdingstad, E. J., Fjerdingstad, E., and Fredens, K.: Heavy metal content in subdivisions of the rat hippocampus (zinc, lead and copper). *Brain Res, 112*:442, 1976.
4. Brun, A. and Brunk, U.: Histochemical indications for lysomal localisation of heavy metals in normal rat brain and liver. *J Histochem Cytochem, 18*:820, 1970.
5. Choie, D. D. and Richter, G. W.: Lead poisoning: Rapid formation of intracellular inclusions. *Science, 177*:1194, 1972.
6. Hsu, F. S., Krook, L., Shively, J. N., Duncan, J. R., and Pond, W. G.: Lead inclusion bodies in osteoclasts. *Science, 181*:447, 1973.
7. Murakami, M. and Hirosawa, K.: Electron microscope autoradiography of kidney after administration of ^{210}Pb in mice. *Nature (Lond): 245*, 1973.
8. Sabbioni, E. and Marafante, E.: Identification of lead-binding components in rat liver: In vivo study. *Chem Biol Interact, 15*:1, 1976.
9. Walton, J. R.: Intranuclear inclusions in the lead-poisoned cultured kidney cell. *J Path, 112*:213, 1974.
10. Walton, J. R.: Granules containing lead in isolated mitochondria. *Nature (Lond), 243*:100, 1973.
11. Malone, C., Koeppe, D. E., and Miller, R. J.: Localisation of lead accumulated by corn plants. *Plant Physiol, 53*:388, 1974.
12. Hamilton, E. I. and Minski, M. J.: Comments upon the inorganic constituents present in DNA and RNA. *Sci Total Environ, 1*:104, 1972.
13. Wester, P. O.: Trace elements in RNA from beef heart tissue. *Sci Total Environ, 1*:97, 1972.
14. Sabbioni, E. and Girardi, F.: Metallobiochemistry of heavy metal pollution: Nuclear and radiochemical techniques for long term—low level exposure (LLE) experiments. *Sci Total Environ, 7*:145, 1977.

15. Irving, H. and Williams, R. J. P.: The stability of transition-metal complexes. *J Chem Soc*, 3193, 1953.
16. Pedersson, K. J.: The effect of metal ions on the rate of decomposition of nitro-acetic acid. *Acta Chem Scand*, 3:676, 1949.
17. Shaw, W. H. R.: Cation toxicity and the stability of transition-metal complexes. *Nature (Lond)*, 192:754, 1961.
18. Williams, R. J. P.: Catalysis by metallo-enzymes: The entatic state. *Inorganica Chemica Acta Reviews*, 5:137, 1971.
19. Williams, R. J. P.: Electron transfer and oxidative energy. Fifth Keilin Memorial Lecture. *Biochemical Society Transactions*, 1:1, 1973.
20. Wood, J. M.: The biochemistry of toxic elements in aqueous systems. In Malins, C. and Sargent, J. R. (Eds.): *Biochemical and Biophysical Approaches to Marine Biology*. New York, Academic Press, 1976.
21. Wood, J. M.: Biological cycles for elements in the environment. *Die Naturwissenschaften*, 8:357, 1975.
22. Koeman, J. H., Peeters, W. H. M., Koudstaal-Hol, C. H. M., Tjioe, P. S., and Goeij, J. J. M.: Mercury-selenium correlations in marine mammals. *Nature (Lond)*, 245:385, 1973.
23. Hamilton, E. I.: Review of the chemical elements and environmental chemistry—strategies and tactics. *Sci Total Environ*, 5:1, 1976.
24. Pentreath, R. J.: The accumulation of mercury by the thornback ray, Raja clavata. *J Exp Mar Biol Ecol*, 25:131, 1976.
25. Pentreath, R. J.: The accumulation of inorganic mercury from seawater by the plaice, Pleuronectes platessa. *J Exp Mar Biol Ecol*, 24:103, 1976.
26. Pentreath, R. J.: The accumulation of organic mercury from seawater by the plaice, Pleuronectes platessa. *J Exp Mar Biol Ecol*, 24:121, 1976.
27. Hulstaert, C. E., Molenaar, I., Goeij, J. J. M., and van Pijpen, P. L.: Selenium in vitamin-E deficient diets and the occurrence of myopathy as a symptom of vitamin E deficiency. *Nutr Metab*, 20:91, 1976.
28. Schwarz, K. and Pathak, K. D.: The biological essentiality of selenium, and the development of biologically active organoselenium compounds of minimum toxicity. *Chemica Scripta*, 8A:85, 1975.
29. *Medical and Biological Effects of Environmental Pollutants: Selenium*. Washington, D.C., Nat Acad of Sci, 1976.

Chapter 11

THE FUTURE

PREDICTING THE FUTURE, even when armed with an abundance of valid information concerning past experiences, is at the best a risky pursuit. Nevertheless, as far as the chemical elements and man are concerned, an objective has to be the prediction of possible harmful or beneficial effects of the elements upon the health of man, together with those constituents of the environment which support mankind. In the past, man has been exposed to a very large number of the elements of the Periodic Table, often at very high concentrations, but yet mankind has survived and longevity has been significantly prolonged. Today the multifactorial nature of many diseases associated with acute or chronic exposure to the elements complicates any effort directed towards identifying a particular element as the cause of a particular state of disease; today the need is to consider inter-element and compound effects set in a complex natural and technological environment.

Although this monograph is biased towards furthering our understanding between disease and elements directly attributable to human activities, beyond the pathways through which the elements reach man there is the more difficult task of identifying their participation in biochemical processes taking place in cells which give rise to various types of diseased states. Superimposed upon what may be termed academic pursuits there is also the practical need to consider economic and political implications, even when evidence is available which indicates that a particular element or compound does present a hazard to the health of man. Interspersed with progressive discovery is the random identification of a hazard such as one associated with an accident. In such instances the pendulum of public opinion often swings too far in the direction of calling for immediate remedial measures and the imposition of bans which are often not practical and which, in many instances, have little significance beyond the local areas in which the accident took place. There are other instances where it is known that, through human activity, elements and compounds have become widely dispersed throughout the global environment and levels have progressively increased with time, for example, lead present in airborne particulate matter much of which is derived from the automobile. It is right that concern should be expressed over such matters, but often any real detrimental effects (for example, those affecting large populations) are difficult to identify, while well-documented evidence of dose-response relationships for past exposures are not available. On the other hand there are various types of diseases which are known to be associated with the elements but yet they still exist and affect large populations, for example endemic goitre and cretinism.

One of the major controls on the proliferation of the human race, globally or regionally, is that of climate and its variability. Over long periods of time climate determines element availability to crops and livestock and what type of species can proliferate in a particular area. Man cannot predict future changes in climate, although he can indicate probable trends; he is virtually powerless, in the long term, to undertake any required remedial action. Major food-producing areas in the northern hemisphere are particularly vulnerable, and quite sizeable catastrophes are likely to occur; for example, the cereal crop may fail for a few seasons. It is against such a background that it seems important to develop hardy crops, with a high yield, but if—in the process of applying modern methods of plant genetics—hybrids can be produced which will discriminate against potentially harmful elements, this would be an added bonus. In the fields of agriculture and veterinary practice significant progress is now being made and often levels of elements such as selenium, copper, and cobalt give rise to more concern than those caused by infectious diseases. In relation to biogeochemistry, clearly the close association between crops and livestock and soils is an advantage; while man is more remote from natural element associations, many similarities in the performance of biochemical systems involving elements are apparent.

The future should improve our knowledge for essential requirements for dietary compositions; in the rush to identify the nature of an essential diet for large populations we tend to forget that people of different countries use different foods and that individuals have different metabolic patterns and requirements for a particular type of diet: for example, Pygmies inhabiting the dense forests of Ruandi have, as a result of forest clearance schemes, been rehabilitated in adjacent but environmentally different areas; when provided with a typical Bantu diet, serious nutritional deficiencies appear. Such a situation is also possible in a developed country if natural traditional diets are supplemented or replaced with new foodstuffs such as soya bean or artificial proteins to the partial exclusion of essential roughage.

The air we breathe is heavily contaminated or polluted by various technological products; individuals at most risk are urban dwellers although the distribution of technological debris in the air is of global dimensions. Insufficient attention is being paid to the abundance and chemical form of the elements in urban areas. Considerable attention is paid to the effluents from individual smelters; while lead emitted from some smelters is of local interest, we tend to forget that urban dwellers are exposed to this element at levels some one hundred times higher than those found in rural communities. The domestic home is a veritable mine for many elements, and exposure to some elements may be quite considerable. Far more attention and research is required for the chemical composition of in-house environments with particular emphasis being paid to exposure to children and babies. The latent period for many diseases can be long; while so-called subtoxic effects are often difficult to identify, it seems reasonable to consider that exposure to some elements, early in life within homes, could have a bearing upon the health of adults. In the future, increased use will be made of fossil fuels; these materials are often highly enriched in many elements which become concentrated in urban communities. Many parts of the world are supplied with excellent geochemical maps for a large number of elements but similar maps are not available for towns and cities which contain a significant number of the total population.

With an ever-increasing demand for raw energy, the use of nuclear power is likely to become important in the future. The effluents derived from this industry contain radioactive species and it is the ionizing radiation which is harmful, rather than the chemical properties of the elements. There is room for considerable public education in this field; at present, projected catastrophes seem unlikely, although local accidents will occur and there is also the new dimension of terrorist involvement in the availability of significant amounts of fissionable materials. The high record of safety in the nuclear industry is quite outstanding, but sincere concern is expressed by many that an element such as plutonium is released into the natural environment. It is generally accepted that a linear relationship exists between exposure to ionizing radiation and effects, hence all radiation is harmful. Ionizing radiation is a natural phenomena and an individual residing on chalklands will receive less dose than one on granitic terrains, yet obvious effects attributable to these extremes of exposure are not apparent. Nevertheless, in recent years there is some evidence that exposure to very low doses of ionizing radiation may produce effects, although this matter is open to considerable controversy. In the future, hopefully, radiation associated with an element such as plutonium will be placed in perspective by comparison with a naturally occurring emitter having similar radiation properties; a useful candidate would be polonium which is possibly one of the most hazardous alpha emitters.

Perhaps containment of radionuclides in natural materials, especially the Oklo Phenomenon (see page 9), which includes a total system including rocks, organic compounds and water, may provide practical solutions to the containment of man-produced radioactive materials?

The overall status of the chemical elements and human health may be likened to the phoenix rising from the ashes; today we live in times when at last the problems surrounding technical measurements of elements and associated biological activity have been mainly surmounted. We can now make a new start with some confidence that future actions can be based upon reliable measurements. We need to reexamine existing theories and hypotheses in the light of technical innovations, not through the activity of a single discipline but rather through inter- and multidisciplinary approaches. In order for this to be accomplished, we need to develop a common language and establish practical objectives. This is most readily accomplished by the support and operation of relatively small working groups of interested individuals provided with a common language. Large international organisations have an important part to play, but—because of remoteness of parts of such structures, the lack of a common language, and usually the dominance of one discipline—much of the potential is lost.

The future requires a very careful appraisal of the possible significance of the chemical elements and human health; if we are not careful massive amounts of data on environmental levels will be accumulated but not used in any objective evaluation. Instead, points of view will become accepted which are likely to be more akin to the substance of the "Emperor's New Clothes" than reality of modern life. Today many new hypotheses are being suggested; although they stimulate the scientific community for short periods of time, for example the story of polywater, some have little substance. The system of Aristotle formed the basis for scientific speculation for nearly two thousand years; supported by Taoist philosophies and through Arabian alchemists, much of modern chemistry and related technical applications was born. Has the discovery of the structure of the atom vindicated one of the oldest alchemical theories? Today we can transform matter but by different methods and purposes than those proposed by the alchemists; recently other methods for element transformation have been suggested, such as through the presence of all matter in the neutrino field. Absurd as some hypotheses may seem when they are first suggested, all have to be considered and, even if disproved, often give rise to interesting fallout products while extending our philosophical thinking beyond material requirements. At present each disease tends to be treated separately, although collectively some are associated with a particular biological system. Perhaps the future will identify some fundamental similarities responsible for the initial onset of a diseased state as distinct from details concerning the results of the presence of a particular disease.

INDEX

(Page numbers in **bold type** indicate whole chapters or sections; in *italic type* indicate figures; with suffix 't' indicate the entry occurs in a Table.)

A

Activation analysis *(see* Radioactivation analysis)
Air-conditioning system, 161, 164-6
  for ultra-clean laboratory, *164*
Alpha particle(s)
  autoradiography of, 288, *292,* 293, 297
  Backscatter Analyser (APBSA), 357-8
  track analysis, 302-7, *302-5*
Aluminium
  abundance ratios, 37t, 40t
  analysis of, 196, 196t
  concentration of, in, air particulates, 31t, 34t, 57t, 163t
    blood, 75t, 79t, 172t, 369t, 456t
    bone, 88, 390t, 391t
    brain, 379t
    car exhaust effluent, 44t
    coal, 50t
    DNA, 440t
    drinking water, 71t, 73, 74t, 83t, 97t
    excreta containers, 141t
    house dust, 55t, 57t, 59, *63*
    kettle fur, 79t
    kidney, 385t, 457t
    liver, 382t, 457t
    lung, 27t, 386t
    lymph nodes, 394t
    milk, 115t
    muscle, 388t, 457t
    ovaries, 397t
    RNA, 440t
    seal tissues, 456-7t
    sugar, 125t
    testes, 396t
  daily intake of, 101t
  disease and, 422
Americium-241, 297, 302, *303*
Amino acids, formation of, 15
Ammonia, 15
  analysis of, 196t, 210
Anodic stripping voltammetry, 214-7, *215, 216*
Antimony, xiit
  analysis of, 241
  concentration of, in, air particulates, 31t, 33t, 56t
    blood, 75t, 78t, 172t, 368t, 376t, 456t
    bone, 390t
    brain, 378t
    car exhaust effluent, 44t, 45
    coal, 48t, 49t
    DNA, 440t, 443t
    drinking water, 74t; particulates, 82t
    enzymes, 444t
    house dust, 56t
    kettle fur, 78t
    kidney, 384t, 457t
    liver, 382t, 383t, 457t
    lung, 27t, 386t
    lymph nodes, 394t
    milk, 114t
    muscle, 388t, 457t
    RNA, 440t, 442t
    seal tissues, 456-7t
    sugar, 125t
    testes, 396t
  daily intake of, 101t
Arsenic, xiit, 25, **104-8**, *106, 107*

469

abundance ratios, 40t
analysis of, 196, 241
concentration of, in, air particulates, 31t, 33t, 56t, 163t
blood, 75t, 79t, 172t, 369t, 456t
brain, 379t
car exhaust effluent, 44t, 45
coal, 48t, 49t, 50t
DNA, 440t, 443t
drinking water, 97t; particulates, 81, 82t, 86
enzymes, 444t
excreta containers, 141t
house dust, 55, 55t, 56t, 57, 59, 64
kettle fur, 79t
kidney, 385t, 457t
liver, 382t, 383t, 457t
lymph nodes, 394t
milk, 114t
muscle, 388t, 457t
RNA, 440t, 441t, 442t
seal tissues, 456-7t
testes, 396t
daily intake of, 101t
Arteriosclerotic heart disease, 69, 124
Ash content of tissues, 142t
Ashing
dry, 148-51
in furnace, limitations of, 148, *149*
low temperature, 148-50, *150*, 151t
by heating under vacuum, 150
Schöniger technique, 150-1
wet, 151-5
closed systems, 154-5, *154*, *155*
Atomic absorption spectroscopy, **231-46**, *233*, *240*, *242*, *243*, *244*, *245* 248-50
applications of, 381
atomisation, flame, 236-8, *237*
electrothermal, 238-41, *240*, *242*, *243*
detection limits, 248-9, 249t
double beam spectrometer, 245-8, *245*
limitations of, 239-41
light source, 234-6
Auger spectroscopy (AES), 356
Automatic analysis
for biological samples, 283-5, *283*, *284*, *285*, *286-7*
precision of, 290-1t
Autoradiography
alpha-particle, 288, *292*, 293
beta and gamma emitter, 307, *308-10*

B

Barium, xiit
analysis of, 197
concentration of, in, air particulates, 31t, 33t, 56t, 163t
blood, 75t, 78t, 368t, 456t
bone, 390t, 391t
brain, 378t, 380t
car exhaust effluent, 44t
coal, 48t, 50t
DNA, 440t, 443t
drinking water, 74t, 76, 97t; particulates, 82t
excreta containers, 141t
hair, 401t
house dust, 56t, 401t
kettle fur, 78t
kidney, 384t, 457t
liver, 382t, 457t
lung, 27t, 386t
lymph nodes, 394t
milk, 114t
muscle, 388t, 457t
ovaries, 397t
RNA, 440t, 441t, 442t
seal tissues, 456-7t
sugar, 125t
testes, 396t
daily intake of, 100t
Barium-130, 7
Beryllium, xiit, 6
analysis of, 201, 256
concentration of, in, air particulates, 31t, 34t, 57t, 163t
blood, 79t, 456t
car exhaust effluent, 44t, 45
coal, 48t, 49t, 50t
DNA, 440t
house dust, 57t
kettle fur, 79t
RNA, 440t
seal tissues, 456-7t
sugar, 125t
daily intake of, 101t

Index

Bismuth, xiit
  analysis of, 196t, 197, 201, 216, 238, 241, 257
  concentration of, in, air particulates, 31t, 32t, 56t, 163t
    blood, 172t, 241, 370t
    bone, 390t, 391t
    brain, 378t
    car exhaust effluent, 44t
    coal, 48t, 51t
    excreta containers, 141t
    house dust, 56t
    kettle fur, 78t
    kidney, 384t
    liver, 382t
    lung, 27t, 386t
    lymph nodes, 394t
    milk, 114t
    muscle, 388t
    testes, 396t
  daily intake of, 100t
Blood, human
  concentration of elements in, 75t, 78-9t, 140t, 172t, 368-9t, 370t, 372, 372-6, 373, 374, 375, 376t, 456t
  compared with crustal rock, 20, *20*, *21*
  compared with seawater, 21-2, *22*, *23*
  determination of elements in, 372-6
  sample contamination of, 138, 139t, 374
  sampling, methods, 138-9, 372
  U.K. sites, *371*
Bone
  concentration of elements in, 387, 389-91, 390t, 391t
  concentration of lead in, *389*
Boron, xiit, 6
  concentration of, in, air particulates, 31t, 34t, 57t
    blood, 79t, 172t
    bone, 390t, 391t
    car exhaust effluent, 44t
    coal, 48t, 49t, 50t
    DNA, 440t
    drinking water, 71t, 74t, 97t; particulates, 83t
    hair, 400t
    house dust, 57t
    kettle fur, 79t
    milk, 115t
    RNA, 440t
    seal tissues, 456-7t
    sugar, 125t
  daily intake of, 101t
Brain, human
  concentration of elements in, 376-7, 378-9t, 380t, 381
Brain, rat
  hippocampus of, lead levels in, 428, *429, 430*
Bromine, xiit
  abundance ratios, 40t
  analysis of, 196t, 263
  concentration of, in, air particulates, 31t, 33t, 43, 56t, 163t
    blood, 75t, 79t, 140t, 172t, 369t, 456t
    bone, 391t
    brain, 378t
    car exhaust effluent, 44t, 45
    DNA, 440t, 443t
    drinking water, 74t, 76, 97t; particulates, 81t
    house dust, 56t
    kettle fur, 79t
    kidney, 384t
    liver, 382t, 383t, 457t
    lung, 27t, 386t
    lymph nodes, 394t
    milk, 114t
    ovaries, 397t
    RNA, 440t, 441t, 442t
    seal tissues, 456-7t
    testes, 396t
  daily intake of, 101t

C

Cadmium, xiit
  analysis of, 201, 216, 216t, 239, 240, 248
  behavior of, in seawater, 455
  concentration of, in, air particulates, 31t, 33t, 56t, 163t
    blood, 75t, 78t, 172t, 370t, 376t

bone, 390t, 391t
brain, 378t
car exhaust effluent, 44t
coal, 48t, 50t
DNA, 440t, 443t
drinking water, 74t, 97t; particulates, 82t
enzymes, 444t
excreta containers, 141t
hair, 400t
house dust, 56t
kettle fur, 78t
kidney, 384t, 457t
liver, 382t, 383t, 457t
lung, 27t, 386t
lymph nodes, 394t
milk, 114t
muscle, 388t, 457t
ovaries, 397t
RNA, 440t, 441t, 442t
seal tissues, 456-7t
sugar, 125t
testes, 396t
daily intake of, 101t
disease and, 422
metabolism of, 442-3, *445*, 445
Calcium, xiit, 69
abundance ratios, 37t
analysis of, 196, 196t, 203, 210, 211, 256, 357
concentration of, in, air particulates, 31t, 34t, 57t
blood, 75t, 76, 79t, 140t, 172t, 369t, 376t, 456t
brain, 379t
car exhaust effluent, 44t
coal, 50t
DNA, 440t
drinking water, 71t, 74t, 83t, 97t; particulates, 81t, 82t, 83t
excreta containers, 141t
house dust, 55t, 57t
kettle fur, 79t
kidney, 385t
liver, 382t, 457t
lymph nodes, 394t
milk, 114t
muscle, 388t, 457t
ovaries, 397t
RNA, 440t
seal tissues, 456-7t
sugar, 125t
testes, 396t
daily intake of, 101t
deficiency (Urov Disease), 120
Californium, 6, 302, *303*
Cancer, 419-23
geographical distribution of, 421
Carbon, xiit, 5
analysis of, 257, 357
concentrations of, in, drinking water particulates, 81t
house dust, 55t
Carcinogens, 420, 421
Cardiovascular disease, 70, 71t, 416-20
Cerium
concentration of, in, air particulates, 31t, 33t, 56t, 163t
coal, 48t, 50t
excreta containers, 141t
house dust, 56t
kettle fur, 78t
lymph nodes, 394t
milk, 114t
RNA, 441t, 442t
testes, 396t
Cesium
abundance ratios, 37t
concentration of, in, air particulates, 31t, 33t, 56t, 163t
blood, 75t, 78t, 172t, 368t, 376t, 456t
bone, 390t
brain, 380t
car exhaust effluent, 44t
DNA, 440t, 443t
drinking water, 74t
enzymes, 444t
excreta containers, 141t
house dust, 56t
kettle fur, 78t
kidney, 384t, 457t
liver, 383t, 457t
lymph nodes, 394t
milk, 114t
ovaries, 397t
RNA, 440t, 441t, 442t
seal tissues, 456-7t

sugar, 125t
testes, 396t
daily intake of, 101t
Cesium-137, 307, *309*, 311, 405
Chemicals toxic, influence of, on evolution, 19
Chemiluminescence, 203
Chlorine, xiit
  abundance ratios, 40t
  analysis of, 196t, 353
  concentration of, in, air particulates, 31t, 34t, 43, 57t, 163t
    blood, 75t, 76, 79t, 140t, 172t, 369t
    brain, 379t
    coal, 50t
    drinking water, 73, 74t, 97t; particulates, 81t
    house dust, 55t, 57t
    kettle fur, 79t
    kidney, 385t
    liver, 382t, 383t, 457t
    lung, 27t, 386t
    lymph nodes, 394t
    milk, 115t
    muscle, 388t, 457t
    ovaries, 397t
    seal tissues, 456-7t
  daily intake of, 101t
Chlorophyll, fossil evidence of, 15
Chromatography, 185
  applications of, 282, 445 (see also Gas chromatography)
Chromium, xiit, 124
  analysis of, 203, 256
    as chromate, 196t
  concentration of, in, air particulates, 31t, 33t, 56t, 163t
    blood, 75t, 79t, 172t, 369t, 370t, 456t
    brain, 379t
    car exhaust effluent, 44t
    coal, 48t, 49t, 50t
    DNA, 440t, 443t
    drinking water, 74t, 76, 83t, 97t; particulates, 82t, 83t
    excreta containers, 141t
    hair, 400t, 401t
    house dust, 55t, 56t, 401t
    kettle fur, 79t

kidney, 385t, 457t
liver, 382t, 383t, 457t
lung, 27t, 386t
lymph nodes, 394t
milk, 114t
muscle, 388t, 457t
ovaries, 397t
RNA, 440t, 441t, 442t
seal tissues, 456-7t
sugar, 125t
testes, 396t
  daily intake of, 101t
Climate, 463
  effect of, on health, 29
  supernovae on, 5-6
Coal
  as source of elements, 47-9, 48t, 49t
  consumption trends, *54*
Cobalt, xiit
  abundance ratios, 40t
  analysis of, 196, 196t, 216
  concentration of, in, air particulates, 31t, 33t, 163t
    blood, 172t, 370t, *375*
    car exhaust effluent, 44t
    coal, 48t, 49t, 50t
    DNA, 443t
    drinking water, 97t; particulates, 82t
    enzymes, 444t
    excreta containers, 141t
    liver, 383t
    RNA, 441t, 442t
    seal tissues, 456-7t
  daily intake of, 102
Cobalt-60, 311
  decay of, *274*
Colorimetry, 192-9, *194, 195*
Colostrum, 112, 118
Concentration techniques, 184-6
Contamination, environmental, 28-59, 411, 414-16, 452, 464
  sample, 185
    from grinding, 143t, 145-8, 323
    from sample containers, 140, 141t, 183
    from sampling method (blood), 138, 139t
    from reagents, 153

474 The Chemical Elements and Man

in anodic stripping, 217
in X-ray fluorescence, 323, 324-5
Copper, xiit
  abundance ratios, 40t
  analysis of, 196t, 196, 201, 203, 216, 216t, 240, 353
  concentration of, in, air particulates, 31t, 33t, 56t, 163t
    blood, 75t, 79t, 140t, 172t, 369t, 370t, *373*, 376t, 456t
    bone, 391t
    brain, 379t, 380t
    coal, 48t, 49t, 50t
    DNA, 440t, 443t
    drinking water, 73, 74t, 76, 83t, 97t; particulates, 81t, 82t, 83t, *84*
    enzymes, 444t
    excreta containers, 141t
    hair, 400t
    house dust, 55t, 56t, 59, *65*
    kettle fur, 79t
    kidney, 385t, 457t
    liver, 382t, 383t, 457t
    lung, 27t, 386t
    lymph nodes, 394t
    milk, 101t
    muscle, 388t, 457t
    ovaries, 397t
    RNA, 440t, 441t, 442t
    seal tissues, 456-7t
    sugar, 125t
    testes, 396t
Corn (*see* Maize (*Zea mays*))
Cyanide, analysis of, 196t, 210

D

Data
  evaluation of, 363-4
  obtaining accurate, 365-6, 370
  retrieval, 360-2, 434
Dental caries, 103, 392-3, 395
Detectors
  neutron, use of, 288
  radiation, 275-7, *276* (*see also* Geiger-Müller counter; Ge(Li) detector; Na(Tl) detector)
Deuterium, 4, 6, 263

Diet, **95-126**, 464
  arsenic in, 104-5, 108
  cancer and, 420
  composition of, 99t, 100
  elements in, 97t, 98, 100-1t, 102-5, 108-16, 114-15t, 124-6, 125t, 126t
  mercury in, 108-9
  tin in, 109-13
  uranium in, 109, 110t, 111t
Disease, **411-23**, 462, 463
DNA
  element concentrations in, 439, 440t, 433t
  formation of, 15
Dust, household
  element concentrations in, 55t, 56-7t, 401t
  structure of, 58, **58**
Dysprosium
  concentration of, in, air particulates, 31t, 32t, 56t
  bone, 390t
  drinking water, 74t
  house dust, 56t
  milk, 114t

E

Earth
  age of, 8
  atmosphere of, 15, 26, 27
  formation of, 10-14
  heavy elements present on, 8-9
  ice ages on, 6
  origin of, 3-6
  ozone layer of, 6
Electrochemical techniques, 204-19
Electron microprobe analyser (EMA), 350-2, *351*
Electron microscope, scanning (SEM), 350
Electron spectroscopy (SCA), 355-6
Element(s)
  abundance, of, on earth, 6-7, *7*
    ratios, 37t, 40t
  availability of, in food, 24-5
  in seawater, 455
  coffee, present in, 98
  concentration of, in, air particulates,

Index

31t, 32-5t, 163t
blood, human, 75t, 78-9t, 140t, 172t, 368-9t, 370t, 372, 372-6, *373*, *374*, *375*, 376t, 389t, 456t; seal, 456t
bone, human, 390t
brain, human, 380t, 389t
car exhaust effluent, 44t
DNA, 440t, 443t
drinking water, 71t, 74t, 83t; hard, 74t; soft, 74t, 97t; particulates, 81t, 82t, 83t
enzymes, 444t
excreta containers, 141t
fetal tissues, 109-11, 397-9
fossil fuels, 48t, 49t, 50t
hair, 400t, 401t, 401-2
house dust, 54-5, 55t, 56-7t, 57-9
kettle fur, 97t
kidney, human, 384-5t, 398t, 457t; seal, 457t
liver, human, 382t, 383t, 398t, 457t; seal, 457t, 458t
lung, human, 27t, 386t
lymph nodes, human, 394t, 398t
maize (*Zea mays*), 126t
milk, 114t, 116
muscle, human, 388t, 398t; seal, 456t
newborn, tissues, 111, 397-9; vertebrae, 391t
ovaries, 397t
placenta, 109-11
RNA, 440t, 441t, 442t
seal tissues, 456-7t; liver, 458t
soya bean, 126t
sugar, 125t
testes, 396t, 398t
detection of, by colour methods, 196-9, 196t, *197*, 198
distribution of, in man, determination of, 427t, 427-8
environmental levels of, local, effects of, 25, 103-9, 120-3
essential, xi, xiit, xiiit, 364-5
geochemical classification of, 36
loss of, after low temperature ashing, 151t
natural levels of, variability of, 41-3

sensitivity for, by neutron activation analysis, 273t
separation and concentration of, prior to analysis, 184-7, 186t
synthesis of, 3-5
tea, present in, 98
toxic, xi, xiiit, 25, 104-9, 455
Encephalopathy, 90-1
Endoplasmic reticulum, 435, 448
Environment,
levels of toxic elements in, 25
laboratory, **156-69**
Enzyme(s), 118-19
determination of elements in, 439, 441
heavy metal content of, 441t
metals, associated with, 449-50
unusual, explanation for, 19-20
primeval, 15-17
Erbium
concentration of, in, air particulates, 31t, 32t, 56t
bone, 390t
coal, 50t
house dust, 56t
kettle fur, 78t
milk, 114t
Europium
concentration of, in, air particulates, 31t, 32t, 56t
bone, 390t
coal, 50t
DNA, 443t
drinking water, 74t
house dust, 56t
kettle fur, 78t
milk, 114t
Evolution, 14-15, *16*, 19
Excitation sources, used in emission spectroscopy, 222-5
Excreta containers,
concentration of elements in, 141t

F

Faeces
collection containers for, element concentration in, 141t
infant, concentration of tin in, 113, 116t

sampling methods for, 140-1
Faraday's Laws, 205-6
Fertilisers, artificial, as source of elements, 103
Fetal tissues, 397-9
 concentration of elements in, 398-9
Filtration techniques, 445-8, *446*, *447*, *448*, *449*
Fission track analysis, 293-7
 applications of, 389
Flame photometry, 227-31, *230*
Fluid intake, in adults, 92t, 96, 98
Fluorescence spectroscopy, 246-8, *247*
 detection limits, 249t
Fluoride,
 analysis of, 196t, 210, 211
 concentration of, in, drinking water, 71t
  teeth, *392*, 393
 dental caries and, 395
Fluorine, xiit
 analysis of, 263
 concentration of, in, air particulates, 31t, 34t, 57t, 163t
  blood, 75t, 79t, 172t, 369t, 456t
  brain, 379t
  car exhaust effluent, 44t, 45
  dental caries and, 103, 393, 395
  DNA, 440t
  house dust, 57t
  kettle fur, 79t
  kidney, 385t, 457t
  liver, 382t, 457t
  lung, 27t, 386t
  lymph nodes, 394t
  milk, 115t
  muscle, 388t, 457t
  ovaries, 397t
  RNA, 440t
  seal tissues, 456-7t
  sugar, 125t
  testes, 396t
 daily intake of, 101t
Food
 additives, 124
 in intensive livestock rearing, 103-4
 consumption of, 99t
 element concentrations of, local variations in, 103

methods of analysis of, 102
sampling of, 99-100, 102
as source of elements, 95-6
Fossil fuels
 consumption trends of, 54
 elements present in, 43, 50-1t (*see also* coal)

G

Gadolinium
 concentration of, in, air particulates, 31t, 32t, 56t
  bone, 390t
  coal, 50t
  house dust, 56t
  kettle fur, 78t
  milk, 114t
Gallium
 abundance ratios, 37t
 analysis of, 216, 238
 concentration of, in, air particulates, 31t, 33t, 56t, 163t
  blood, 172t
  brain, 378t
  car exhaust effluent, 44t
  coal, 49t, 50t
  DNA, 440t, 443t
  drinking water, 74t, 76
  house dust, 56t
  kidney, 385t, 457t
  liver, 382t
  lung, 27t, 386t
  lymph nodes, 394t
  muscle, 388t, 457t
  ovaries, 397t
  RNA, 440t
  seal tissues, 456-7t
  sugar, 125t
  testes, 396t
Gamma ray spectrometers (*see* Ge(Li) detector; Na(Tl) detector)
Gas chromatography, 186
 linked to microwave plasma, 263-4, *264*
Gaussian distribution, 134
Ge(Li) detector, *276*, 279, *280*, *282*
Gieger-Müller counter, *276*, 277-8
Germanium
 abundance ratios, 37, 37t

Index   477

concentration of, in, air particulates,
    31t, 33t, 56t, 163t
  blood, 75t, 79t, 172t
  car exhaust effluent, 44t, 45
  coal, 49t, 50t
  DNA, 440t
  house dust, 56t
  kettle fur, 79t
  RNA, 440t
  sugar, 125t
  daily intake of, 101t
Gold
  concentration of, in, air particulates,
    31t, 32t, 56t
  blood, 172t, 368t, 376t
  bone, 390t
  brain, 380t
  DNA, 443t
  enzymes, 444t
  house dust, 56t
  RNA, 441t, 442t
  daily intake of, 101t
Granules, intracellular (see Inclusions
  (granules), intracellular)

H

Haemodialysis, water quality for, 87-93
  (see also Renal dialysis)
Hafnium
  abundance ratios, 37, 37t
  concentration of, in, air particulates,
    32t, 56t
  DNA, 443t
  house dust, 56t
Hair, 399-402
  concentration of elements in, 440t,
    401t, 401-2
Helium, 4, 5
Hemochromatosis, 422
Holmium
  concentration of, in air particulates,
    31t, 32t, 56t
  bone, 390t
  coal, 50t
  house dust, 56t
  kettle fur, 78t
  milk, 114t
Homogenisation, sample, **142-8**
  by grinding, 144-8, *146, 147*

for mass spectrometry, 342
for X-ray fluorescence, 323-5
wet, 143, *144-5*
Hydrogen, xiit
  analysis of, 357
  importance of, in organic evolution,
    15
  origin of, 3-4
Hydrogen sulphide, analysis of, 196t
Hypothalamus, 29

I

Inclusions, (granules), intracellular 433,
  435, *436, 437, 438,* 439
Indium
  abundance ratios, 40t
  analysis of, 216
  concentration of, in, air particulates,
    31t, 33t
  car exhaust effluent, 44t
  DNA, 443t
Infant food, concentration of elements
  in, 112-3, 114t, 115-6, 115t, 116t
Instrumentation, **190-362**
Intercalibration exercises (see Interlaboratory comparisons)
Interlaboratory comparisons, 171-6, 174-
  5t, 176-7t, *178-80,* 182
Iodide
  analysis of, 203, 210
  concentration of, in drinking water,
    71t
Iodine, xiit
  abundance ratios, 40t
  analysis of, 196t, 263
  concentration of, in, air particulates,
    31t, 33t, 56t, 163t
  blood, 75t, 78t, 172t, 368t, 456t
  bone, 391t
  brain, 378t
  car exhaust effluent, 44t
  DNA, 440t
  drinking water, 74t, 97t
  house dust, 56t
  kettle fur, 78t
  kidney, 384t, 457t
  liver, 382t, 457t
  lung, 27t, 386t
  lymph nodes, 394t

milk, 114t
muscle, 388t, 457t
ovaries, 397t
RNA, 440t
seal tissues, 456-7t
sugar, 125t
testes, 396t
daily intake of, 101t
Iodine-131, 41
  in thyroid of sheep, 42
Ion-exchange resins, 185
  chelating, 185
    applications of, 281, 282-3, 284
    recovery values for, 289t
Ion-microprobe analyser (IAM), 352
Ion-protein bonding, 448-51
Ion scattering spectrometer (ISS), 355
Ion-selective electrodes, 206-7, 210-2, 207, 208, 209
Iridium
  analysis of, 257
  concentration of, in, air particulates, 32t
  DNA, 443t
  daily intake of, 257
Iron, 20
  abundance ratios, 40t
  analysis of, 196, 196t, 201, 203, 216, 256, 353
  concentration of, in, air particulates, 31t, 33t, 56t
    blood, 75t, 79t, 140t, 172t, 369t, 370t, 376t, 456t
    brain, 379t, 380t
    car exhaust effluent, 44t
    coal, 50t
    DNA, 440t, 443t
    drinking water, 74t, 83t, 97t; particulates, 80, 81t, 81, 82t, 82, 83t, 86
    enzymes, 444t
    excreta containers, 141t
    hair, 400t
    house dust, 55t, 56t, 59, 62
    kettle fur, 79t
    kidney, 385t
    liver, 382t, 383t, 457t
    lung, 27t, 386t
    lymph nodes, 394t

milk, 114t
muscle, 388t, 457t
ovaries, 397t
RNA, 440t, 441t, 442t
seal tissues, 456-7t
sugar, 125t
testes, 396t
daily intake of, 101t
disease and, 422
Iron-56, 7
Irving-Williams order, 450
Isolator system, trace element free, 117, 118

K

Kettle fur
  concentration of elements in, 78-9
  sampling and analysis of, 77
Kidney
  concentration of elements in, 381, 383, 384-5t, 457t
  lead in, 433, 434t

L

Laboratory, analytical
  contamination sources in, 165, 165t
  design of, 158-64, 159, 160, 161, 162
  types of, 156-7
  working conditions in, 157-8
Lanthanum
  concentration of, in, air particulates, 31t, 33t, 56t, 163t
  bone, 390t
  brain, 380t
  coal, 48t, 50t
  DNA, 440t, 443t
  drinking water, 74t; particulates, 82t
  excreta containers, 141t
  house dust, 56t
  kettle fur, 78t
  liver, 382t, 383t
  lymph nodes, 394t
  milk, 114t
  RNA, 440t, 441t, 442t
  testes, 396t
Laser probe analysers, 353-5, 354
Lead, xiit, 25
  abundance ratios, 40t

Index   479

analysis of, 196t, 197, 201, 216, 216t, 238, 348, 353
behavior of, in seawater, 455, 458
concentration of, in, air particulates, 31t, 32t, 43, 51-4, 52t, 56t, 163t
   blood, 74t, 78t, 368t, 370t, 375, 423, 456t; determination of, 177, 181-3, 239-40, *244*
   bone, 390t, 391t
   brain, 378t, *430*
   car exhaust effluent, 44t, 45
   coal, 47-9, 48t, 49t, 51t
   DNA, 440t
   drinking water, 74t, 76, 83t, 97t; particulates, 80, 81t, 81, 82t, 83t, 85
   excreta containers, 141t
   hair, 400t, 401t
   house dust, 55t, 56t, 59, *64*, 401t
   kettle fur, 78t
   kidney, 384t, 457t
   liver, 382t, 457t
   lung, 27t, 386t
   lymph nodes, 394t
   milk, 114t
   muscle, 388t, 457t
   ovaries, 397t
   RNA, 440t
   seal, tissues, 456-7t; liver, 458t
   sugar, 125t
   testes, 396t
daily intake of, 100t, 102, 115
disease and, 422
distribution of, in rat tissues and intracellular components, *433*, 434t, 435t, 435-6, *436*, *437*
tissue levels of, 24
Lead-203, -204, 307
Lead-210, 403, 405
Life, development of, 14-17 (*see also* Evolution)
time required for, 17
Lithium, xiit, 6
abundance ratios, 37t
analysis of, 256
concentration of, in, air particulates, 31t, 34t, 163t
   blood, 75t, 79t, 172t, 369t, 370t, 456t
   brain, 379t
   car exhaust effluent, 44t
   coal, 48t, 50t
   DNA, 440t
   drinking water, 74t, 97t
   excreta containers, 141t
   hair, 400t
   kettle fur, 79t
   kidney, 385t, 457t
   liver, 382t, 457t
   lung, 27t, 386t
   lymph nodes, 394t
   milk, 114t
   muscle, 388t, 457t
   ovaries, 397t
   RNA, 440t
   seal tissues, 456-7t
   sugar, 125t
   testes, 396t
daily intake of, 101t
Liver, human
concentration of elements in, 381, 382t, 383t, 457t
cirrhosis of, 422
heavy metals in, 430
Liver, rat
distribution of lead in, 434t, 435-9, *436*, *437*
fractionation of, *431*, 435
lead binding components of, *432*, 435
Lung
concentration of elements in, 26, 27t, 383, 386t, 387
functioning of, 414
Lutetium
concentration of, in, air particulates, 31t, 32t
bone, 390t
coal, 50t
milk, 114t
Lymph nodes
concentration of elements in, 394t, 395
Lysosomes, 448
uptake of metals by, 430

M

Magma, 12, *13*
Magnesium, xiit, 5, 69

abundance ratios, 37t
analysis of, 196, 203, 353, 357
concentration of, in, air particulates, 31t, 34t, 57t, 163t
  blood, 75t, 172t, 369t, 456t
  brain, 379t
  car exhaust effluent, 44t
  coal, 50t
  DNA, 440t
  drinking water, 71t, 74t, 97; particulates, 81t, 82t
  house dust, 55t, 57t
  kettle fur, 79t
  kidney, 385t, 457t
  liver, 382t, 383t, 457t
  lung, 27t, 386t
  lymph nodes, 394t
  milk, 114t
  muscle, 388t, 457t
  ovaries, 397t
  RNA, 440t
  seal tissues, 456-7t
  sugar, 125t
  testes, 396t
daily intake of, 101t
Maize (*Zea mays*), 125
concentration of elements in, 126t
precipitation of lead in, 439
Manganese, xiit, 20
abundance ratios, 40t
analysis of, 196t, 353
concentration of, in, air particulates, 31t, 33t, 56t
  blood, 75t, 76, 79t, 172t, 369t, 370t, 374, *374*, 376t, 456t
  brain, 379t
  car exhaust effluent, 44t
  coal, 48t, 49t, 50t
  DNA, 440t, 443t
  drinking water, 71t, 74t, 76, 83t, 97t; particulates, 81t, 82t, 83t
  excreta containers, 141t
  hair, 400t
  house dust, 55t, 56t
  kettle fur, 79t
  kidney, 385t, 457t
  liver, 382t, 383t, 457t
  lung, 27t, 386t

  lymph nodes, 394t
  milk, 114t
  muscle, 388t, 457t
  ovaries, 397t
  RNA, 440t
  seal tissues, 456-7t
  sugar, 125t
  testes, 396t
daily intake of, 101t
Mass spectrometry, 332-48
isotope dilution analysis by, 344-5, *346*
of liquids, 347-8
of lead, 348
spark source, 337-44, *338*, *340-1*, 366
  applications of, 381, 439
  detection limits for human tissues, 367
thermal ionisation, 334-7, *335*
Mercury, xiit, **108-9**
analysis of, 197, 241, 248
concentration of, in, air particulates, 31t, 32t
  blood, 75t, 172t, 368t
  bone, 390t
  coal, 48t, 51t
  DNA, 443t
  drinking water particulates, 82t
  enzymes, 444t
  hair, 400t, 401t
  house dust, 401t
  liver, 383t
  RNA, 441t, 442t
  seal, tissues, 453, 455; liver, 458t
daily intake of, 100t
disease and, 422
Metabolic pathways, xv, 2, 311
Meteorites, 8
carbonaceous chondrites, 14
Methane, 15
Methods, analytical, xvi, xvii, **130-362**
problems associated with, xii-xvii, 130-3
standardisation of, 169-84
(*see also* individual techniques *eg.* Anodic stripping voltammetry; Atomic absorption spectroscopy; Automatic analysis; Autoradiog-

Index 481

raphy; Chromatography; Colorimetry; Electrochemical techniques; Fission track analysis; Flame photometry; Fluorescence spectroscopy; Mass spectrometry; Optical emission spectroscopy; Polarography; Radioactivation analysis; Spectrophotometry; X-ray fluorescence spectrometry; *etc*)
Methylmercury, 108, 453-4
Microorganisms, 452-3
Microprobe optical laser examiner (MOLE), 359
Milk
  concentration of, elements in, 114t, 116
    tin in, 113, 115t
    zinc in, 112
Mitochondria, 448
  lead granules in, in rat liver, 435-6 *436, 437*
Mitochondrial membranes, distribution of lead in, 434t, 435t
Molybdenum, xiit
  concentration of, in, air particulates, 31t, 33t, 56t
    blood, 75t, 78t, 172t, 368t
    bone, 391t
    car exhaust effluent, 44t
    coal, 48t, 49t, 50t
    DNA, 440t, 443t
    drinking water particulates, 82t
    enzymes, 444t
    excreta containers, 141t
    house dust, 56t
    kettle fur, 78t
    liver, 382t, 383t
    lung, 27t, 386t
    milk, 114t
    muscle, 388t
    RNA, 440t, 441t, 442t
  daily intake of, 101t
Molybdenum-92, 7
Muscle
  concentration of elements in, 387, 388t, 457t

N

Na(T1) detector, *276*, 277, 279, *280*
Neodymium
  analysis of, 257
  concentration of, in, air particulates, 31t, 32t, 56t
    bone, 390t
    coal, 50t
    house dust, 56t
    kettle fur, 78t
    milk, 114t
Neon, 5
Nernst Equation, 205
Neutron activation analysis (*see* Radioactivation analysis)
Neutron irradiation
  analysis (*see* Radioactivation analysis)
  artificially induced, 268
  naturally occurring, 9
Nickle, xiit
  abundance ratios, 40t
  analysis of, 196t, 196, 201, 216, 256
  concentration of, in, air particulates, 31t, 33t, 56t, 163t
    blood, 172t, 370t
    car exhaust effluent, 44t
    coal, 48t, 49t, 50t
    DNA, 440t
    drinking water, 74t, 83t, 97t; particulates, 81t, 82t, 83t
    excreta containers, 141t
    hair, 400t, 401t
    house dust, 55t, 56t, 401t
    kettle fur, 79t
    RNA, 440t
    seal tissues, 456-7t
  daily intake of, 101t
Niobium
  concentration of, in, air particulates, 31t, 33t, 56t
    blood, 75t, 78t, 368t, 456t
    bone, 390t
    car exhaust effluent, 44t
    coal, 48t
    house dust, 56t
    kettle fur, 78t
    kidney, 384t, 457t

liver, 382t, 457t
lung, 27t, 386t
lymph nodes, 394t
milk, 114t
muscle, 388t, 457t
seal tissues, 456-7t
sugar, 125t
testes, 396t
daily intake of, 101t
Nitrate
analysis of, 196t, 210
Nitrite
analysis of, 196t, 203
Nitrogen, xiit
concentration of, in, drinking water particulates, 81t
house dust, 55t
Nucleic acids
formation of, 15
(see also DNA; RNA)
Nuclear structure, 10

O

Oklo phenomenon, 9, 456
Optical emission spectroscopy, 220-1, 222-7, *223*, 343, 366
detection limits for, 226t
with plasma source, 253-66
Osmium
concentration of, in, air particulates, 32t
DNA, 443t
daily intake of, 100t
Osteodystrophy, effect of water quality on, 88
Ovaries
concentration of elements in, 395, 397t
Oxygen, xiit, 5, 6, 15
Ozone layer, 6

P

Palladium
concentration of, in, air particulates, 31t, 33t
DNA, 443t
Particulates
airborne, **28-66**
carcinogenic, 421

composition of, 29-36
concentration of elements in, 31t, 32-4t, 56t, 163t
element enrichment in, 43
global transport of, 41
in house dust, 54-9, 58
in lung, 26, 414-16, 421
particle size distribution in, 30, 31
production of, 28-9
sampling methods for, 29
in water,
sampling, methods, 77, 80-1; problems, 68-9
Phosphorus
analysis of, 263, 353
concentration of, in, air particulates, 31t, 34t, 57t, 163t
blood, 75t, 140t, 172t, 369t, 376t, 456t
brain, 379t
car exhaust effluent, 44t
coal, 49t, 50t
DNA, 440t
drinking water, 74t, 83t, 97t; particulates, 81t, 83t
enzymes, 444t
excreta containers, 141t
house dust, 55t, 57t
kidney, 385t
liver, 382t, 457t
lung, 27t, 386t
lymph nodes, 394t
milk, 115t
muscle, 388t, 457t
ovaries, 397t
RNA, 440t
seal tissues, 456-7t
sugar, 125t
testes, 396t
daily intake of, 101t
Photosynthesis, evolution of, 15
Placenta, 110
elements found in, 110, 397-9
Plasma emission sources, **252-66**
capacitance-coupled, 261-2
coupled to optical emission spectrometer, 253-5, *255*
detection limits, 256-7
inductively coupled, 258-61, *258*

detection limits, 259, 260t
microwave, 262-6, *264*, *265*
Platinum
  concentration of, in, air particulates, 32t
  DNA, 443t
  daily intake of, 100t
Plutonium, 9
Plutonium-239, analysis of, 297, *298-301*
Poisson distribution, 134
Polarography, 212, *213*, 214
Pollution *(see* Contamination, environmental)
Polonium-210, 297, *302*, 306, 403
Potassium, xiit, 36-7
  abundance ratios, 37t
  analysis of, 196t, 210, 211, 357
  concentration of, in, air particulates, 31t, 34t, 57t, 163t
    blood, 75t, 79t, 140t, 172t, 456t
    brain, 379t
    car exhaust effluent, 44t
    coal, 50t, 51t
    DNA, 440t
    drinking water, 74t, 97t; particulates, 81t
    house dust, 55t, 57t
    kettle fur, 79t
    kidney, 385t
    liver, 392t, 383t, 457t
    lung, 27t, 386t
    lymph nodes, 394t
    milk, 114t
    muscle, 388t, 457t
    ovaries, 397t
    RNA, 440t
    seal tissues, 456-7t
    sugar, 125t
    testes, 396t
  daily intake of, 101t
Potassium-40, 403
Praseodynium
  analysis of, 257
  concentration of, in, air particulates, 31t, 32t, 56t
    bone, 390t
    coal, 50t
    house dust, 56t

kettle fur, 78t
milk, 114t
Precipitation techniques, 184-5
Promethium-147, 307, *310*
Protein(s)
  determination of elements in, 439, 441-2
  formation of, 15
  -ion bonding, 448-51
  metal binding, 442

R

Radiation, ionizing
  detection of, 275-7, *276*
  effects of, 465
Radioactivation analysis, 185, **267-314**, 366
  by neutron irradiation, 268-72, *268*, 288, 293, 306, 307-8, 310
  detection systems, 275-7, *276*, 279-88 *(see also* Autoradiography)
  applications of, 374, 380, 381, *404*, 441-5
Radioactive decay, **272-5**, *274*, 277
  half-life, calculation of, 272-3, *274*
  use of, to determine age of earth, 8
Radioelements, 403-5 *(see also* Radionuclides)
Radionuclides
  analysis of, xiv-xv, **275-311**
  naturally occurring, 310-1
  nuclear fallout, 311
Radium-226, 403
Rare-earths, 39
  abundance ratios, 39t
Rat
  distribution of lead in tissues of, 434-5t
  brain, lead levels in hippocampus of, 428, *429*, *430*
  liver, fractionation of, 431
    lead binding components of, *432*, 435
Reagents
  contamination from, 153
  purification of, 152-3, *153*, 217, *218*
Reference materials, 171
Renal dialysis, 87-93
  element levels in water for, 97t

preparation of water for, 91, *91*
salt mixture, 89
Renal failure, causes, 87-8
Rhenium
  concentration of, in, air particles, 32t
    coal, 51t
  daily intake of, 100t
Rhodium, concentration of, in air particulates, 31t, 33t
RNA
  concentration of elements in, 439, 440t, 441t, 442t
  formation of, 15
Rocks
  crustal, elements in, compared with human blood, *20, 21*
  origin of, 10-14, *11, 13*
  sedimentary, formation of, 12
  fossil record in, 14
Rubidium
  abundance ratios, 37, 37t
  concentration of, in, air particulates, 31t, 33t, 163t
    blood, 75t, 79t, 140t, 172t, 368t, 376t, 456t
    bone, 391t
    brain, 378t, 380t
    car exhaust effluent, 44t
    coal, 48t, 49t, 50t
    DNA, 443t
    drinking water, 74t, 97t; particulates, 81t
    house dust, 55t
    kettle fur, 79t
    kidney, 384t
    liver, 382t, 383t, 457t
    lung, 27t, 386t
    lymph nodes, 394t
    milk, 114t
    muscle, 388t, 457t
    ovaries, 397t
    RNA, 441t, 442t
    seal tissues, 456-7t
    testes, 396t
  daily intake of, 101t
Ruthenium
  concentration of, in, air particulates, 31t, 33t, 163t
    DNA, 443t

S

Samarium
  concentration of, in, air particulates, 31t, 32t, 56t
    bone, 390t
    coal, 50t
    house dust, 56t
    kettle fur, 78t
    milk, 114t
    RNA, 442t
Sample preparation, **142-87**, 186t
  biological material, for radioactive counting, 282 (*see also* individual techniques *eg* Ashing, Concentration techniques, Homogenisation, Separation techniques *etc*)
Sampling
  constant, 134
  methods, **130-141**
    for air particulates, 29, 53, 57, 59
    for deposits in household water supplies, 80
    for lead budget in coal, 47-8
    for liquids, 138-42
    for tissues, 135-8
    water, problems associated with, 68-9
Scandium
  abundance ratios, 40t
  concentration of, in, air particulates, 31t, 34t, 57t
    blood, 79t, 369t
    car exhaust effluent, 44t
    coal, 50t
    DNA, 443t
    house dust, 57t
    kettle fur, 79t
    liver, 383t
    RNA, 441t, 442t
    sugar, 125t
    water, 97t
Seal
  correlation between, mercury and selenium levels in, *453,* 455
  sulphur and selenium levels in, *454,* 455
  element levels in, tissues of, 456-7t
  liver of, 458t

Index

Seawater
  elements in, compared with human blood, 21-2, *22*, *23*
  metals in, behavior of, 455, 458
Sedimentation, 12, 35
Selenium, xiit, 37-9, 459
  abundance ratios, 37-8
  analysis of, 203, 238, 241, 257
  concentration of, in, air particulates, 31t, 33t, 56t, 163t
    blood, 75t, 79t, 172t, 368t
    brain, 379t, 380t
    car exhaust effluent, 44t
    coal, 50t
    DNA, 440t, 443t
    drinking water, 97t; particulates, 82t
    hair, 400t
    house dust, 56t
    kettle fur, 79t
    kidney, 385t
    liver, 382t, 383t
    lung, 27t, 386t
    lymph nodes, 394t
    muscle, 388t
    ovaries, 397t
    RNA, 440t, 441t, 442t
    seal, tissues, *453*, *454*, 455; liver, 458t
    sugar, 125t
    testes, 396t
  daily intake of, 101t
Selenium-74, 7
Senile dementia, 422
Separation techniques
  chemical, 184-7, 186t, 204
  radiochemical, 277-9, *281*
  (*see also* Chromatography; Ion-exchange resins; Precipitation techniques; Solvent extraction)
Signal response, 190-2
  typical curve, *191*
Silica
  analysis of, 196t
  concentration of in drinking water, 71t
Silicon, xi, xiit
  abundance ratios, 37, 37t
  analysis of, 197

concentration of, in, air particulates, 31t, 34t, 57t
  blood, 75t, 79t, 172t, 369t
  brain, 379t
  car exhaust effluent, 44t
  DNA, 440t
  excreta containers, 141t
  house dust, 55t, 57t, 59, *62*
  kettle fur, 79t
  kidney, 385t, 457t
  liver, 382t, 457t
  lung, 27t, 386t
  lymph nodes, 394t
  milk, 114t
  muscle, 388t, 457t
  ovaries, 397t
  RNA, 440t
  seal tissues, 456-7t
  testes, 396t
daily intake of, 101t
Silver
  analysis of, 201, 210, 216
  concentration of, in, air particulates, 31t, 33t, 43, 56t
    blood, 75t, 78t, 172t, 368t, 456t
    bone, 390t, 391t
    brain, 378t, 380t
    car exhaust effluent, 44t
    coal, 48t, 50t
    DNA, 440t, 443t
    excreta containers, 141t
    hair, 400t
    house dust, 56t
    kettle fur, 78t
    kidney, 384t, 457t
    liver, 382t, 457t
    lung, 27t, 386t
    lymph nodes, 394t
    milk, 114t
    muscle, 388t, 457t
    ovaries, 397t
    RNA, 440t, 441t, 442t
    seal tissues, 456-7t
    sugar, 125t
    testes, 396t
  daily intake of, 101t
Skin, concentration of elements in, 402-3
Sodium, xiit, 5
  abundance ratios, 40t

486   *The Chemical Elements and Man*

analysis of, 210, 211, 353, 357
concentration of, in, air particulates, 31t, 34t, 57t, 163t
    blood, 172t, 369t, 376t
    car exhaust effluent, 44t
    coal, 50t
    DNA, 440t
    drinking water, 71t, 74t, 97t
    house dust, 55t, 57t
    liver, 383t
    RNA, 440t
daily intake of, 101t
Solvent extraction, 185
Soya bean, 124-6
    concentration of elements in, 126t
Spectrophotometry, 185-6, 192-3, **199-203**
Spectroscopy, emission, 220-66
    atomic, 220-50
    plasma, 252-66, *254, 255*
    (*see also* Atomic absorption spectroscopy; Flame photometry; Fluorescence spectrometry; Optical emission spectroscopy)
Staining techniques
    determination of element tissue distribution by, 428-30, *429*
Standard(s)
    additional analysis, 229-31, *230*
    environmental, establishment of, 181t
    international, 173t
    solutions, 183-4
Standardisation of techniques, 311-3
Stars, hot, 4
Strontium
    abundance ratios, 37t
    concentration of, in, air particulates, 31t, 33t, 56t, 163t
        blood, 75t, 79t, 172t, 368t, 370t, 376t, 456t
        bone, 390t, 391t
        brain, 378t
        car exhaust effluent, 44t
        coal, 48t, 50t
        DNA, 440t, 443t
        drinking water, 74t, 97t; particulates, 81t
        excreta containers, 141t

        house dust, 55t, 56t
        kettle fur, 79t
        kidney, 384t, 457t
        liver, 382t, 457t
        lung, 27t, 386t
        lymph nodes, 394t
        milk, 114t
        muscle, 388t, 457t
        ovaries, 397t
        RNA, 440t
        seal tissues, 456-7t
        sugar, 125t
        testes, 396t
    daily intake, 101t
    substitution of, for calcium, 120
Sugar
    concentration of elements in, 125t
Sulphur, xiit
    abundance ratios, 37-8
    analysis of, 263
        as sulphate, 196t, 203
        as sulphide, 196t, 203, 210
    concentration of, in, air particulates, 31t, 34t, 57t, 163t
        blood, 79t, 140t 172t 369t, *372*, 456t
        brain, 379t
        car exhaust effluent, 44t
        coal, 50t
        DNA, 440t
        drinking water, 74t, 83t, 97t; particulates, 81t, 83t, *85*
        excreta containers, 141t
        house dust, 55t, 57t, 59, *63*
        kettle fur, 79t
        kidney, 385t, 457t
        liver, 382t, 457t
        lung, 27t, 386t
        lymph nodes, 394t
        milk, 114t
        muscle, 388t, 457t
        ovaries, 397t
        RNA, 440t
        seal tissues, 456-7t
        sugar, 125t
        testes, 396t
    daily intake of, 101t
Supernovae, 5-6

Surface composition determined by analysis of impact radiation (SCANIIR), 357
Symbiotic unions, in evolution, 19
System modelling, use of, in health care, 413

T

Tantalum
  analysis of, 257
  concentration of, in, air particulates, 32t
    DNA, 443t
    kettle fur, 78t
  daily intake of, 100t
Teeth
  concentration of elements in, 391-2
  fluoride concentration in, 392-3, *392*, *393*, 395
Tellurium
  concentration of, in, air particulates, 31t, 33t, 56t
    blood, 172t
    house dust, 56t
Terbium
  concentration of, in, air particulates, 31t, 32t, 56t
    bone, 390t
    coal, 50t
    drinking water, 74t
    house dust, 56t
    kettle fur, 78t
    milk, 114t
Testes
  concentration of elements in, 395, 396t
Thallium, xiit
  abundance ratios, 37, 37t
  analysis of, 201, 216
  concentration of, in, air particulates, 31t, 32t, 56t
    blood, 172t, 368t
    bone, 390t
    brain, 378t
    house dust, 56t
    kettle fur, 78t
    kidney, 384t
    liver, 382t
    lung, 384t
  daily intake of, 100t
Third world, diseases of, 413-4
Thorium
  abundance ratios, 37, 37t
  analysis of, 257, 403
  concentration of, in, air particulates, 31t, 32t, 45, 56t
    blood, 172t, 368t, *404*
    bone, 309t, 391t, *404*
    coal, 48t, 51t
    DNA, 443t
    house dust, 56t
    kettle fur, 78t
    lymph nodes, 394t
    milk, 114t
    urine, *404*
  daily intake of, 100t
Thulium
  concentration of, in, air particulates, 31t, 32t
    bone, 390t
    coal, 50t
    drinking water, 74t
    kettle fur, 78t
    milk, 114t
Tin, xi, xiit, 109-13, 115
  analysis of, 196t, 197, 201, 216
  concentration of, in, air particulates, 31t, 33t, 56t, 163t
    blood, 75t, 78t, 172t, 368t, 456t
    bone, 390t, 391t
    brain, 378t
    canned foodstuffs, 113
    car exhaust effluent, 44t, 45
    coal, 48t, 49t, 50t
    drinking water, 74t, 97t; particulates, 82t
    DNA, 440t, 443t
    excreta containers, 141t
    hair, 401t
    house dust, 56t, 401t
    infant excreta, 113
    kettle fur, 78t
    kidney, 384t, 457t
    liver, 382t, 383t, 457t
    lung, 27t, 386t
    lymph nodes, 394t

488 *The Chemical Elements and Man*

milk, 113, 114t, 115t
muscle, 388t, 457t
ovaries, 397t
RNA, 440t
seal tissues, 456-7t
sugar, 125t
testes, 396t
daily intake of, 101t
  for baby, 113, 116t
-induced encephalopathy, 90-1
Titanium
analysis of, 201
concentration of, in, air particulates,
  31t, 34t, 56t, 163t
blood, 172t
brain, 379t
car exhaust effluent, 44t
coal, 49t, 50t
DNA, 440t
drinking water particulates, 81t
excreta containers, 141t
house dust, 55t, 56t
kidney, 385t
liver, 382t, 457t
lung, 27t, 386t
muscle, 388t, 457t
RNA, 440t
seal tissues, 456-7t
sugar, 125t
daily intake of, 101t
Toxicology, 411
Trace-element free isolator system, 117, *118*
Tungsten
analysis of, 257
concentration of, in, air particulates,
  31t, 32t, 56t
blood, 172t
car exhaust effluent, 44t
DNA, 443t
drinking water, 97t
house dust, 56t
RNA, 441t, 442t
daily intake of, 100t

U

Ultra-centrifugation, 445
Uranium, 8-9, **109**

abundance ratios, 37, 37t
analysis of, 288, 293-7, *294*
concentration of, in, air particulates,
  31t, 32t, 45, *46*, 47, 56t, 163t
blood, 172t, 368t
bone, 390t, 391t
coal, 48t, 51t
DNA, 443t
drinking water, 74t
house dust, 56t
kettle fur, 78t
liver, 382t
lung, 27t, 386t
lymph nodes, 394t
milk, 114t
muscle, 388t
daily intake of, 100t
Urine
collection containers, element concentration in, 141t
infant, concentration of tin in, 113, 116t
sampling methods for, 140-1
Urov disease, 120

V

Vanadium, xiit
analysis of, 196t
concentration of, in, air particulates,
  31t, 33t, 56t, 163t
blood, 79t, 172t
brain, 379t
car exhaust effluent, 44t
coal, 48t, 49t, 50t
DNA, 440t
drinking water, 74t, 97t
excreta containers, 141t
hair, 400t, 401t
house dust, 56t, 401t
kettle fur, 79t
liver, 382t, 383t, 457t
lung, 27t, 386t
lymph nodes, 394t
muscle, 388t, 457t
RNA, 440t
seal, tissue, 456-7t; liver, 458t
sugar, 125t
testes, 396t

Index   489

daily intake of, 101t
Volcanic activity, in formation of Earth, 12, *13*

W

Water, 68-93
  content of tissues, 142t
  deionised, use of for renal dialysis, 88-90
  drinking, and health, 69-70, 71t
    sources of, 70, 72
    treatment of, 72-3
  elements in hard and soft, 69t
  preparation of clean, for laboratory supply, 166-9, *167, 168, 169*
  sampling problems, 68-9
  supply, deposits in household, 80-1, 81t, 82t, *83*

X

X-ray fluorescence spectrometry, **314-31**, *315, 316, 321, 324, 327, 328, 329*, 366
  analytical methods, 322-31
    tablet technique, 323-5
    thin film technique, 325

Y

Ytterbium
  concentration of, in, air particulates, 31t, 32t, 56t
  bone, 390t
  coal, 50t
  house dust, 56t
  milk, 114t
Yttrium
  analysis of, 256
  concentration of, in, air particulates, 31t, 33t, 163t
  blood, 75t, 79t, 368t, 456t
  bone, 390t
  brain, 378t
  car exhaust effluent, 44t
  coal, 50t
  DNA, 440t
  drinking water, 74t, 97t
  excreta containers, 141t
  kettle fur, 79t

kidney, 384t, 457t
liver, 382t, 457t
lung, 27t, 386t
lymph nodes, 394t
milk, 114t
muscle, 388t, 457t
ovaries, 397t
RNA, 440t
seal tissues, 456-7t
sugar, 125t
testes, 396t
daily intake of, 101t

Z

Zinc, xiit, 110-12
  abundance ratios, 40t
  analysis of, 194, 195, 196t, 197, 201, 203, 216, 216t, 240, 248
  concentration of, in, air particulates, 31t, 33t, 56t, 163t
  blood, 75t, 79t, 140t, 172t, 369t, 370t, 373t, 456t
  bone, 390t, 391t
  brain, 379t, 380t
  car exhaust effluent, 44t
  coal, 48t, 49t, 50t
  DNA, 440t, 443t
  drinking water, 73, 74t, 76, 83t, 97t; particulates, 81t, 82t, 83t, *84*
  enzymes, 444t
  excreta containers, 141t
  hair, 400t
  house dust, 55, 55t, 56t, 59, 65
  kettle fur, 79t
  kidney, 385t, 457t
  liver, 382t, 383t, 457t
  lung, 27t, 386t
  lymph nodes, 394t
  milk, 112, 114t
  muscle, 388t, 457t
  ovaries, 397t
  RNA, 440t, 441t, 442t
  seal, tissues, 456-7t; liver, 458t
  sugar, 125t
  testes, 396t
daily intake of, 101t

Zirconium
  abundance ratios, 37, 37t
  concentration of, in, air particulates, 31t, 33t, 56t, 163t
  blood, 75t, 78t, 172t, 368t, 456t
  bone, 390t, 391t
  brain, 378t
  car exhaust effluent, 44t
  coal, 48t, 49t
  DNA, 440t, 443t
  drinking water, 97t
  excreta containers, 141t
  house dust, 56t
  kettle fur, 78t
  kidney, 384t, 457t
  liver, 382t, 457t
  lung, 27t, 386t
  lymph nodes, 394t
  milk, 114t
  muscle, 388t, 457t
  ovaries, 397t
  RNA, 440t
  seal tissues, 456-7t
  sugar, 125t
  testes, 396t
  daily intake of, 101t